# GRID Systems and FORMATS Sourcebook

# GRID systems and FORMATS Sourcebook

Ready-To-Use Material
For Print, Projected,
And Electronic Media

## Ernest Burden

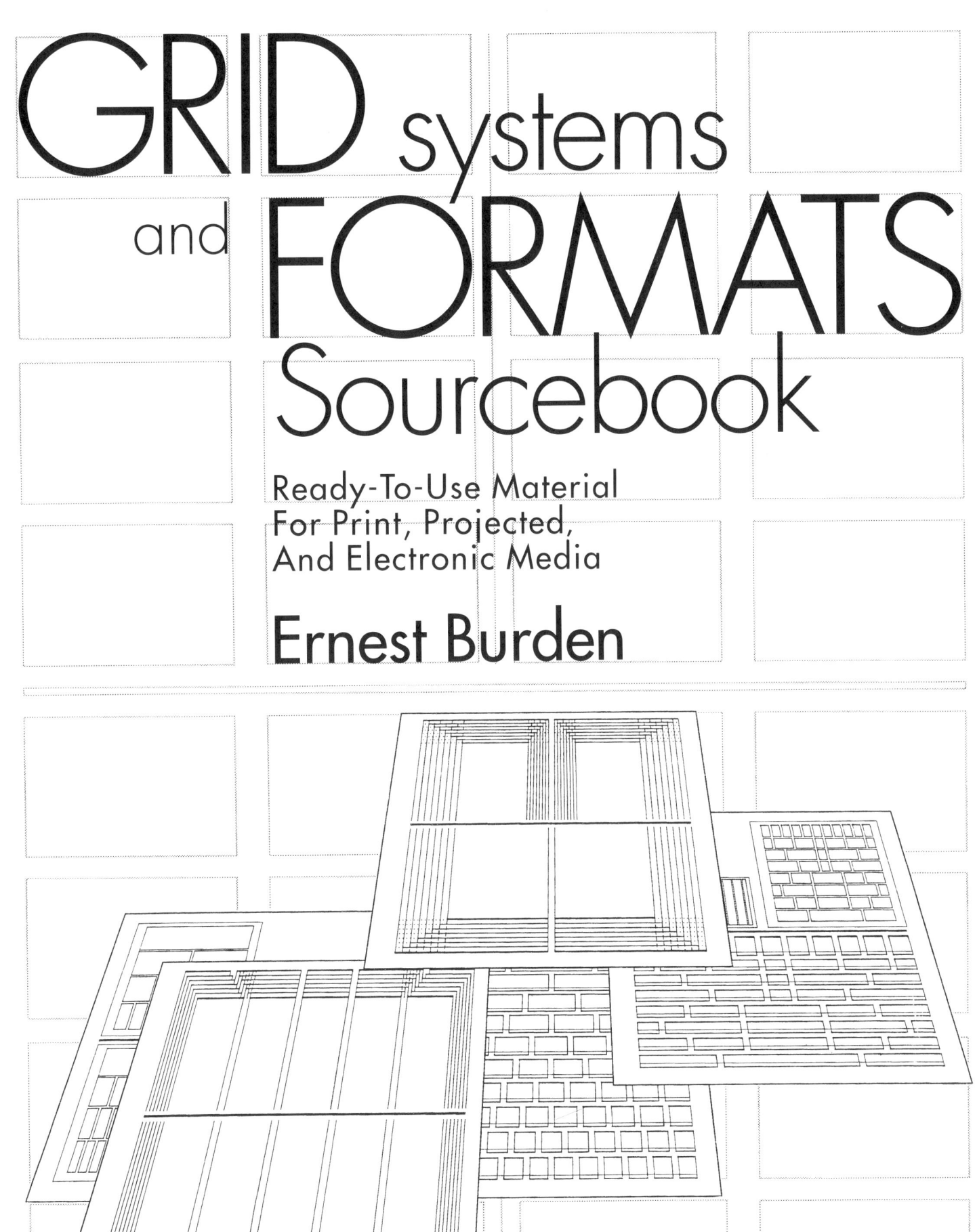

Copyright © 1992 by Van Nostrand Reinhold

Library of Congress Catalog Card Number
ISBN 0-442-01174-1

All rights reserved. No part of this work covered by
the copyright hereon may be reproduced or used in any
form by any means—graphic, electronic, or
mechanical, including photocopying, recording, taping,
or information storage and retrieval systems—without
written permission of the publisher.

Printed in the United States of America.

Van Nostrand Reinhold
115 Fifth Avenue
New York, New York 10003

Chapman and Hall
2-6 Boundary Row
London, SE1 8HN, England

Thomas Nelson Australia
102 Dodds Street
South Melbourne 3205
Victoria, Australia

Nelson Canada
1120 Birchmount Road
Scarborough, Ontario MIK 5G4, Canada

16 15 14 13 12 11 10 9 8 7 6 5 4 3 2 1

Library of Congress Cataloging-in-Publication Data

Burden, Ernest E., 1934-
   Grid systems and formats sourcebook: ready-to-use materials for
print, projected, and electronic media / Ernest Burden.
      p.    cm.
   ISBN 0-442-01174-1
   1.  Engineering graphics.  2.  Computer graphics.  I.  Title.
TA357.B87   1992
604.2' —dc20                                         92-15795
                                                         CIP

# CONTENTS

Preface vii

■ PART ONE
1 Corporate ID 1
2 Anatomy of Formats 13

■ PART TWO
3 Generic Grids 75
4 Case Study Formats 115
5 Appendix 189
   Credits 230
   Bibliography 232

## ACKNOWLEDGMENTS

The material for this book was produced by a combination of manual and electronic techniques. Many of the manual techniques could have been produced electronically, of course. The reverse is also true, as you can use this material either manually, or by inputting it into your computer and using it electronically.

I would like to thank my editor, Wendy Lochner for her encouragement throughout the long process to produce this book, and for her assistance with the final text. Monika Keano, art director, provided valuable suggestions along the way for making this book appropriate and useful to graphic designers, as well as architects and engineers. Kurt Andrews oversaw the production to insure a quality job, along with others in the production department.

Particular thanks go to my son Ernest III for developing the section on generic grids on his CADD system. These grids were plotted by the computer and its dimensioning capabilities were particularly useful here.

Thanks to Joy Arnold who convinced me to give up my phototypesetter in favor of electronic tools. Her input and processing the text was appreciated very much. Many thanks also to Arthur Spaet, Mary Lawrence, and Roy Sokolowski for the use of their Mac, Microsoft and Pagemaker. Thanks to Creighton Nolte, Architect, for his contribution of electronic and CADD images in the Anatomy Of Formats section.

A very special thanks to Joy for her unending encouragement over the entire design and production process, and for her belief in the usefulness of this publication.

# PREFACE

There is nothing new about the grid. It has been a useful tool in nearly every art form for centuries. The Egyptians drew grids on stone to enlarge their drawings, sculptures, and other architectural details. Early papyrus scrolls had lettering and inscribed forms called hieroglyphics, which were arranged within partitioned margins on the top, bottom, and sides. The early Sumerians used a form of pictograph writing called cuneiform using wedge-shaped marks on clay tablets. These were applied along horizontal lines or rules, similar to the Egyptian markings. When the cuneiform markings were used on buildings or sculptured monoliths, the markings were placed at eye level. Other ancient civilizations had also developed similar pictorial forms of communication using clay tablets, until the use of parchment was developed by the Romans.

Medieval manuscripts were painstakingly lettered by hand, along guidelines that were ruled onto the parchment. These early forms of page layout grids are the predecessor of our most sophisticated forms of electronically produced type, and contain some of the same characteristics. The first letter in the column of type was enlarged and often decorated with flourishes, color, or gold leaf to call attention to it. Other enlarged letters were used throughout the text to provide emphasis or stopping points. These enlarged letters broke into the margins, and the type ran around it. This is a popular device used today in nearly all desktop publishing programs.

The development of typography followed these early examples, replacing hand lettering with wooden type and later with cast-metal type. Printers created the columns of type and provided the guidelines for calculating the length of the material. These were grid systems controlled by the printer. The first system to offer an alternative to this method was photo-typesetting, wherein letters were exposed directly onto film. Now the designer and not the printer controlled the elements of the page layout. Advances in technology has given designers the ability to enlarge and reduce type electronically, to "cut and paste" and import images into a publication through electronic means. We are now able to see and alter on-screen images that replicate the printed page. These images can be manipulated and perfected until printed out with high-quality output devices.

Everything in a publication contributes to its image, and that image reflects the image of its designer. It is a statement about his or her knowledge, understanding, and ability. The most effective way to assure a conceptually integrated work is to use a grid as an underlying organizing element. A grid is simply a framework of lines forming rectilinear, angular, or circular patterns. A grid system is a related group of elements, interrelated ideas, or procedures that are grouped together for classification. Grid systems exist throughout science and art.

The complex design of our universe with its intricate movements of planets and satellites whirling within galaxies is a system that we have only begun to explore. On our own planet there are a myriad of elements to examine in microcosm, which reveals natural structure and organization within matter which is endless. The closer we look, the more that we discover how much the part is similar to the whole. Electronic tools are opening the doors to discoveries of these relationships, and at the same time making it possible to apply them to our artistic endeavors.

In the arts, there are a myriad of examples and applications of the organizing principle of the grid and grid systems. Nearly all design begins with some basic organizing structure. Writers make outlines and plot out the adventures of their characters. Poets use the meter as a grid and an underlying element; influencing the verse and the rhythm of their work. Musicians decide on the overall structure of their compositions. Artists use the grid in a variety of ways, and architects and city planners have long understood the necessity for organizing the elements of urban and physical space.

In art the grid has been used to enlarge study sketches for paintings, murals, or mosaics. Perspective grids have been set up by Renaissance artists to depict urban scenes and interior spaces. Many compositions and paintings have underlying geometric grids and guidelines. In music, the grid is expressed by the staff upon which a musical score is written. The lines represent a formal spatial relationship to the sounds produced by the instruments. Another grid is expressed in the rhythm, or cadence of the piece. This is the grid of motion, and it determines how fast or slow the piece will be played. Within these elements are the musical notes themselves, which are more like a format than a grid.

In architecture, the grid is an indispensable element for organizing the complexities of a building. The intricate elements of the Gothic cathedrals, as expressed in plan and elevational view, were all strictly designed according to elaborate grid and proportion considerations. Some of the finest architectural work of the modern age uses the grid in both two-dimensional and three-dimensional modes. In dance, the pattern traced by the choreography represents a grid for that dance, and the interpretations within that grid become the format.

Whereas all other arts must be constantly aware of the limitations of materials, graphic images are easily arranged and manipulated. Working within a system of grids, various formats can be developed quite easily for any publication. The format includes everything from the organization of material to the arrangement within a specific production. It also describes the material form of a layout for a publication.

There are many definitions of formats that apply to other material covered in this book. In proposals, for example, there are both horizontal and vertical formats. In photography, format refers to the size and proportion of the image as produced by the camera. In audiovisual terms, it refers to the projected image, or multiple images. In video, format refers to the width of the magnetic tape. In electronic imaging terms, format refers to a specific software program or configuration. Each is designed to perform specific electronic functions according to the programmed specifications.

The images in this book explore both grids and formats in a wide variety of applications. The opening section focuses on the importance of developing an image through a corporate identity program. The section on the anatomy of various formats explores standard formats for newsletters, magazines, and annual reports. It demonstrates formats for charts, diagrams, and schedules, including audio-visual, video, and electronic imaging. The generic grids offer a number of variations on the traditional two, three, and four column grids, as well as showing combinations using up to twelve columns. The section on case-study formats illustrates designs that have already been published in brochures, newsletters, and annual reports. These designs can be used directly, rearranged, or combined with others. They are intended to serve as a guide and an inspiration for new designs. These designs can be enlarged to actual size by photocopying them, and they can be used as an underlayment for tracing or developing a new grid. They can also be scanned into a desktop publishing program and coped directly or manipulated to suit any particular need. By utilizing these grids you will be able to enhance your conceptual designs and develop graphic page layouts much more effectively. Your design will demonstrate a deeper understanding of the underlying organizing principles of your material.

# PART ONE

# 1 Corporate ID

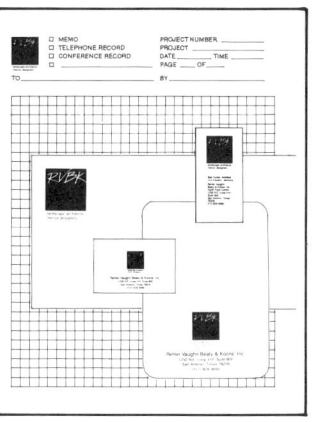

## Overview

### Goals and Objectives

The overall goals of any firm should be identified in terms of its potential growth within a market or markets, new markets to explore potential income growth, and an increase in the number of personnel required to sustain that growth. The immediate challenges of any firm are to identify and clarify the targeted clients within each market, to develop a marketing plan to address the overall direction of the firm, and to specify a direction for each market segment. Two goals that marketing plans address are: to increase awareness among past, current and potential clients in the firm's markets, and to instill pride in the firm among its employees and prospective employees.

### The Marketing Plan

The written marketing plan is the guide behind the marketing communications, in terms of research, design and production. There should also be a written communications plan that lists specific collateral material for promotional use. Once a firm has identified the markets to pursue, the plan identifies how to reach those markets effectively with the appropriate material and message. It also identifies what format to use and what should be highlighted in each piece of marketing material. Each item in the marketing plan is given a priority and specific relationship to all the others.

### Comunications Plan

The communications plan should contain a schedule and a budget for all items to be developed. The schedule should outline when all the pieces are to be prepared and how long it will take to produce them. An integral part of the schedule is the budget. The communications budget does not include marketing salaries but should include public relations retainers, advertising, and ad funds for writers, photographers and graphic artists. It should also include production, printing and mailing costs of all items. A comprehensive corporate communications plan includes, among other things, a corporate identity program, an advertising and public relations program, and a direct selling campaign.

### Corporate Identity Program

The graphic representation of a firm to a variety of publics which, at the same time, fulfills regulatory, operational and marketing needs. It is applied to the entire package of material used by the firm or company.

### Advertising Program

Individual advertisements placed in business, trade or general-interest publications. A campaign usually consists of three ads with a creative related concept.

### Communications Program

A comprehensive communications program using a variety of media in a single campaign or in the firm's total marketing program, showing a consistency in the development of advertising, brochures, newsletter, media placement, and direct mail campaigns.

### Annual Report

Reports made annually for distribution externally for publicly held or privately owned companies. Usually focuses on accomplishments of the company, or firm, during the past year, progress, future expectations and a financial report. The financial information is usually in a totally separate section from the general information.

### Direct Mail Campaign

One or more packages of marketing materials developed to reach a specific target audience and to elicit a response. It can include circulars, fliers, form letters, pamphlets, response mechanisms, announcements, newsletters and market specific brochures.

### Special Events Piece

A printed piece or unusual item representing a one-time effort for a special event. This can include anniversaries, a new name, new office, a holiday, staff promotions and relocations, groundbreaking and dedications.

### Company Brochure

A general corporate publication giving introductory information on the firm's total capabilities.

### Special Market Brochure

A publication representing a one-time effort to market a particular service, discipline, office, or joint venture arrangement to a targeted market segment.

### Service Brochures

These are intended to introduce the firm or a specific part of the firm to a potential client. They outline specific services to specific markets.

### Category Brochures

These are usually devoted to a specific building type with text related to the depth of experience in the project type.

### Magazine

An external publication with liberal use of photos and art, and usually employing a more interpretive writing style. The format is feature oriented.

### Newsletters

A high frequency external or internal publication with few pages, employing a more concise writing style. The format is news or news feature oriented.

### Project Page

A single sheet printed on one or both sides that can be arranged and bound in any sequence, or used loose in a pocket folder, usually focusing on the visual aspect of a project.

### Pocket Folder

A folded sheet of heavy stock designed with a pocket on one or both sides of the inside panels. The pockets are glued to provide a method of holding supplemental loose material. Brochures can be designed with a pocket folder on the last inside page, which is usually die-cut for the placement of a business card.

## Corporate Identity

Corporate identity is the image a firm projects to the outside world. A corporate identity program uses a graphic representation of the firm to a variety of publics, which fulfills all operational and marketing needs. This graphic image is applied to the entire package of print and visual materials used by the firm. The first visual contact a client may have with any firm is its logo and other business items such as letterheads. The logo should be appropriate to the majority of clients served. The communications program should creatively express the firm's operation, display its unique character and capabilities, and provide a workable format for the design of collateral promotional material.

## Advertising and Publicity

The second element of corporate communications program consists of individual advertisements or campaigns placed in business, trade or general-interest publications. Another form of general public contact is where material is developed for display and disseminated at regional or national professional or market-based trade shows. Here a firm can market its capabilities, develop visibility, or introduce a specific service. Publicity and public relations programs previously covered everything a firm did to promote itself. Today promotion is more highly specialized and brings awareness to potential clients through many diverse items such as organized tours, seminars and lectures.

## Direct Mail Programs

The third element of a corporate communications program involves direct mail programs, wherein marketing materials are developed to reach a specific target audience and to elicit a response. A direct mail campaign can include a form letter, pamphlets, response mechanisms, announcements, and market-specific brochures. A direct-mail campaign gets specific messages to targeted audiences. There is a side range of items that can be used as direct-mail items, such as brochures and fliers, newsletters and magazines, posters and announcements, and reprints of articles from publications.

## Special Events Pieces

Two items round out the promotional materials list, one at each end of the spectrum. The special events piece usually represents a one-time effort for a special occasion or event. This can include anniversaries, a new name, new location of office, staff promotion, or client oriented announcements, such as groundbreakings and dedications. On the other end of the scale is the annual report. These are for distribution externally for publicly held or privately owned companies. They usually focus on accomplishments of the firm during the past year, progress, future expectations and a financial report.

---

## Project Fact Sheets

One sided sheets focusing on project data such as project name, client, square footage, and other technical data to explain the functional aspects of a project. They sometimes contain a floor plan or project site plan but rarely includes photographs as the main focus.

## Case Studies

These concentrate on a single project that has specific significance and application to other projects of its type.

## Resume Pages

Usually consists of a single sided sheet, with a resume of an individual, with or without a photograph.

## Client Lists

The mainstay of any design firm is its list of clients. The listings can be categorized in several different ways, either by project type or by client type. Clients within each listing can be arranged by order of importance, by order of date of completion or alphabetically.

## Article Reprints

Articles which appear either in newspaper or magazines can be used for other promotional purposes. Reprints of articles can be ordered from trade journals and printed with the original magazine's cover or simply with the name of the article and the firm. These can be used as targeted direct mail pieces, bound into proposals or used as a leave-behind in client presentations.

## Proposal Graphics

These usually consist of a combination of customized items such as covers, titles and project related photographs, with off-the-shelf promotional material. This includes photo project pages, project descriptions and resumes.

## Organization Charts

Most charts are customized for each proposal outlining the names, positions, and organization of the proposed project team. They are sometimes reproduced in the proposal as a fold-out, and are used in presentations on overhead transparencies.

## Schedules

These are prepared for each specific job based on the task breakdown and projected duration. They are most often reproduced in proposals as fold-outs, and are used in presentations on overhead transparencies.

## "Process" Diagrams

These are generic diagrams that outline a firm's planning, design or construction process in sequential steps. The diagrams are often reproduced in brochures or folders, or used in presentations where specific schedules are not appropriate.

## Overhead Transparencies

These consist of 8' x 10" transparent sheets of acetate which can be projected to a large image using an overhead projector. Charts and diagrams are best presented in this manner. Computers can be hooked up to the overhead projectors to project live data originating from the PC.

## Electronic Imaging

Electronic formats are used for practically every aspect of design and presentation. They include: word processing, desktop publishing, design and structural analysis, schedules, topographic images, and a wide array of design and drafting formats. Many of these formats can be combined for multi-media applications.

## Electronic Templates

These consist of programs that exist within the software package and allow the user to design layouts according to prearranged designs. New text replaces the generic text within a fixed format of rules, column widths and margins. Many of these elements can be rearranged in any desktop publishing system.

## Corporate AV Presentations

These include both audio-visual and video presentations designed to portray the general capabilities of a firm. They are generic in nature and are often produced in modules relating to specific market segments. Those modules can be combined in different ways for specific applications.

## Project Presentations

Both audio-visual and video are used for these presentations, which are related to a specific project. They include reference to the project location, time schedule, budget and other specific client requirements.

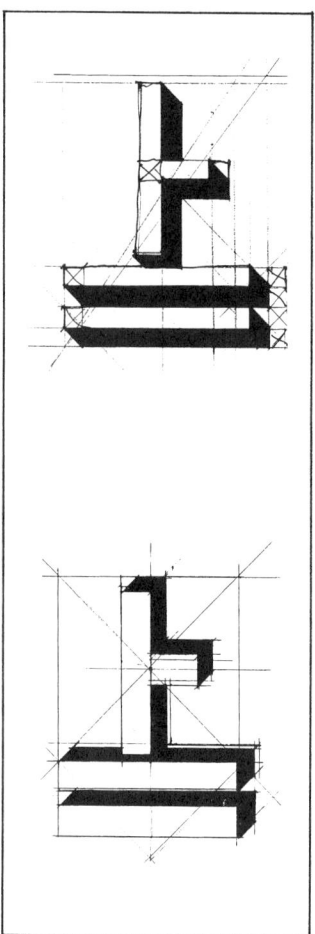
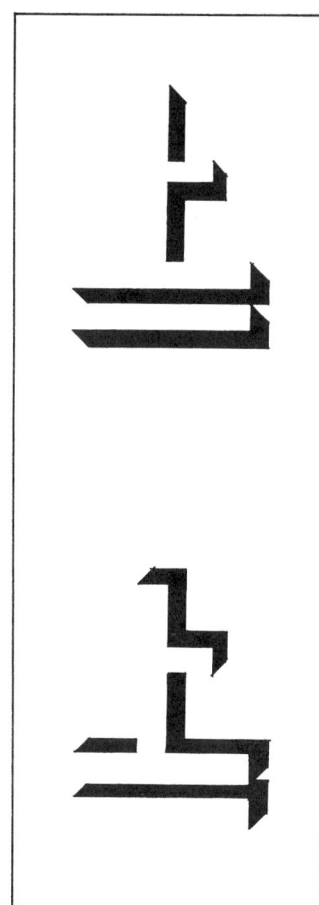

### Logos

Most design firms' logos, unlike corporate logos, are not something that the firm lives with forever. In fact, they change them very often due to personnel changes in the firm. They are usually developed from rough sketches, studied in a few variations, and then developed for use on the material. They are most often done internally, except for the larger firms which either have or hire PR firms to do it for them. The logos are often a graphic device which may be closely related to a firm's general services. Or they may be simply a graphic device representing an abstract idea. Those shown below are typical of logos used by professional design firms.

 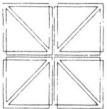  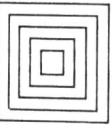    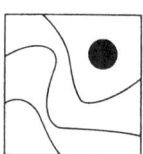

    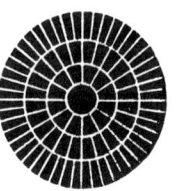    

### Monograms

A monogram is a device which utilizes the initials of the firm or individuals in an abstract graphic form. Shown here are the numerous variations on the letter "G," for example. Most design firms develop monograms using letters designed geometrically rather than using standard fonts. It is the way the letters are combined that makes a distinctive monogram, rather than the letters individually. The monograms shown below are typical of those developed by design firms.

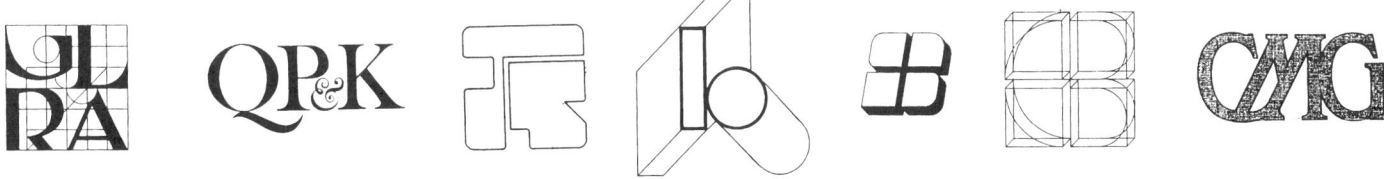

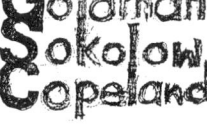

## Logotypes

Usually there is enough variation in typefaces that a design firm can select one that fits its image. This simplifies the graphic design process, while allowing firms to get a distinctive looking logotype. Often a number of sketches helps in the initial selection to narrow the choice down to a select few before actually setting type. The logotypes illustrated here are typical of those used by design firms.

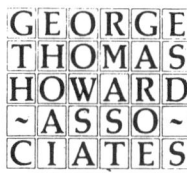

Once the design has been established, the next stage is to study how it will be used on the various pieces, including letters, envelopes and business cards, or course, but also forms, transmittals, labels, and announcements.

Perhaps the most common use is the business card and letterhead. Several typical layouts are shown below, although there are as many variations and possibilities as there are design firms.

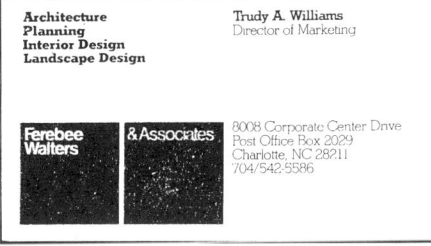
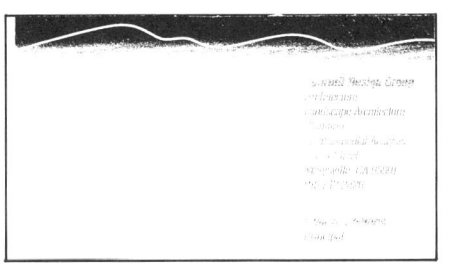
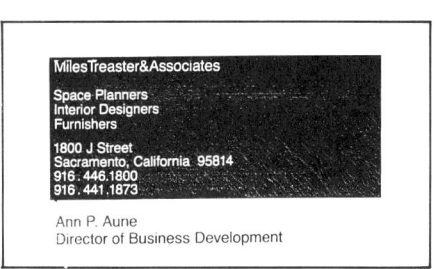

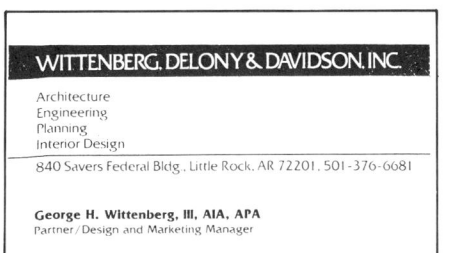
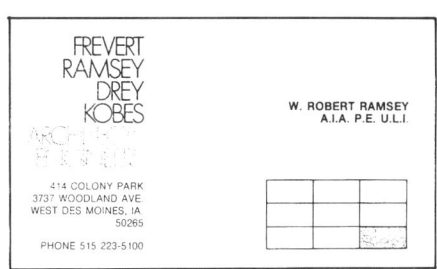

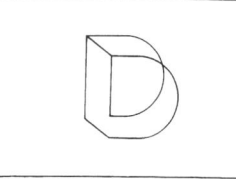

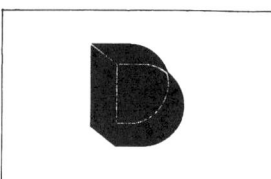

## Letterheads

Some design firms change their logo quite often. This is due to the change in key personnel, partners retiring, and other internal changes. The study of this logo development shows that there is a wide number of choices available to the designer. The logo design studies explored many variations of the two partner's initials (which coincidentally were the same). These studies were carried into the design of the firm's letterhead, which studied the total effect of these combinations. The number of possibilities that were studied is shown on the following page. This study compared both serif and sans-serif typefaces in all caps, and in caps and lower case letters for each combination. They were all studied using the same point size type for consistency in the evaluation.

The logo selected was not one of the graphic play on the two partner's initials, but an abstract representation of the berry, called the dewberry. This exercise shows the extent of study that went into the final selection. The development of a corporate logo for major corporations can sometimes take months, even years.

DEWBERRY/DAVIS ASSOCIATES

Dewberry/Davis Associates

**DEWBERRY/DAVIS ASSOCIATES**

**Dewberry/Davis Associates**

**DEWBERRY/DAVIS ASSOCIATES**

**Dewberry/Davis Associates**

DEWBERRY/DAVIS ASSOCIATES

Dewberry/Davis Associates

DEWBERRY/DAVIS ASSOCIATES

Dewberry/Davis Associates

DEWBERRY/DAVIS ASSOCIATES

Dewberry/Davis Associates

---

DEWBERRY/DAVIS ASSOCIATES

Dewberry/Davis Associates

DEWBERRY/DAVIS ASSOCIATES

Dewberry/Davis Associates

DEWBERRY/DAVIS ASSOCIATES

Dewberry/Davis Associates

DEWBERRY/DAVIS ASSOCIATES

Dewberry/Davis Associates

**DEWBERRY/DAVIS ASSOCIATES**

**Dewberry/Davis Associates**

**DEWBERRY/DAVIS ASSOCIATES**

**Dewberry/Davis Associates**

---

**Dewberry+Davis**

**Dewberry and Davis**

**Dewberry & Davis**

**Dewberry** and **Davis**

**Dewberry & Davis** Associates

---

**DEWBERRY/DAVIS** ASSOCIATES

**Dewberry/Davis** Associates

DEWBERRY/DAVIS ASSOCIATES

Dewberry/Davis Associates

**Dewberry/Davis Associates**

**Dewberry/Davis** Associates

## ID Manuals

Corporate identity manuals are created to establish guidelines for the proper use and placement of the firm's logos and type on various pieces of promotional material and business forms. Consistent and correct adherence to the rules outlined in these manuals is important to the success of the program as a whole. The size of manual may vary from a few printed pages to a large column depending on the size of the firm and the concern for specific control of its identification program. Some of the applications typically illustrated in these manuals are listed below. The illustration here shows a typical page from a corporate identity manual for a large multi-discipline firm.

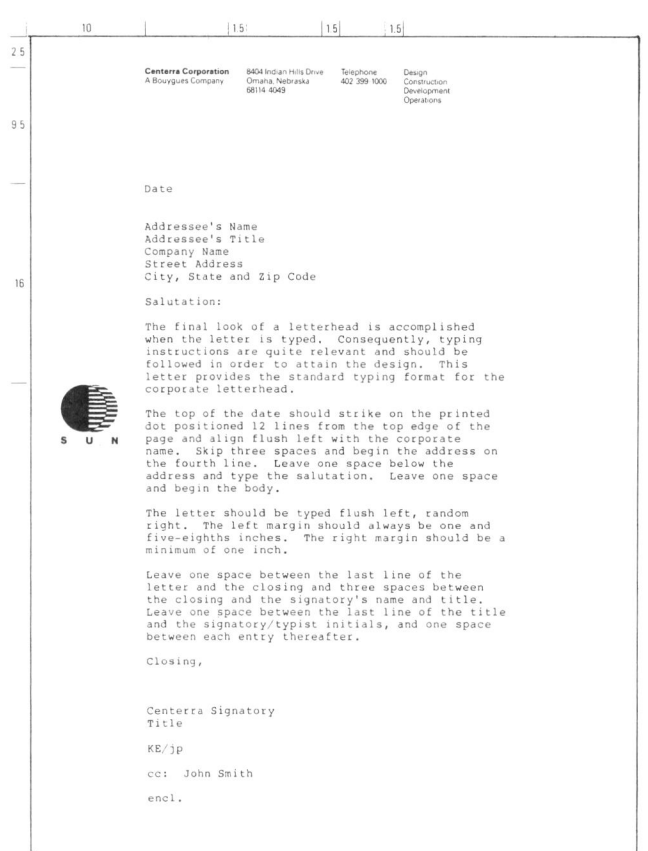

The letterhead illustrated on this page is for use by Level A1, the parent corporation. A typing format which should be used for all letterhead correspondence is shown. Measurements are also provided for placing a typing dot on the artwork to serve as an anchor point for typed copy.

**Size**
8½" x 11"

**Color**
Symbol:
Centerra Bordeaux

All other elements:
Centerra Dark Grey

**Paper Stock**
Strathmore Bond
Fluorescent White,
Wove Cockle
Substance 24

**Typesetting**
Legal name:
8 point Univers 75

All other copy:
8 point Univers 55

All type will be set:
Upper and lower case
Minus ½ unit letterspacing
Flush left, random right
1 point leading

All measurements are shown in picas.

**2.2 Level A1 Letterhead**

The maximum line width for the Identifier, address, telephone and Business Class Designator copy blocks is 8 picas. The distance from the end of the longest line of each copy block to the start of the next copy block equals 1.5 picas.

Symbol centers horizontally within 10 pica margin. Symbol width equals 5 picas.

---

Some of the applications typically illustrated in corporate identification manuals are:

**BASIC STANDARDS**
- Glossary of Terms
- Corporate Symbol/ Grid/Guidelines
- Levels of Identification
- Signature Configurations
- Design Control
- Color Standards/Swatches
- Decorative Use
- Reverse Use
- Supporting Typography

**STATIONERY SYSTEM**
- Introduction
- Letterhead
- Envelope
- Business Card
- Personalized Note Pad
- Mailing Label
- Press Release
- Internal Memorandum
- Transmittal
- Proposal Covers
- Project Report
- Title Page Format
- Purchase Order
- Invoice
- Check/Envelope

**ADVERTISING AND PROMOTION**
- Newsletter
- Brochures
- Binders
- Image Ads
- Calling Card Ads
- Yellow Page Ads

**SIGNAGE SYSTEM**
- Exterior Formats
- Interior Formats
- Construction Sign Formats

**VEHICLE SYSTEM**
- Trucks, Vans, Cars
- Trailer
- Corporate Jet

**TITLE BLOCKS**
- Cover Sheet 8½x11
- Title Blocks

**BASIC ITEMS**
- Hard Hat/Name Tag
- Coaster/Lapel Pin/ Napkin/Cup

The same concern for the use of the firm's logo is shown in this manual. It lists the elements that are covered in the manual. A sample of one of the sub-topics is reproduced here to show the level of detail that goes into such an award-winning program. A sample of the letterhead as described in the manual is shown here. By giving all the print and AV communications an integrated look, the overall image of the firm appears organized. Consistency eliminates the need to start all over each time a new printed piece is needed. By using the guidelines in the manual many of the decisions are already made.

Stationery

**B1** Standard Letterhead

**Corporate Signature** Use reproduction photostat size A, located in section H1 of this manual.

**Typesetting** Use reproduction photostat size A, located in section H1 of this manual for the address and phone numbers. In the event of an address or phone number change, use 8 point Helvetica regular, all capital letters. See section A5 for more typography information.

**Punctuation** The corporate signature uses a comma after Engineering and a period after P and after C. The address punctuation is limited to periods after P and after O in P.O. Box, a comma between the city and state, hyphens in the zip code and phone number, and parentheses around the area code.

**Print Colors** The corporate mark is printed in PMS 535. The corporate name and address are printed in PMS 424. Color samples are located in section H2 of this manual.

**Paper Stock** Simpson Starwhite Vicksburg, 70 lb. text, Tiara Vellum.

**Printing Method** Offset lithography.

**Typing and Format Specifications** Follow guidelines in sample letter shown below (use courier 10 pitch; courier 12 pitch is acceptable when fitting long letters on the page).

**Short Letters** On extremely short letters, the date is centered on the corporate mark. The rest of the letter follows the specifications outlined in the sample below. The objective is to have the letter centered on the page.

**Note:** The sample letter on this page is 62% of the original size, but the position specifications in the margin are the actual full-size measurements.

**Page Size** 8-1/2" x 11"
**Grid Size** 6-1/4" x 9-7/8" (59 picas)
**Basic Unit** 1/4" x 1 pica

The grid system is especially important to a corporate identity program because, along with the corporate signature, colors, and typography, it helps to give all of the printed pieces a cohesive, organized look. Every element on Einhorn Yaffee Prescott's forms is positioned in an exact place on the grid.

**A**
**Basic Identity**
Corporate Mark
Corporate Signature
Basic Standards
Corporate Colors
Typography

**B**
**Stationery**
Standard Letterhead
Second Sheet
Standard Envelope
Executive Letterhead
Executive Envelope
Business Card
Clip-On Card
Large Mailing Label
Small Mailing Label

**C**
**External Forms**
Basic Standards
 The Grid System
 Transmittal Letter
Print Order
Purchase Order
Invoice
Memo Letter
Window Envelope
Project Report
Check
Field Report
Programming
 Questionnaire
Furniture Type Chart
Sketch Sheet
Meeting Notes

**D**
**Internal Forms**
Work Pad
Estimate Sheet
Typing/Xerox Request
Travel Information
 and Advance
 Request
Travel Expense Report
Reimbursement
 Voucher
Mailing List Data
 Entry
Telephone
 Memorandum

**E**
**Presentation Materials**
Proposal and Report
 Cover
Proposal and Report
 Internal Pages
Working Drawing
 Title Block
Presentation Drawings
 Title Block
Title Sheet
Title Slide
Interiors Board

**F**
**Promotional Materials**
News Release
Pocket Folder
Magazine Reprint
Advertising
Photographic
 Enlargements for
 Office Display
Newsletter
Case Study

Corporate ID 11

# 2 Anatomy of Formats

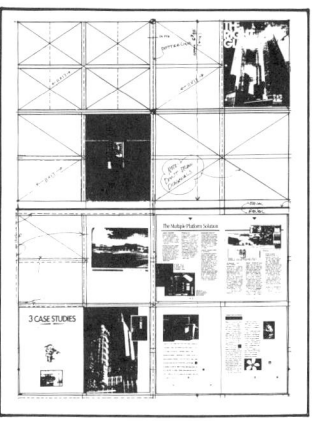

## Anatomy Of A Grid

Graphic designers have been using grids ever since the printed page was divided into more than one column of type. A grid is the placement of one column of type or photograph next to one another throughout a publication. It is an elementary concept, but it can be elaborated into complex arrangements.

Each spread of a publication must be regarded as a single entity since it is viewed together. The first determination of a grid occurs when the margins are drawn in. In the first example below, showing minimal margins, the top, bottom, side and gutter margins are very thin, leaving a wide column inside. The illustration show the variations that are possible changing the head margin first, the foot margin next, and finally each of the side margins. The lower illustration shows two spreads. The one on the left represents a "mirror" grid, where the margins are mirrored around the center of the spread. The one on the right shows a "repeat" grid. The grid on the left page is simply moved over and repeated on the right page in the same aspect.

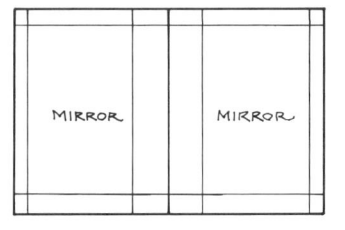

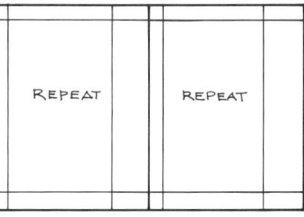

The design of each page must compliment the other. In many cases the design includes both pages, as some photographs go across the gutter margins to the other page. Also, consecutive pages become part of a three-dimensional format as the reader turns from page to page. The majority of examples in this book follow the standard 8 1/2 x 11" format. The subdivisions of the grid is proportioned to this dimension as well as the 11 x 17" open spread. Any brochure, magazine, catalog, newsletter, or project page could be shown in several different sizes and still be consistent. The design of the grid related specifically to the design of the material that was to be shown in this book. Minimum top, bottom, and side margins set the outer boundaries as wide as possible, to display the examples as large as possible. The case study grids were shown as large as possible on the page leaving room for a single column of type. The generic grids were shown actual size using the same margins throughout. The horizontal rule separated the page into two equal spaces so that two examples could be shown on each page.

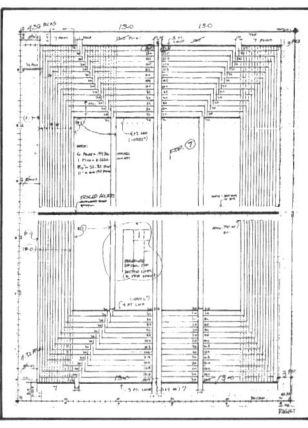

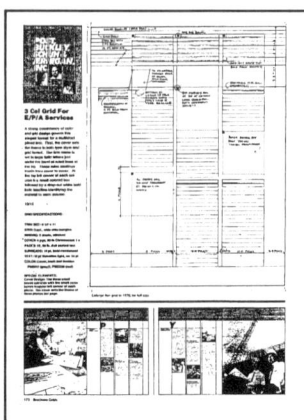

The grids depicted in the case study section of this book were developed from existing material. The headline, text and column outlines are accurate renditions of each page layout. Although the trim size for all case studies is 8 1/2 x 11 inches, the interior dimensions, column widths and margins, are noted in picas and points. Most desktop publishing programs can accommodate either inches, picas or metric measurements. The pica measurement was chosen because it is used most often by graphic designers. The anatomy of a printed page with discussion about type size and style has filled many chapter in other books, which are highly recommended as references. The hand written notes that will be found throughout the case study section are descriptive of the type size, style, and spacing. The type can be set on 1/2 point increments of leading as the chart below indicates. Use this as a rough guide in laying out the spacing. There are special rulers called Haberules for marking leading. These and other rulers for inches, picas, and metric measurements are available in most art supply stores.

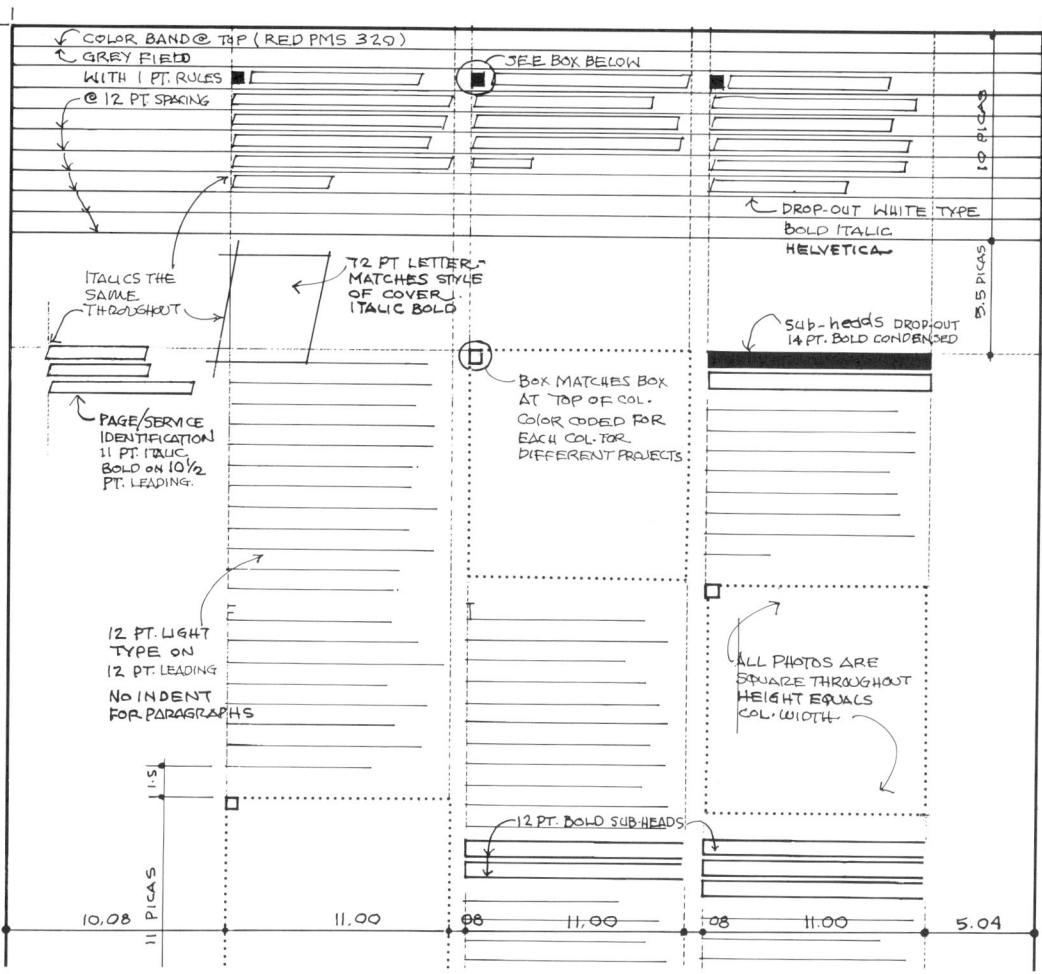

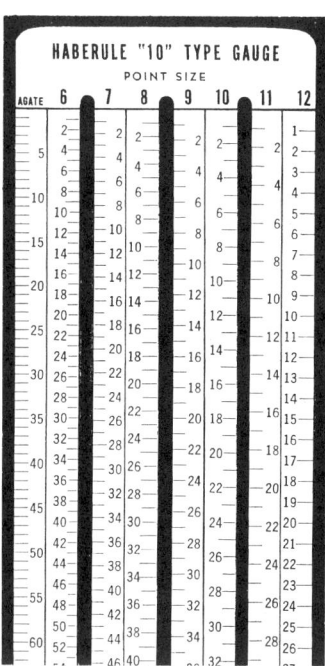

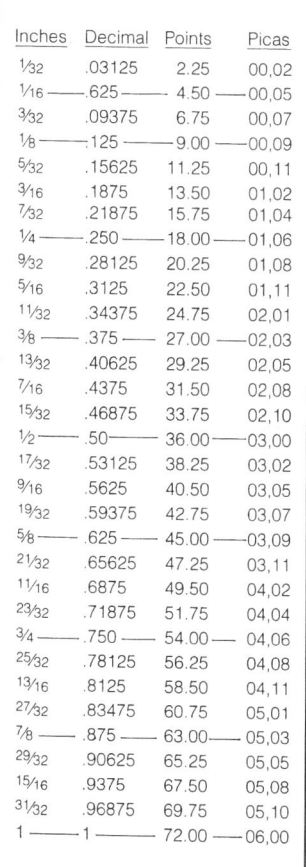

The size of a typeface is defined by its body. Type set on its own body size is said to be "set solid." A 10 point type with 2 points of additional space between the lines is specified as "10 on 12 point leading."

The unit of typographic measurement used to specify column widths and margins can be given in either inches, picas and points, or millimeters. Inch measurements are conveniently close to the point system. One pica is 0.166 of an inch, so 6 picas are only slightly less than one inch. This is particularly useful when calculating margins if the trim size is given in inches.

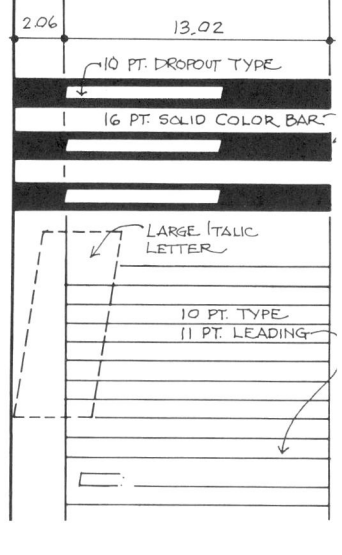

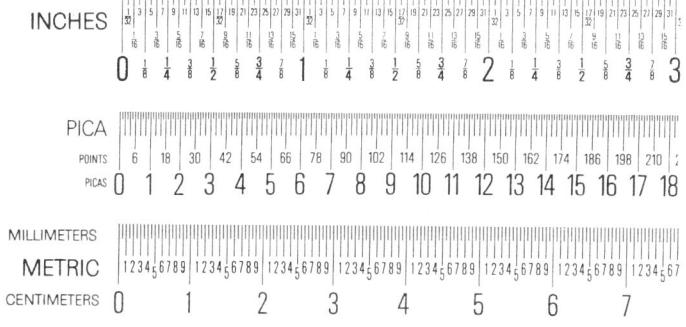

Anatomy Of Formats 15

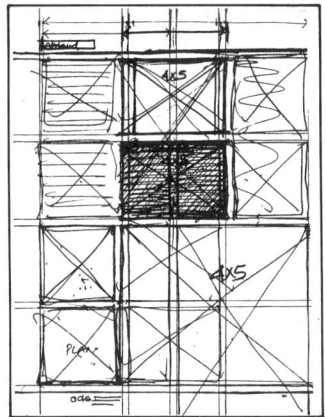 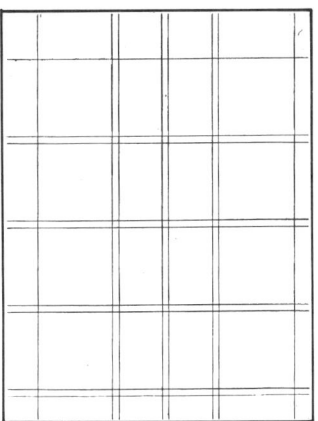

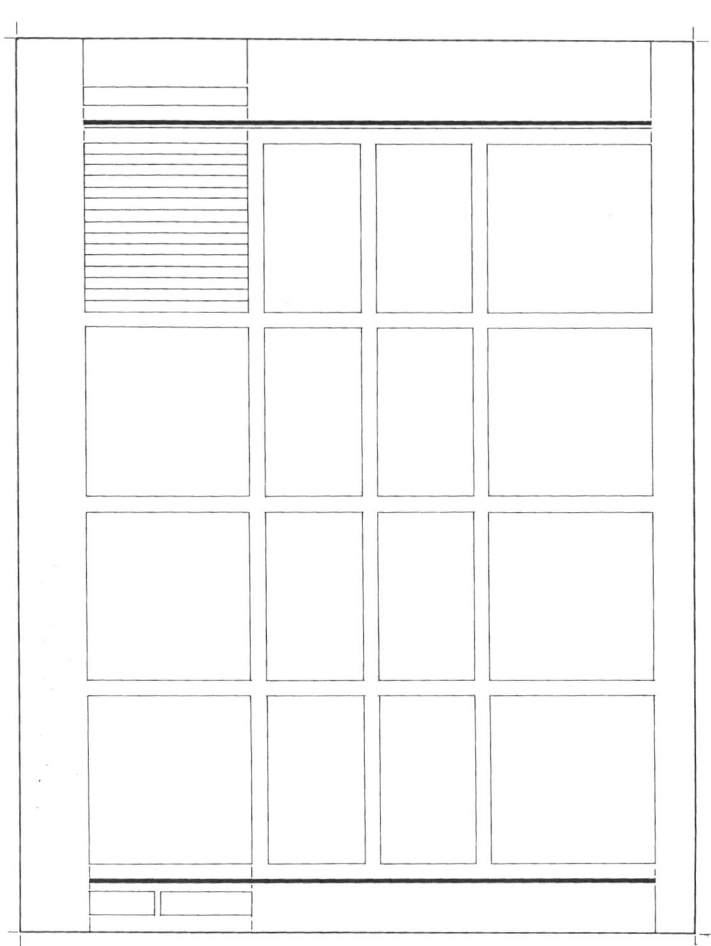

## Designing Grids

Grids can be developed from many different and divergent beginnings. In this case the design firm had an archive of photos which were taken with two different camera formats. One was square, and the other a 4x5 format, which was perfect for making 8x10 color prints for their portfolio. The grid had to accommodate both formats and in the case of the 4x5 format, had to be useable both horizontally and vertically. The full value of the photo had to be used, only minor cropping would be allowed. After many rough sketches were made, a format began to evolve which would work with the combination of these two camera formats. Rules were added to the top and bottom to emulate layouts found in interior design magazines.

 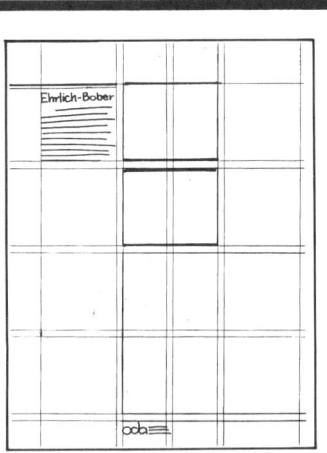 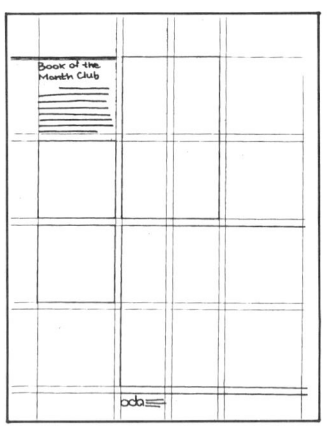 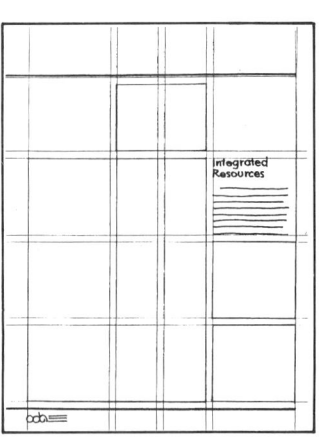

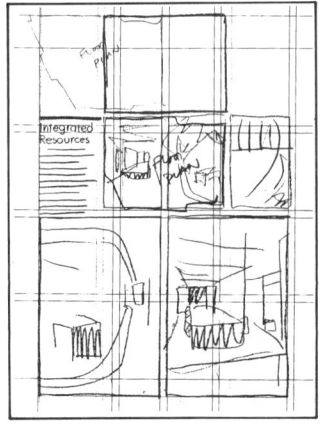 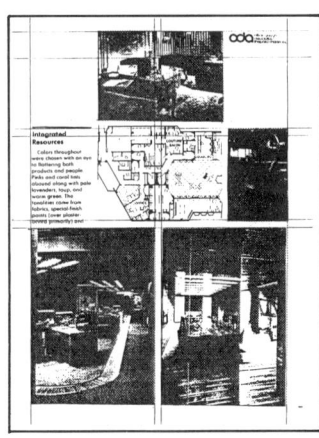

Once the design grid was developed, sketches could then be made to determine how the achieve of photos could be used to create the individual project sheets. After the initial sketches were made the selected photographs were then enlarged or reduced to fit the grid patterns selected. Floor plans were also reproduced in small scale and a composite paste-up of a photo-dummy resulted. Now each page could be studied and altered very easily without a large investment of money.

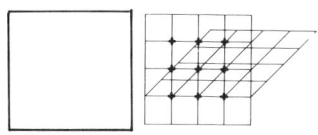

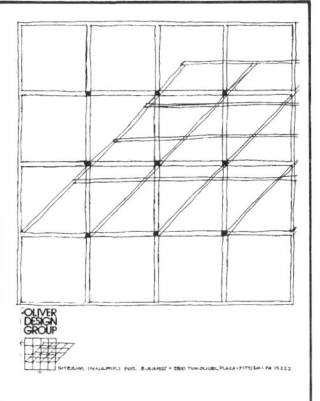

Sometimes there are elements within a design firm that naturally lend themselves to adaptation as a grid design. Such was the case here, wherein the firm's logo was already a representation of a grid-like design. This design was simplified and became the basis for a four unit wide by five unit high square grid module. A six point rule was added to the top of the grid to encompass headings. Early sketches were made of various layouts which included a list of services, a list of representative clients, and other project related pieces. The series of project pages developed from this grid are shown in the case study section.

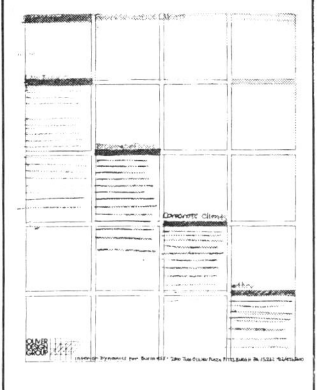

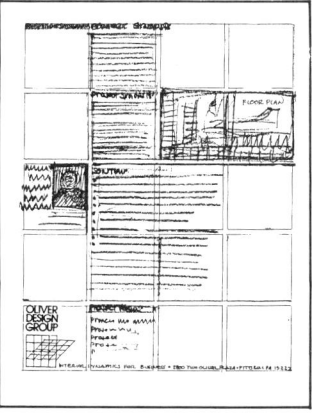

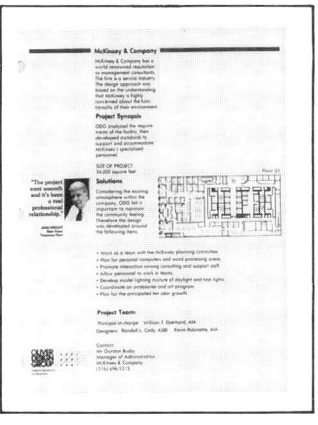

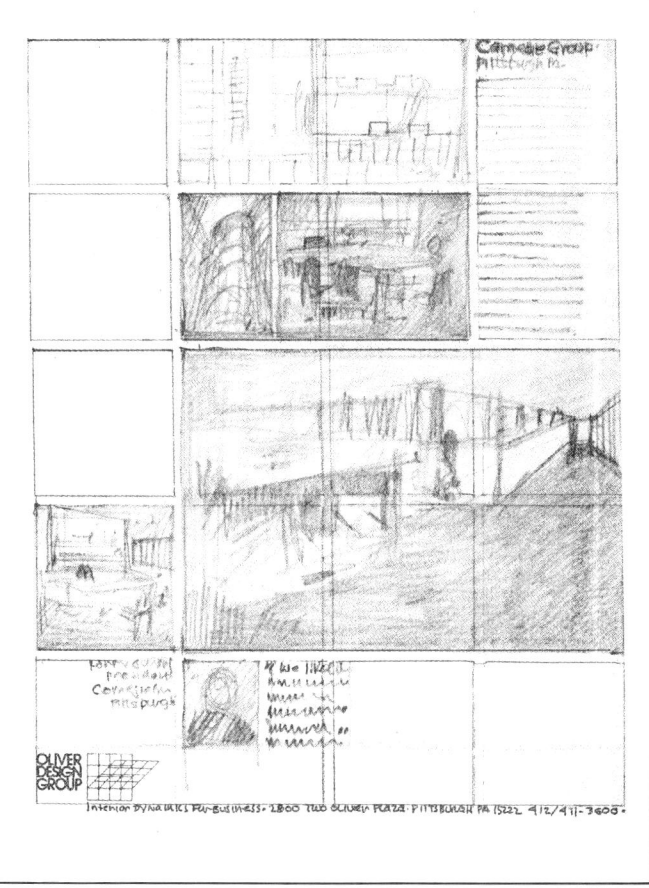

Anatomy Of Formats 17

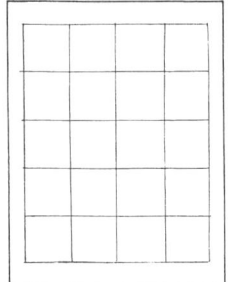
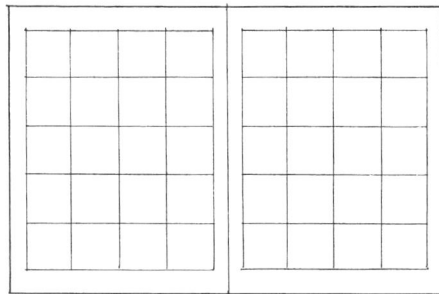
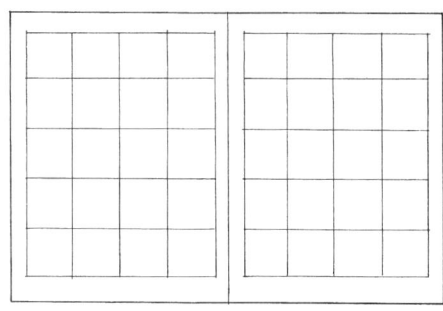

### Mini-Grids

One of the most valuable aids in planning page layouts is the use of mini-grids or thumbnail layouts. These are designed to fit on an 8 1/2" x11" page with the prearranged formats reproduced in miniature on the page. Using these mini-grids can assist in sketching out the cover, inside spreads, and back cover all on the same sheet of paper for easy viewing. The images can be moved and revised much more easily at this stage than on a larger or full size scale.

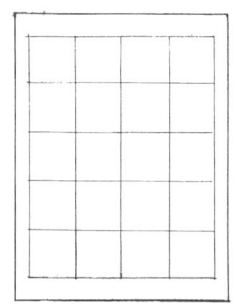
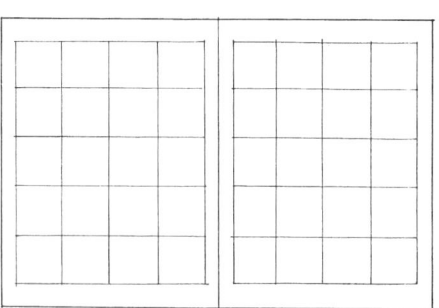
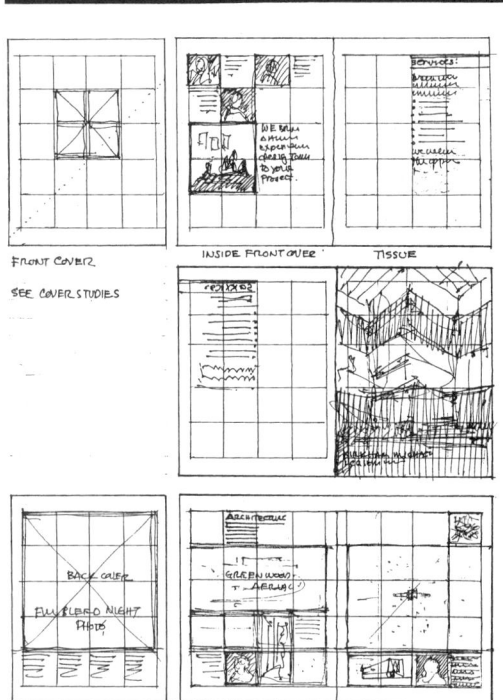
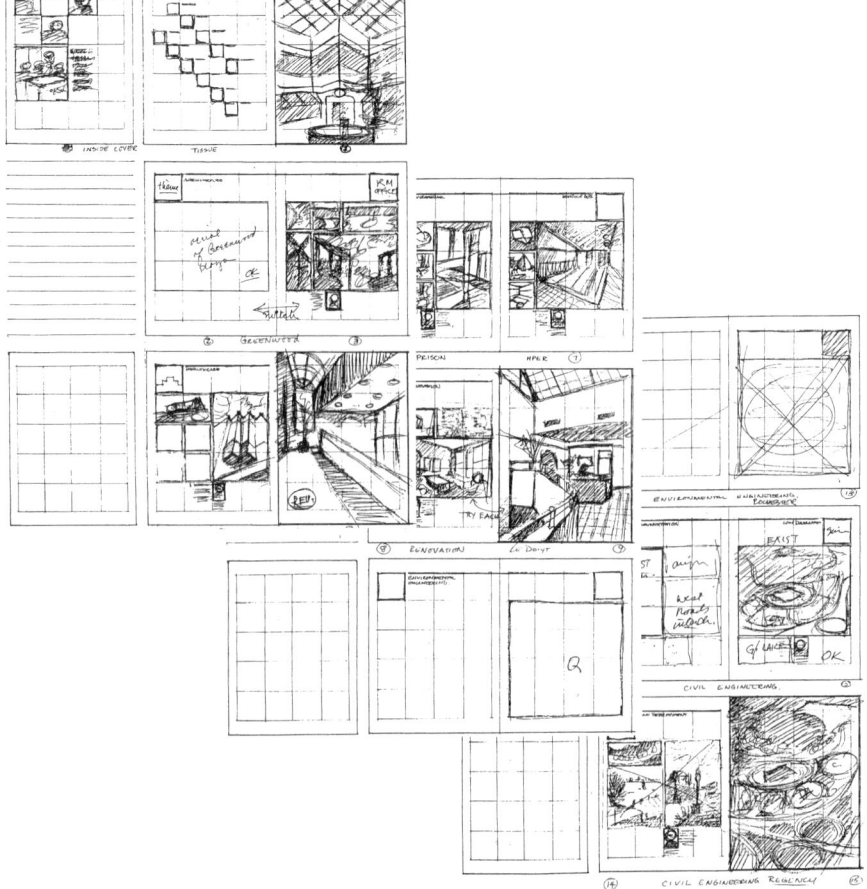

The mini-grid shown on this page is designed especially for the three panel foldout format. It therefore shows the cover and back cover, plus the first and overlay page, then the full three page unfolded layout. Concepts can be sketched out here and then further developed through a sketch dummy or photo dummy stage. The illustrations here compare the original concept sketch and the final photo dummy superimposed on the grid layout.

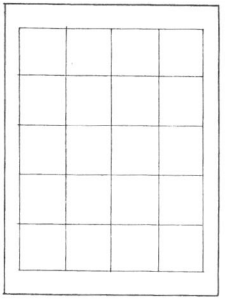
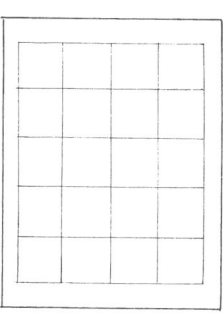
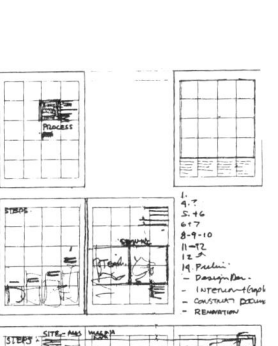
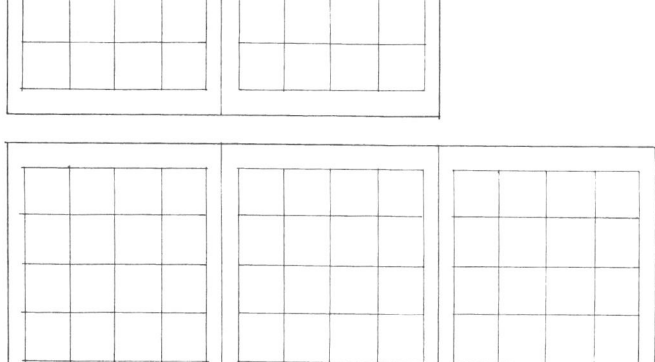
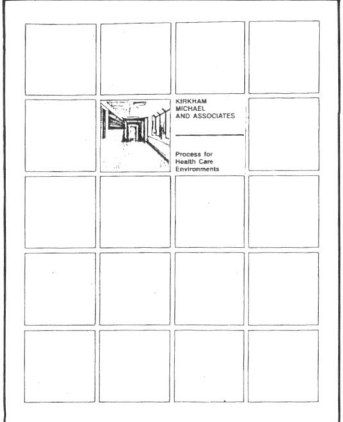
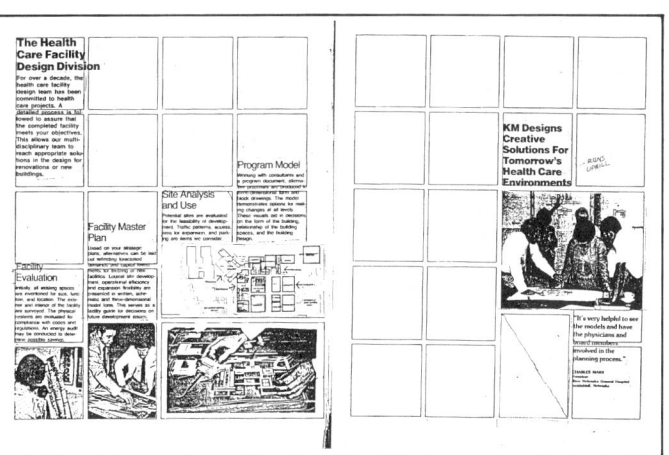
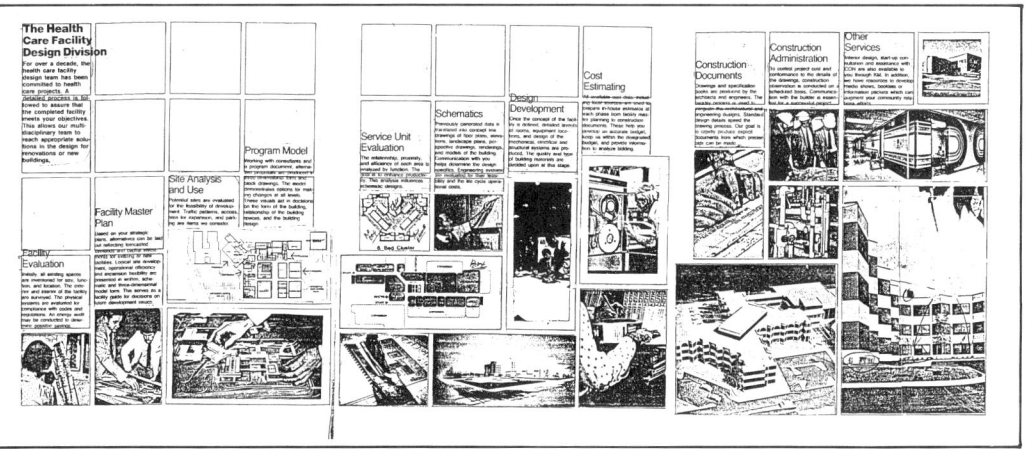

Anatomy Of Formats 19

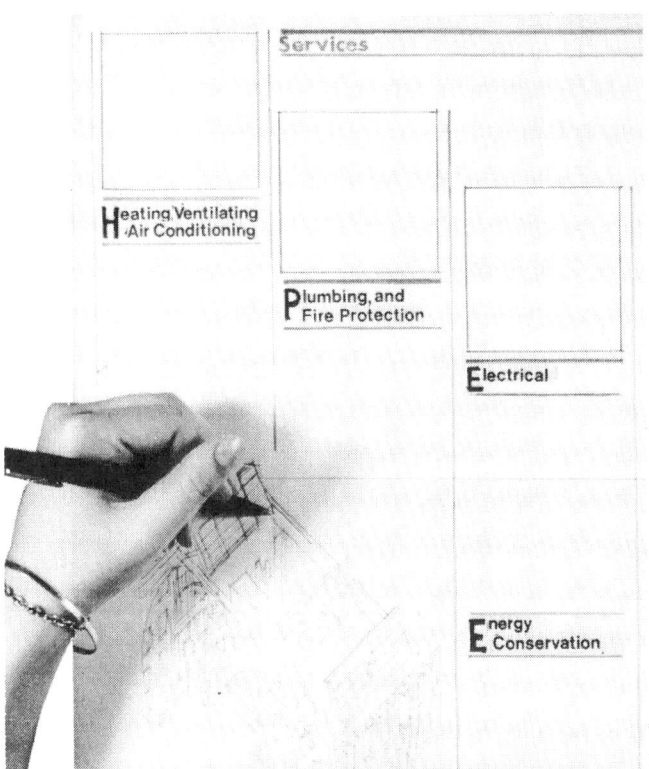

## Sketch Layouts

Few brochures go from concept to completion without numerous stages of design and revisions. Initially, the rough ideas are sketched out on tracing paper, oftentimes with the aid of a grid underlayment. By beginning with a grid, it is easier to establish design parameters, and to work out the placement of photographs. Design study sketches are usually done at the same size as the finished piece. However, if there are certain elements that could be studied more easily at large scale, simply photocopy them to the desired size, do the studies, then reduce them again. This is helpful for studying headline lettering in relation to photographs and other details. The illustrations here show two examples of the rough sketches compared to the final printed work. The first is a spread highlighting a mechanical engineering firm's services. The bottom example shows the three partners of the same firm and their credentials printed beneath the photo. It follows the same stepped design as the main part of the brochure.

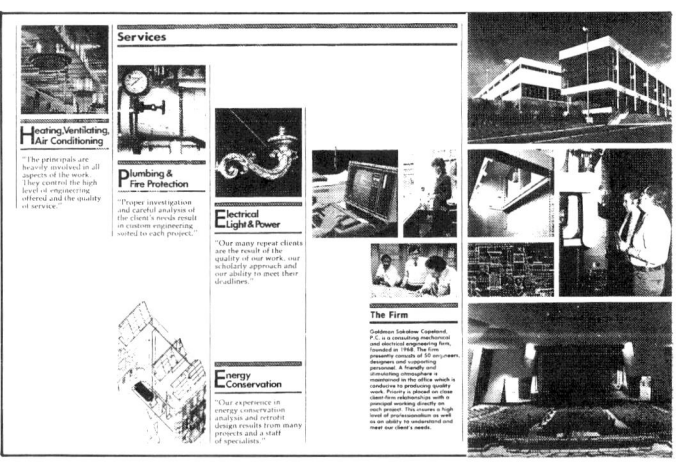

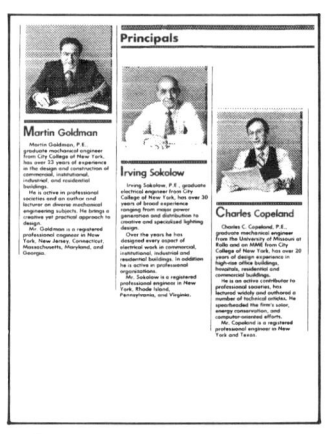

The sketch dummy can be produced from many sources and can incorporate many items into the stages of development. This includes pieces of type, either actual or dummy type or art from other sources. Sketches can be made directly over photographs that exist to quickly see if they fit the desired pattern. Compared here is the initial sketch for a page in a brochure and the final printed version.

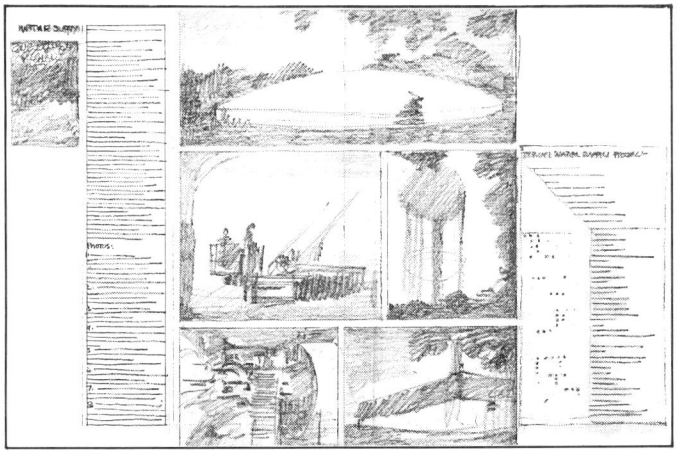 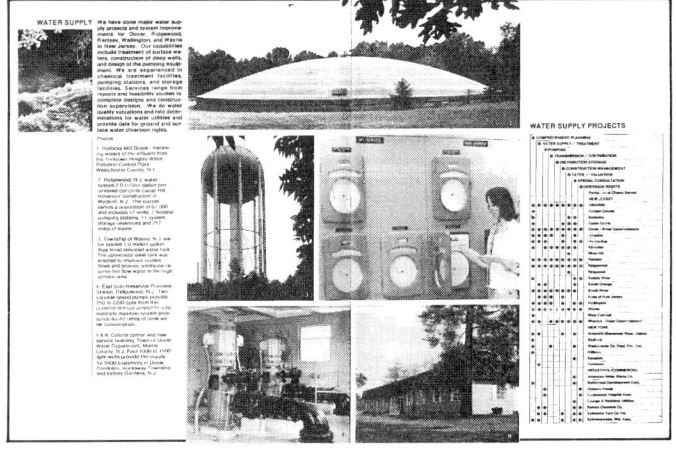

 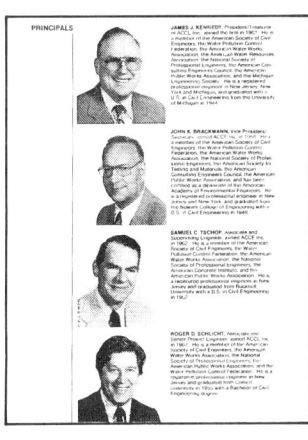

Blocks of type can be denoted by free-hand lines and photos can be quickly sketched in. If the photos exist, they can be enlarged or reduced to the size of the sketch dummy and then traced over directly for exact proportions. If the photos are not yet taken, the sketches must be approximate to the views anticipated. The illustrations shown here compare the original sketch for a spread in an engineering brochure and the opening page with photos of the principals.

Anatomy Of Formats 21

## Collateral

All the promotional print material for this firm was designed on this four column grid. It is based upon a square module and the overall grid is four units wide and five units high. Since all the firm's material was to be interchangeable within pocket folders, the design of the covers was a critical factor. Various cover designs were studied in relation to the square grid module. This was done to ensure that the diagonal flap of the folder did not obscure the name of the firm when it was carried in a folder.

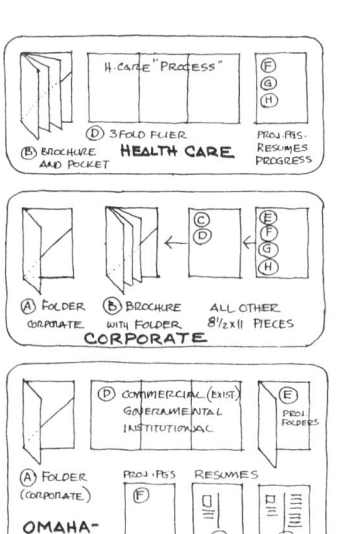

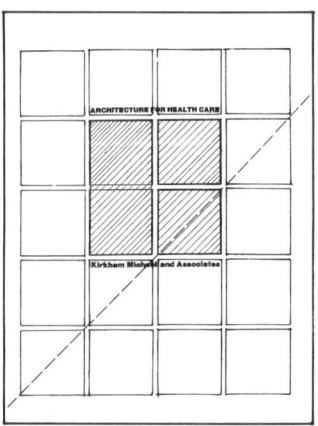
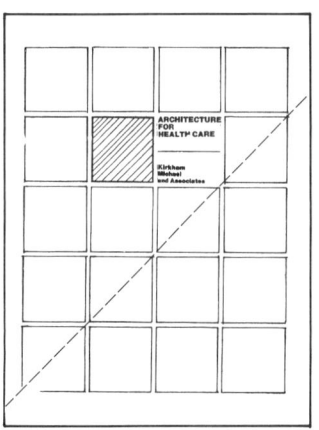
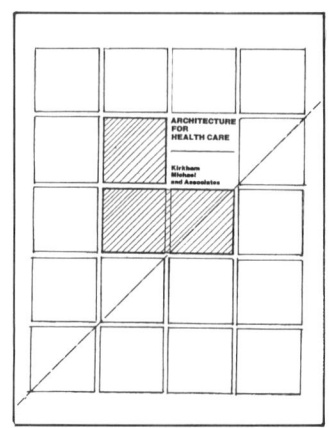
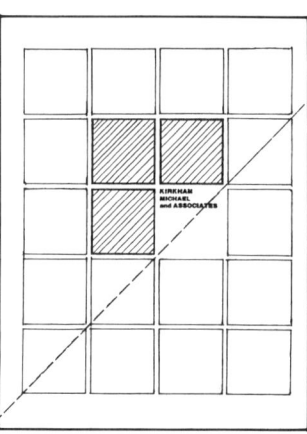

Once the grid format was finalized, it was used for the design of many pieces of the firm's promotional material. This four column format worked very well with their four color newsletter. It provided a base for a dramatic use of photos while still leaving room for the text.

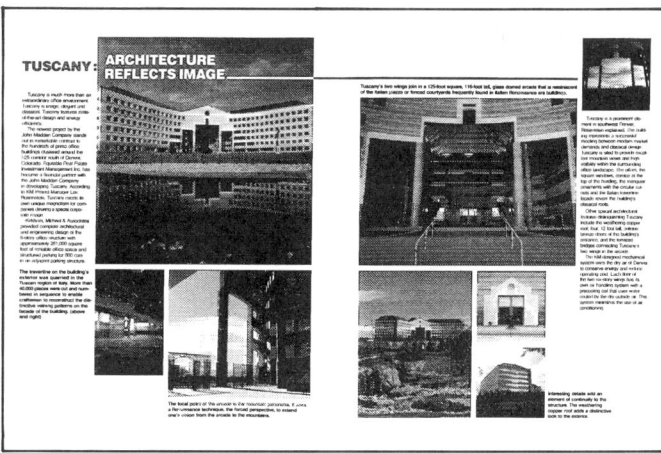

A series of specialty four page folders was also designed on this grid. The cover featured a full bleed black and white photo. Inside the grid allowed a flexible arrangement of both text and photos.

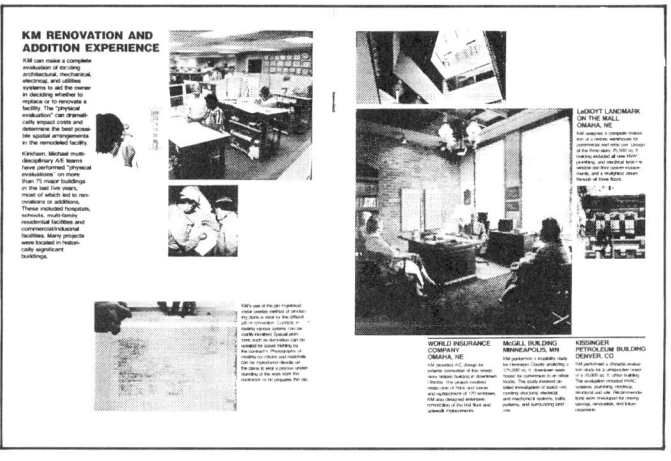

Although this specialty brochure is printed on a black background, the grid is still evident in the strict geometry of the photos. Each project is accompanied by a photograph of the client, ad the project is described in a brief quotation from this client.

Color project pages were also designed on this grid. Although most of them have a liberal amount of photographs, they have minimal descriptive text. These project pages were used in conjunction with pocket folders, or bound into proposal packages.

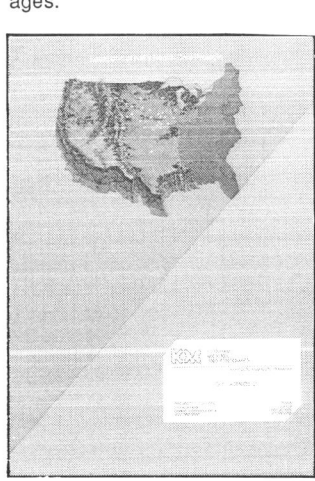

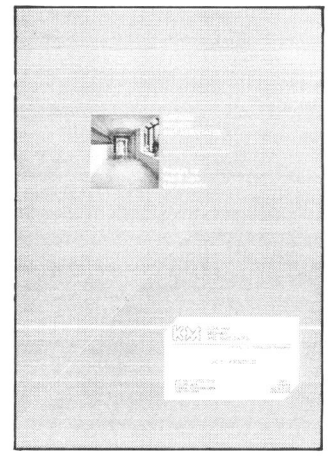

Anatomy Of Formats 23

## Advertisements

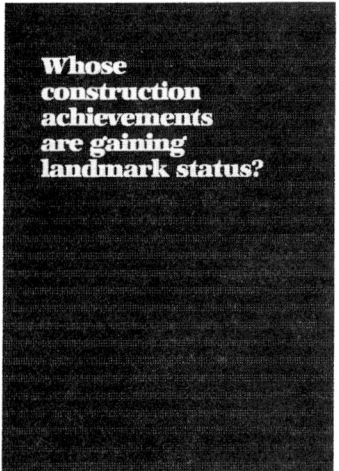

Some advertisements are designed for magazines, some for direct mail and some for both. Reprints as single pages or folders are very common and very effective. Here a non-specific advertisement in a trade journal becomes targeted to a specific person by a direct mail program.

In most cases, advertisements are not designed as single entries but as multiple variations on the same theme, at the very least in groups of three. The majority of the ads shown here are for large construction companies or construction management firms. It is interesting to note the elements that comprise most of the advertisements and to see how they relate to the industry.

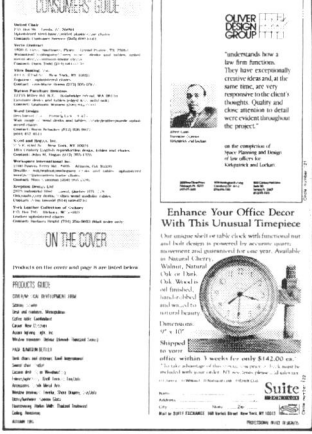

This series of advertisements was developed to be placed in trade journals read by the clients of the design firm. Therefore, the theme was developed around the use of comments from other clients about the design firm's services. A series of informal interviews were scheduled, and tape recorded conversations were transcribed for review by the design firm. Passages were then selected that could be used for these ads, as well as for other promotional purposes. Clearance was obtained for the use of this material, and a design was developed which featured the client, the quote, the name of the project and the office locations of the design firm. They were displayed as one-quarter-page ads and repeated on three consecutive upper right hand corner pages to increase exposure in the same issue. Simple but very effective.

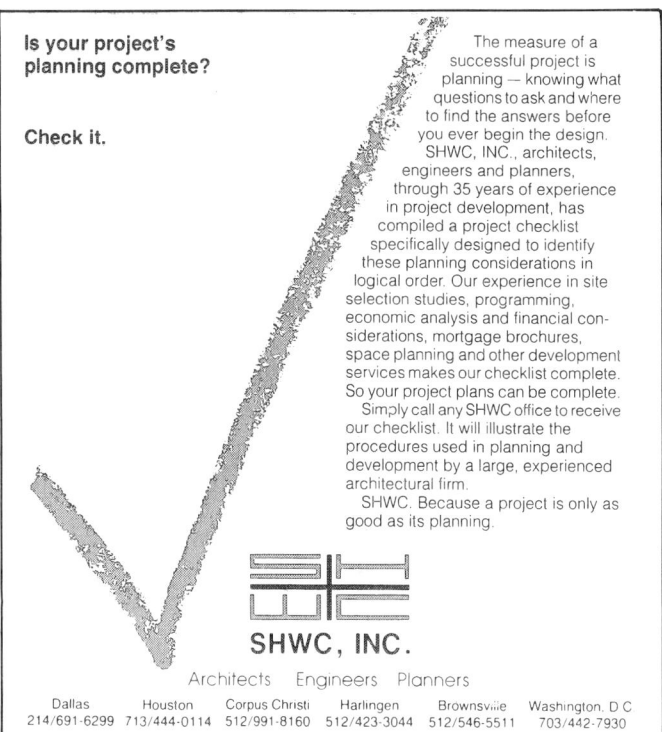

The ads depicted here were developed to attract large corporate clients to this growing design firm. They were designed as full page advertisements for *The New York Times*. Each one featured a partner of the firm standing alongside an enlarged photograph of a model of one of their projects. The copy was very client oriented, yet mentioned the design firm's stand on design issues and its service to clients. These advertisements did a lot for projecting this firm into the big leagues.

These advertisements were designed as quarter pages in client read trade journals. They were all based on creating a message using graphics. In this case a "check mark" was used in bright red, the same color as the firm's initials in the monogram which appeared along with it. Other marks included exclamation points, quotes, and circles. The copy related the mark to the design firm's services in each case.

Shown here are two versions of a similar advertising campaign, showing three examples from each. The first uses the device of cartoon sketches, yet depicts serious concerns of their clients. The copy reassures the reader that these problems will not occur.

The second set uses a humorous approach as well but using photos of real people. The top photo depicts the negative issue, and the bottom portrays the solution. There is a picture of a friendly construction person with a reassuring look that the stated problem will not occur.

Anatomy Of Formats 25

## Announcements

When new people join a professional organization, particularly in the design industry, it is a good opportunity to publicize the event. Although a press release may suffice for some firms, a well designed printed announcement is more effective. They can be quite simple in design; on a single folded card, or even a single card. They usually contain a photograph of the individual joining the firm and a brief accounting of their prior experience.

These announcements are sent out on a regular basis and they all follow the same format. The card is single sided and the number of individuals shown varies each time even if the format does not.

BALTIMORE • DALLAS • WASHINGTON • FORT LAUDERDALE • LOS ANGELES

**RTKL**  With great pleasure, RTKL Associates Inc. announces the naming of four new Principals.

| James R. Baker AIA | Laurent Myers AIA | Michael A. Sichel PE | Paul C. Zugates AIA |
|---|---|---|---|
| *Interiors* | *Interiors* | *Engineering* | *Health Care* |

BALTIMORE • DALLAS • WASHINGTON • FORT LAUDERDALE • LOS ANGELES

**RTKL**  With great pleasure, RTKL Associates Inc. announces the naming of four new Principals.

*James R. Baker AIA* — *Interiors*  
*Laurent Myers AIA* — *Interiors*  
*Michael A. Sichel PE* — *Engineering*  
*Paul C. Zugates AIA* — *Health Care*

This firm chose to personalize this announcement card which measures 4"x 6". It carries the usual photograph, but a very thin rule is used to frame the inside of the picture. This device is similar to the design of the cover to the packet that holds this card. Above the photo is the name of the individual and their signature at the head of the card. This adds a personal touch to the announcement without which it becomes just another card.

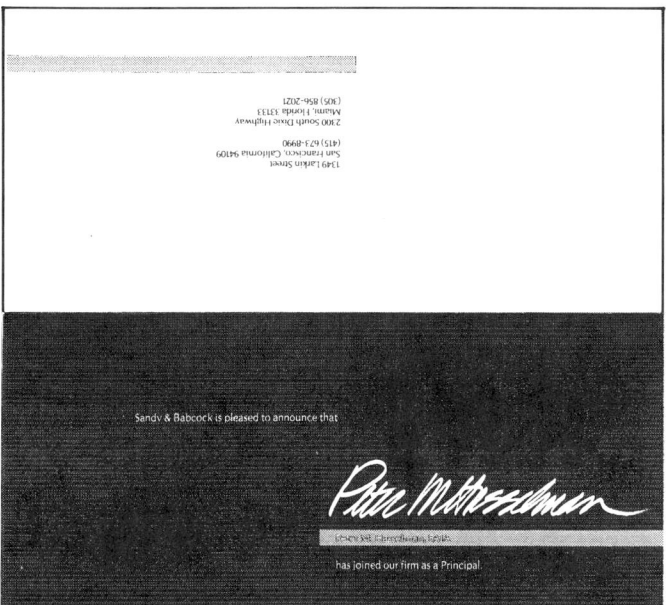

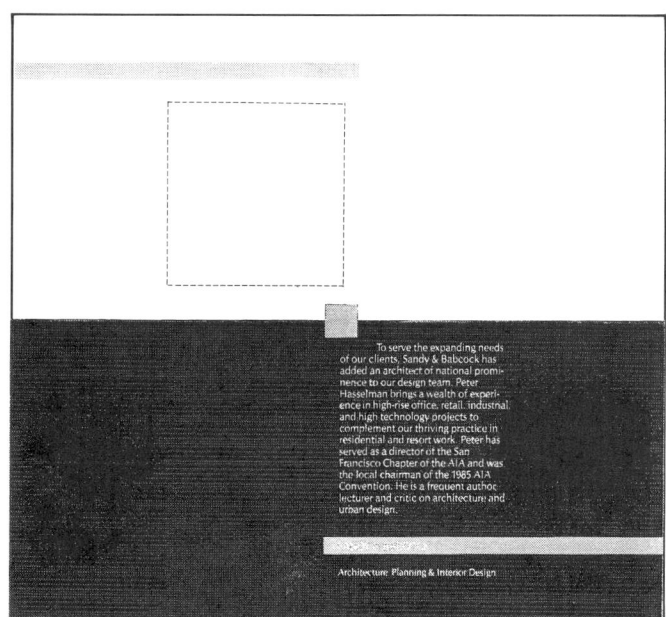

The design of this card is by the same architectural firm as the one above and it uses some of the same elements such as the personal signature. But here it appears on the outside of the card (which fold in half) and the photo is on the inside of the card. The other design elements are repeated inside and outside, such as the positions of the heavy rules.

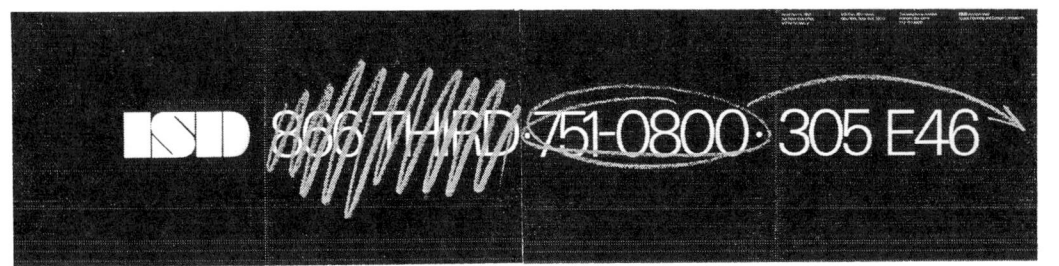

### Announcements

One of the most appropriate times to contact friends and clients is when the firm moves to a new location or holds an open house. The example here is simple and inexpensive. The card by ISD begins as a square and unfolds to the oblong shape shown. It is done simply with white drop-out type from a black background. The card below uses a pop-art theme on the cover, combined with a metaphoric "yellow brick road" leading to a city skyline - the location of their new regional office.

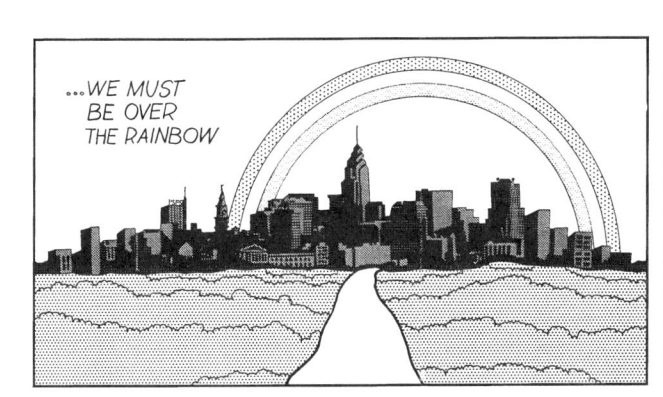

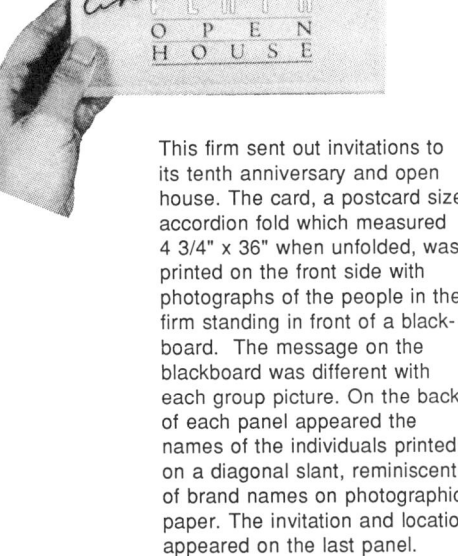

This firm sent out invitations to its tenth anniversary and open house. The card, a postcard size accordion fold which measured 4 3/4" x 36" when unfolded, was printed on the front side with photographs of the people in the firm standing in front of a blackboard. The message on the blackboard was different with each group picture. On the back of each panel appeared the names of the individuals printed on a diagonal slant, reminiscent of brand names on photographic paper. The invitation and location appeared on the last panel.

This unique promotion was designed by a marketing consulting firm just opening the office for the first time. The '"doors now open" theme came with a card, which held another smaller package in a "mail slot" on the large photo of the door. The smaller package opened up in an accordion folded design to a panel 3 1/4"x 25" long. A series of doors, each featuring different messages, unfolded with the name of the new firm at the end.

A simple card with a holiday theme and message about their new location. The blocks spelled out the name of the firm.

One of the most appropriate times to contact friends and clients is when the firm moves to a new location or holds an open house. The examples here are simple and inexpensive. The card begins as a square and unfolds to the oblong shape shown. It is done simply with white drop-out type from a black background. The card below uses a pop-art theme on the cover, combined with a metaphoric "yellow brick road" leading to a city skyline - the location of the firm's new regional office.

Anatomy Of Formats 29

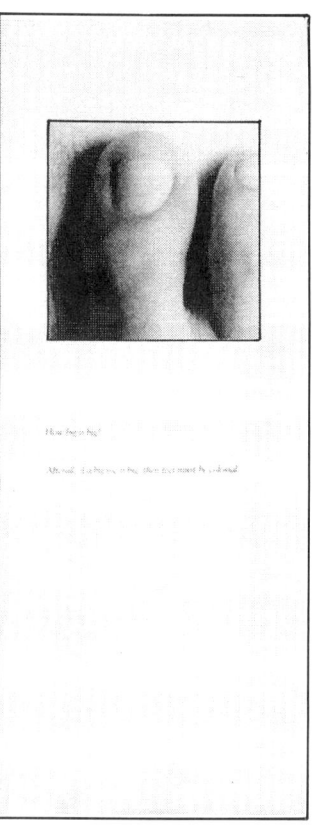

## Accordion Folders

An accordion fold in a mailing piece is a very effective way of keeping your message in front of the viewer in the proper sequence. A single long sheet of paper is simply folded like an accordion after printing. The size of each panel is determined mostly from the preferred size of the envelope, such as the standard letter size. The theme can be quite simple as in this example which features one square picture per panel and just a few lines of type under each picture. This panel unfolded measures 11" x 30".

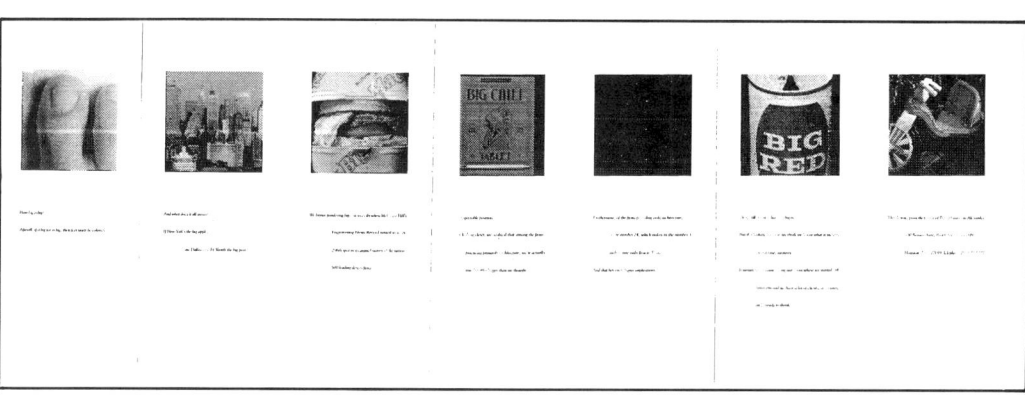

One perfect shape...
the essence of art.

An accomplishment in architectural artistry, 1 North Main is a monument to perfect shape and form. Its reflective glass and bronzed brick exterior harmonizes with the surrounding environment, embodying the cosmopolitan atmosphere of downtown Ann Arbor in a singular combination of corporate, retail and residential suites tailored to the needs of the successful individual.

This series of accordion folded mailers was all designed around a single theme, which is expressed in the opening panel. These were designed to be sent in a standard #10 envelope, so they measure 9 3/8" x 3 5/8" when folded and 9" x 18" when fully opened. By keeping all the images in one panel the message is strongly reinforced. The text must also be cohesive since it is read as a continuous unfolding story.

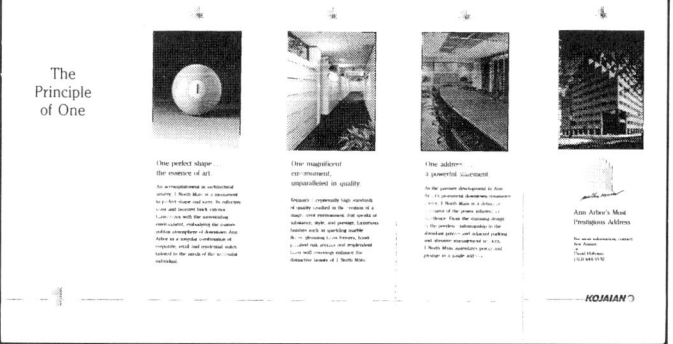
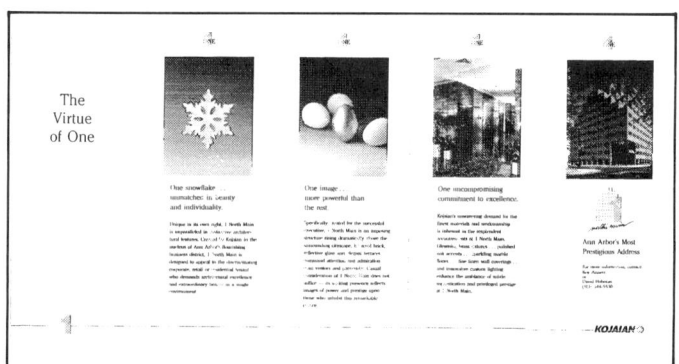

30

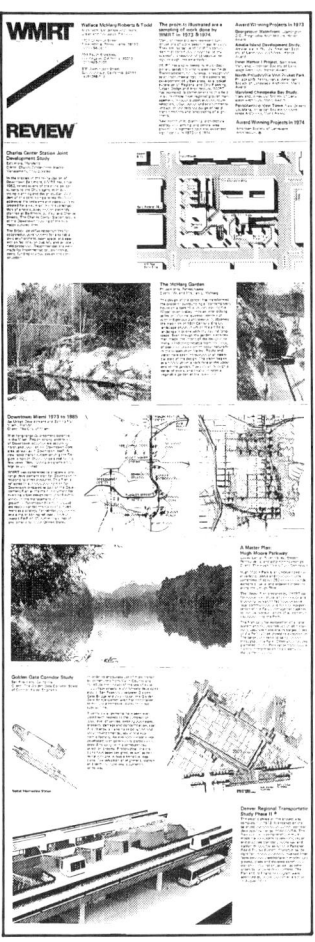

The accordion fold format can be used in a horizontal direction as well, and it unfolds toward the viewer. This is a good format for planners and architects because it allows a flexible arrangement of type and photos. In this example, the layout for each panel is a four column grid. The length of these panels depends upon the maximum paper size that the printer can take and the number of panels desired. They can also be printed on the reverse side.

This design is based on a five column grid, and is split into a ten column grid on the cover to highlight small vignettes which appear inside at a larger size. Each panel is folded on the heavy black rule, and each is a self-contained panel about one particular project. The five column grid allows a lot of flexibility in the layout, even allowing photographs to go outside the grid modules and still maintain a sense of harmony.

Anatomy Of Formats 31

## Fliers

The direct mail flier is a versatile form of promotion. It can be directed to a general audience or a specific audience. Since it is produced by a single folded sheet of paper, it is much less costly than a brochure. The piece is meant to give an overview of a firm's services, rather than present specifics. It is usually printed on 8 1/2" x 11" paper and folded into three panels to go into a standard business envelope. Some fliers contain more than 3 panels and are designed in the gatefold format which opens up to 8 1/2" x 16", but they all fit a standard envelope.

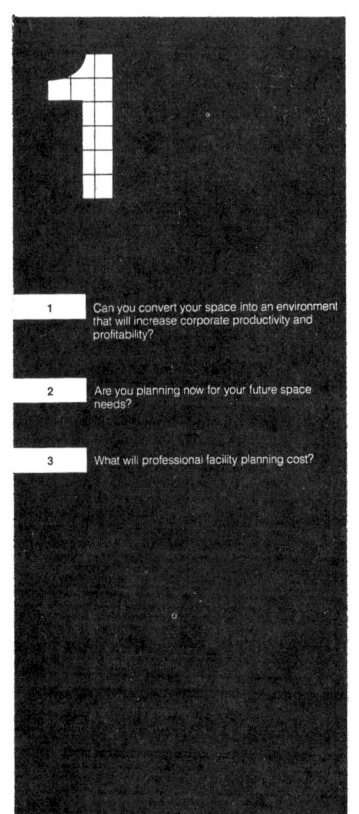
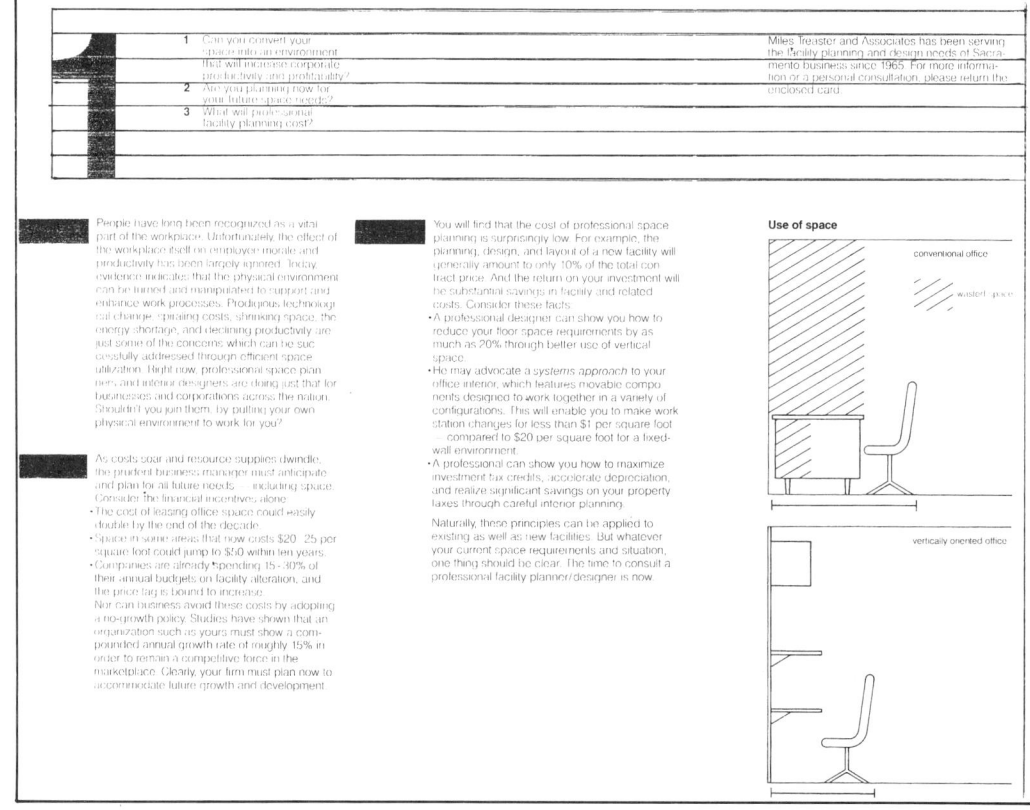

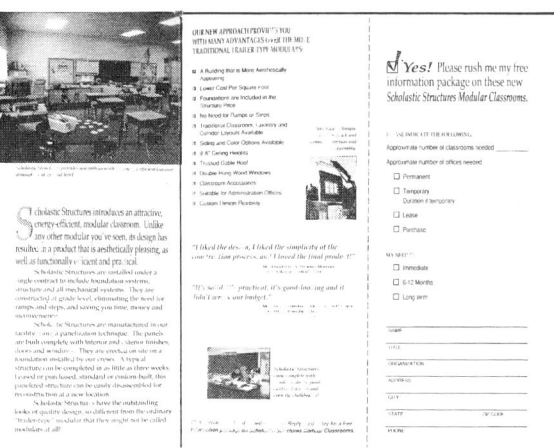

There is a wide range of uses for these pieces, but their most distinctive advantage is the relative cost per piece compared to the bound brochure. Information must be capsulized and summarized, while leaving room for small photographs. Of primary importance is the use of a response mechanism, either on the flier or enclosed in the mailing package.

There are pre-printed papers available today for use in desktop publishing systems wherein the user supplies only the type. The paper is pre-printed with a design and color image so that the only printing required is one plate with black ink. Photos can be included, and the overall effect is a professional looking piece for a minimum investment in time and money.

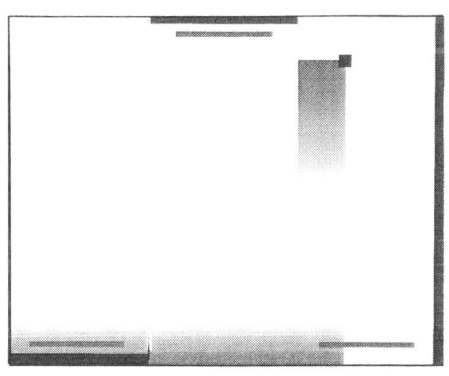

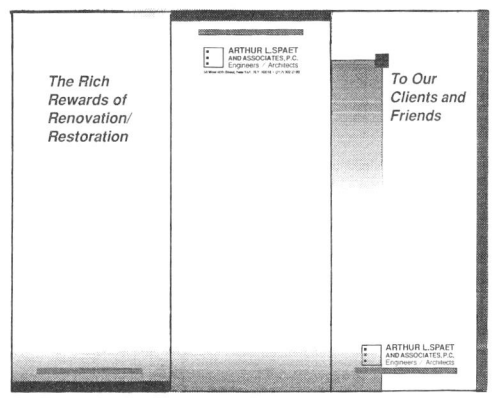

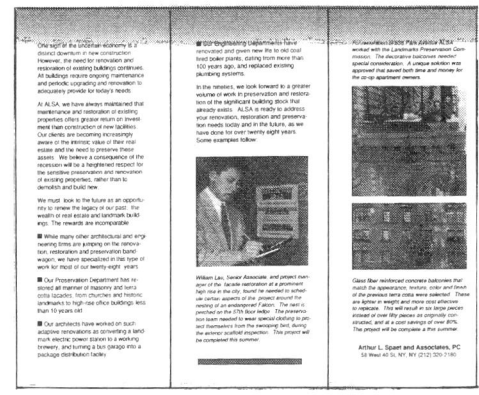

Anatomy Of Formats 33

## Magazines

Many of the larger firms produce magazine-format publications, since the projects featured are large scale and require lengthy coverage. The covers for these publications are varied but usually focus on the large-scale image, using full bleed color photos and dramatic photography. Like their counterpart, newsletters, each has a distinctive name and masthead. Some are extensions of their newsletters into quarterly publications.

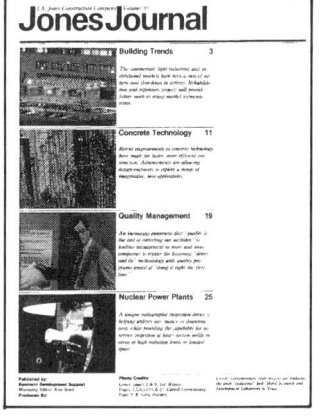
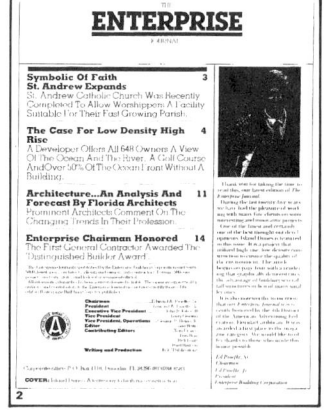

The inside cover page, like consumer magazines, lists the contents of publication, plus a listing of key people and the location of corporate offices. Some contain a brief message from the CEO or Chairman, President, or other officers, often accompanied by a photograph. It personalizes the magazine, just as consumer magazines may have a message from the editor.

On the inside there are numerous photographs throughout the text, just as in other promotional pieces. The majority use a three column format throughout, although some magazines change the format within each chapter or segment. There are numerous examples of these formats in the case study section of this book. The generic grid section provides a number of variations to the three column format as well.

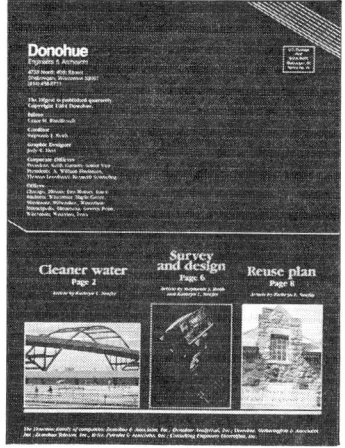

Some formats are less straightforward and employ rules between the columns of type or around the edges of the page. The use of bold headlines and large opening paragraph letters is also common. However, the most distinguishing feature is the use of large dramatic photographs of projects.

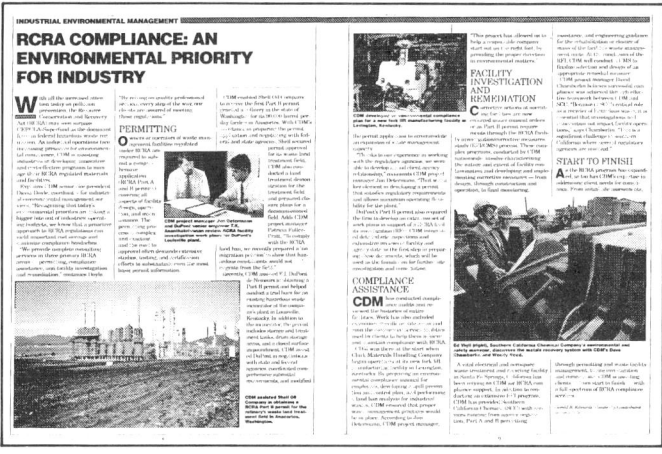

Anatomy Of Formats 35

## Annual Reports

The explicit nature of the annual report is found in the cover. Its primary purpose is to review, explain, tabulate, and review the progress of the company over the past year for its readers. Progress is expressed in many different ways both graphic and photographic. There is no mistaking an annual report for any other publication, since its format must cover, by law, certain aspects of the company's business. It is therefore divided into two distinct sections, one general and one financial.

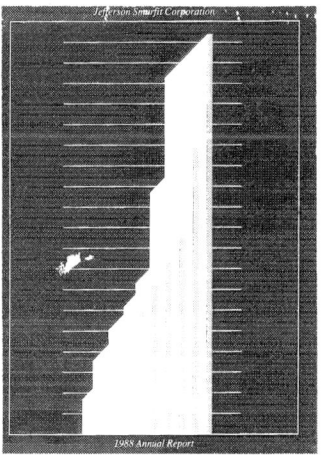

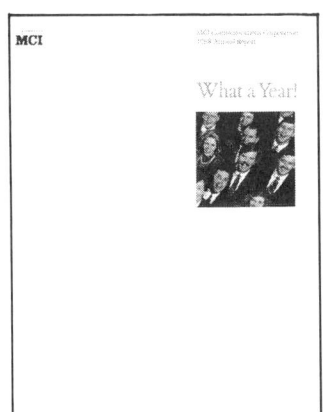 

An important aspect of any company is its management, and it is the purpose of the annual report to show who they are, either in general terms or by specifically listing or naming those individuals pictured. More than half have a signed statement by the CEO along with a carefully posed photograph. Some have adopted less formal ways of showing officers, but it is a very important introduction to the publication.

The general section of the report looks like any corporate brochure, and there are numerous examples in the case study section of this book. The most distinctive aspect of the report is the financial information. This consists of a financial highlights page or section, sometimes covering a ten year period, to indicate growth. It also includes the past year's financial information most often in a single pie chart or bar graph diagram. The trend is towards three-dimensional charts, as they are now easier and less expensive to produce using electronic graphics.

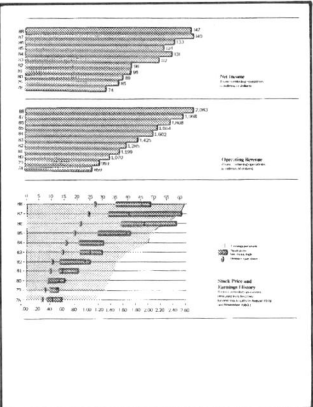
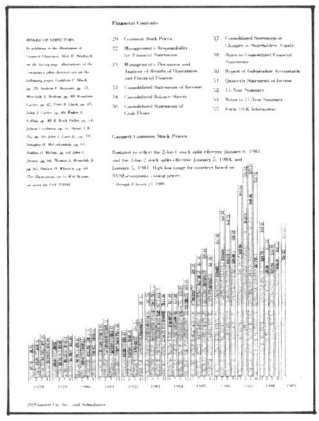
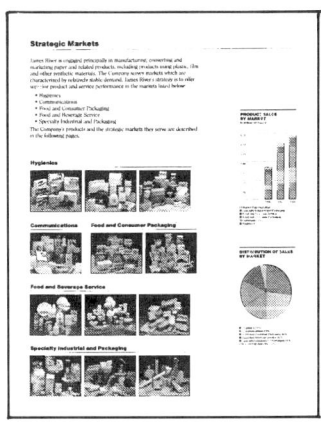
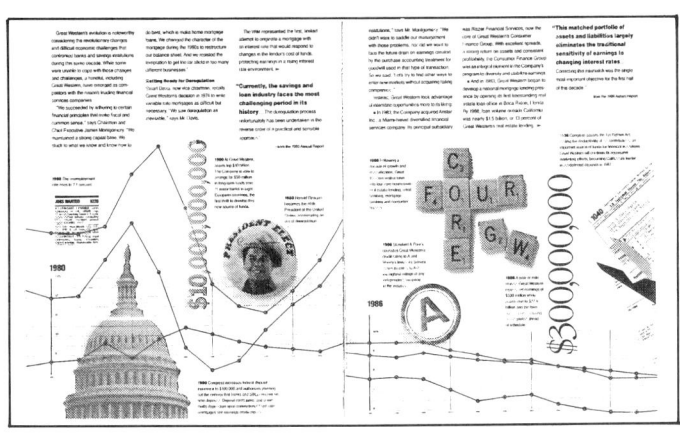
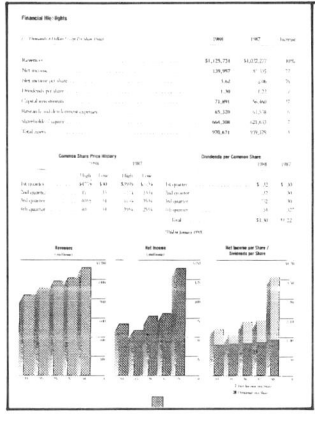
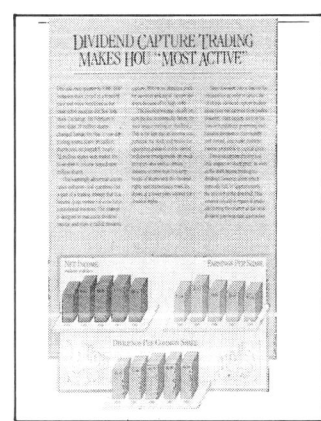

Another specialized area of graphics can be readily found in annual reports. The use of maps is widespread and covers the spectrum of graphic representation. A map of the country or the world is something that we often take for granted. In the annual report, these images display a high degree of graphic ingenuity. The maps show locations of plants, operations, networks, products, markets, or other information that can be graphically illustrated, and tell the reader at a glance the extent of the company's coverage on a national or global scale.

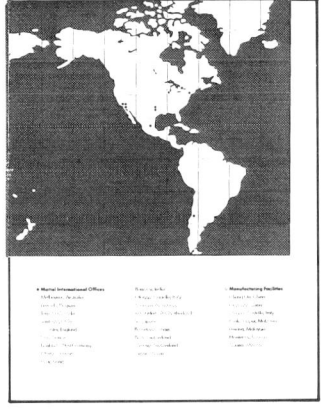
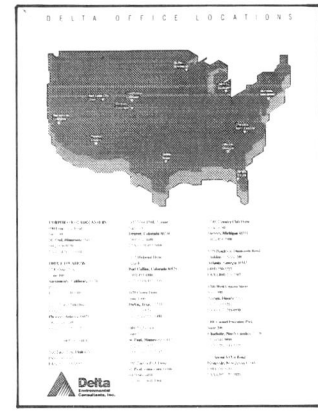
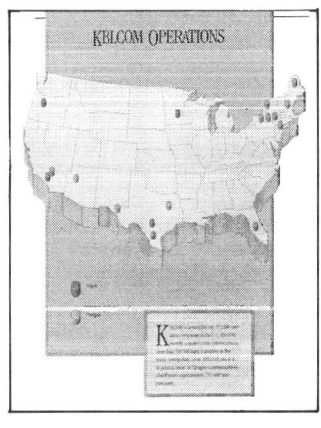
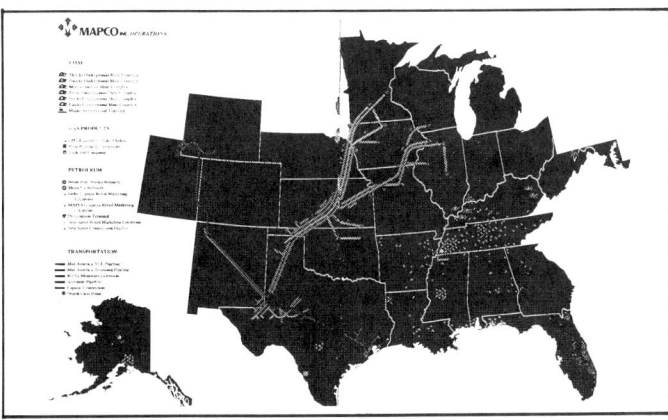
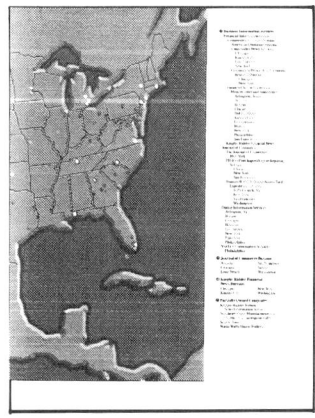
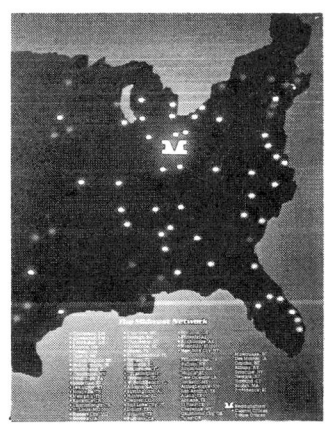

Anatomy Of Formats 37

## Books

The design of a book is like expanding the design of a brochure by ten to twenty times. There is usually one grid which is consistent throughout the entire book and governs the placement of everything in it. The design of a grid for a book is based largely on the use of type, supplemented by photographs. Therefore, the two or three column format is most often used. If the book contains a lot of photographs, the grid will have to accommodate them. There is an additional 1/8" margin on all sides of the grid to accommodate photos that "bleed" off the page. On some book formats there may be pages where bleeds cannot occur, due to the gripper margin required. Using oversized paper will overcome this, but may add to the cost of the printing. Books are being used as a promotional vehicle in instances where landmark projects are restored or renovated, and the story is printed in book form. However, their permanent value cannot be achieved by any other format.

There are certain tools that are helpful in the design and layout of books. The first is a miniature page layout of the grid system used. This can be used to create sketch layouts, or "thumbnails." Notes can accompany the sketches to assist in the production. These sketches can be reduced and used on a storyboard to help plan the page sequence. The storyboard is a series of outlines representing the spreads of the book arranged in series to the desired length. The final spreads are numbered according to the storyboard. This tool is useful in the early stage of planning. It is helpful throughout the design stage and becomes essential as the material nears completion. Reduced repros of the pages can be arranged on the storyboard, and as material is completed it can be replaced by final pages. By using this method, the entire contents of the book are always visible in a miniature format. This can facilitate in both the planning and production process. The actual size of the storyboard is 24" x 36".

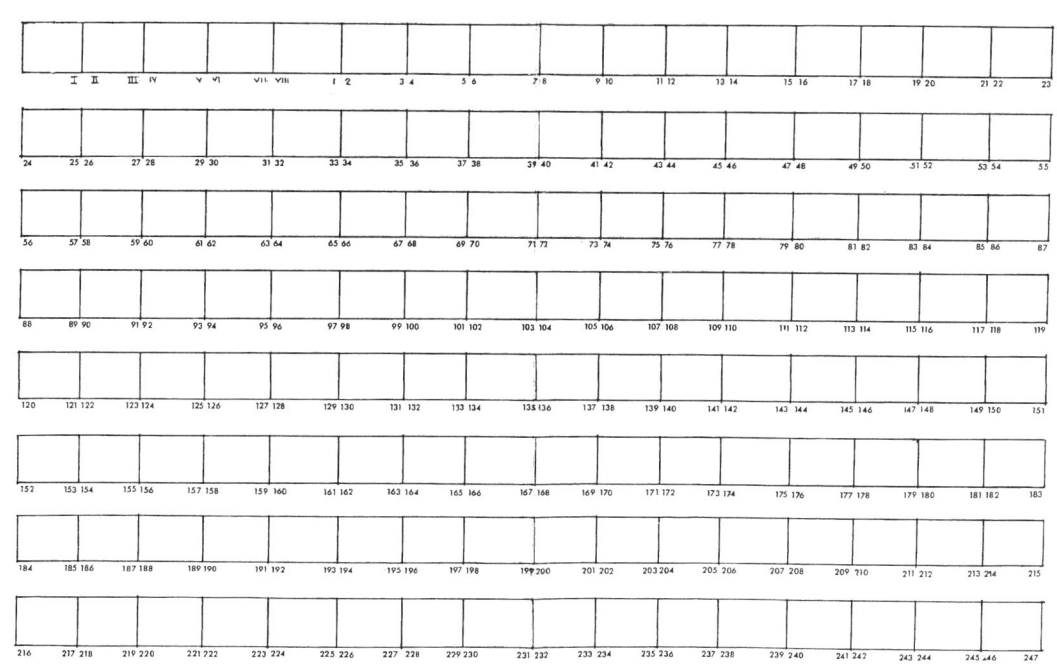

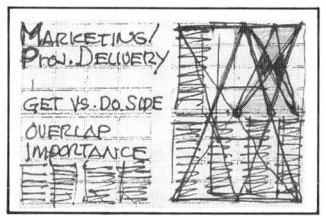

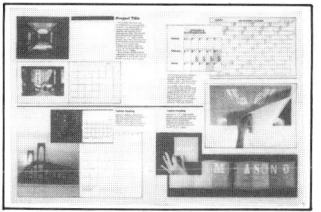

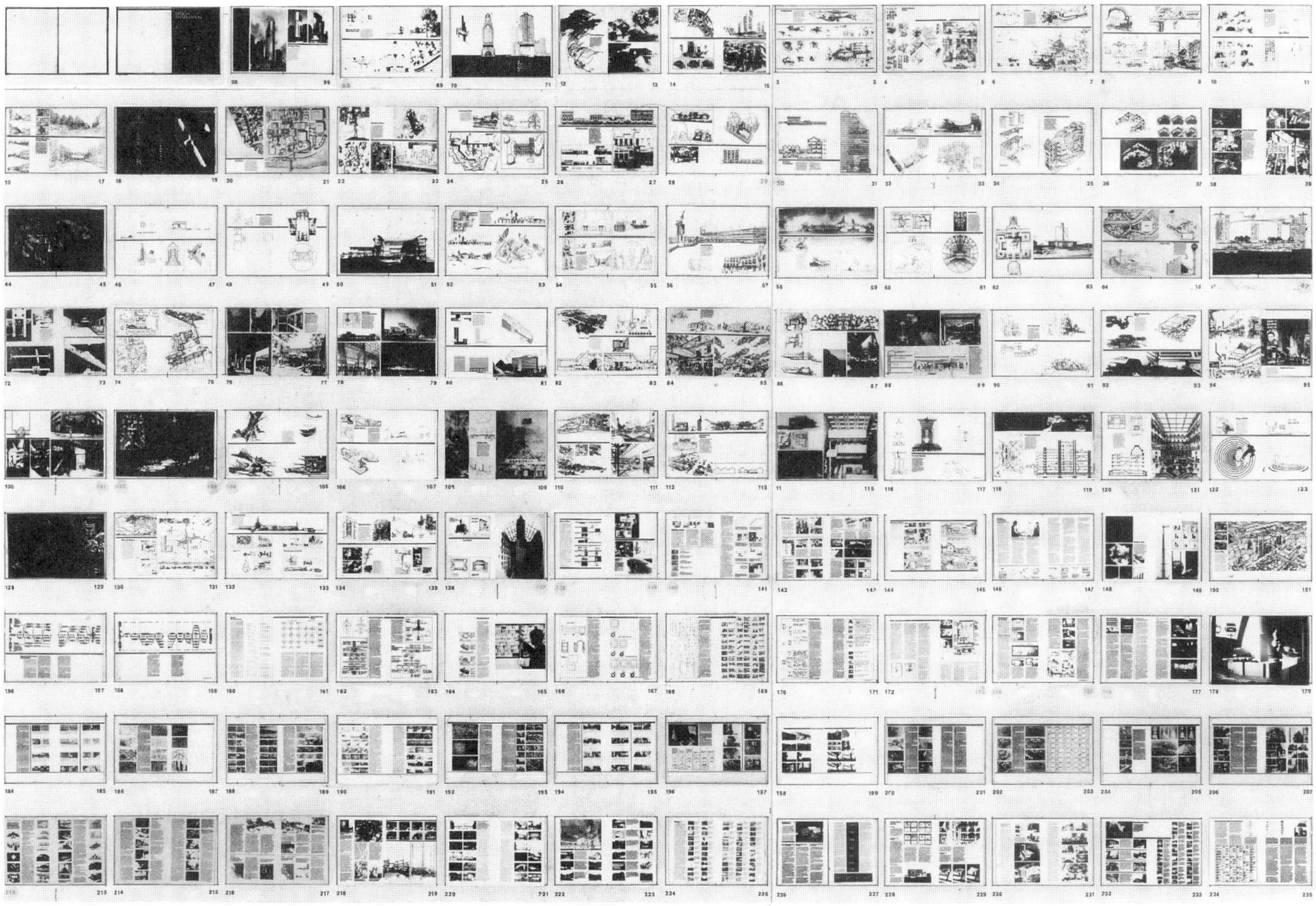

Anatomy Of Formats 39

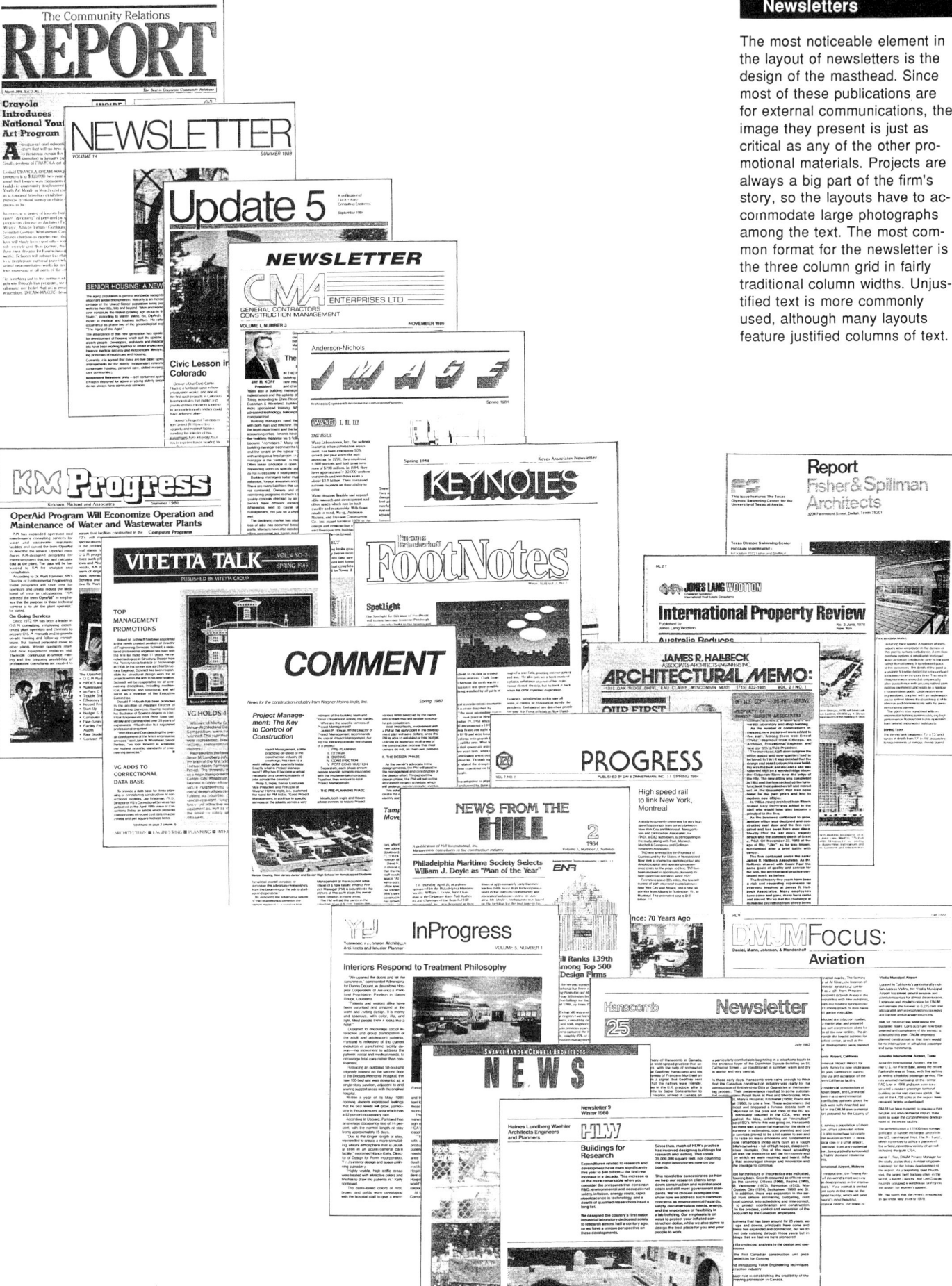

## Newsletters

The most noticeable element in the layout of newsletters is the design of the masthead. Since most of these publications are for external communications, the image they present is just as critical as any of the other promotional materials. Projects are always a big part of the firm's story, so the layouts have to accommodate large photographs among the text. The most common format for the newsletter is the three column grid in fairly traditional column widths. Unjustified text is more commonly used, although many layouts feature justified columns of text.

The section on generic grids shows a wide variety of formats that can be used in the design of newsletters. They are mostly two and three equal column formats, but many newsletters have been designed with four and five columns. The narrow column width is more difficult to set in type, but it provides more possibilities for photo layouts. The text is easy to read in narrow columns. It is interesting to note the similarities among names of the publications shown here. What is equally noticeable is the dissimilarity of the design of the mastheads, as expressed by the selection of different type styles.

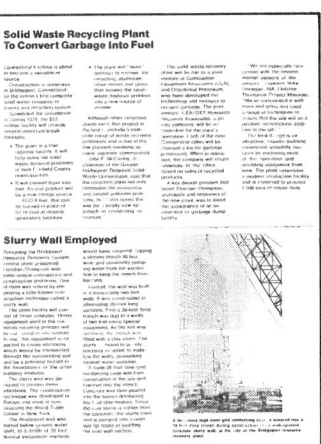

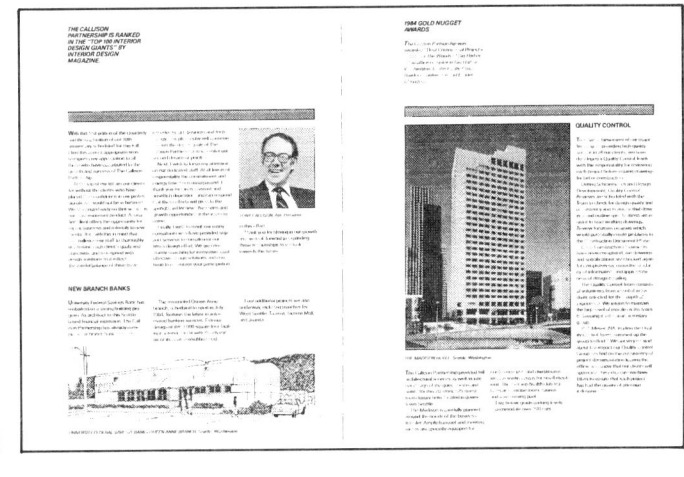

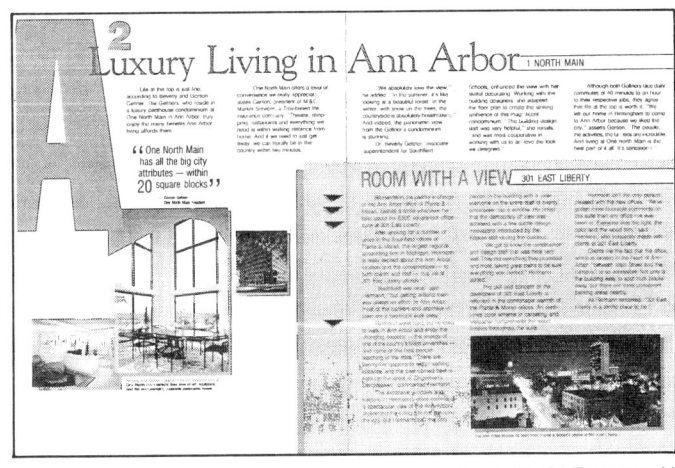

Anatomy Of Formats 41

## Project Pages

The individual photo project page is perhaps the mainstay of many firms' promotional arsenal. It is extremely flexible, particularly when printed on only one side. When bound in a proposal, the blank back of one page is a perfect foil for the other facing project. When printed on both sides, with a spread visible, there can be confusion about the two images. However, if the back side features drawings and other information, and is clearly identified, then confusion is minimal. These pages are usually dominated by large photographs with supporting information usually contained in a single column of type.

The purpose of these is to allow a completely customized package of material to be inserted into a pocket folder, or bound up into a tailored proposal. Some project pages have incorporated the logo of the client into the design, as those illustrated here. Some have put statistical information on the back of the page such as the time schedule for the development of the project illustrated on the front. Although they may appear at the bottom of the list of promotional materials, with the corporate brochure at the top, they are at the top of the list in terms of flexibility. This book contains many layouts that can be adapted to the individual photo project page, both in the generic grid section, and in the case study section. In fact, any page from any brochure can be used as an individual photo project page.

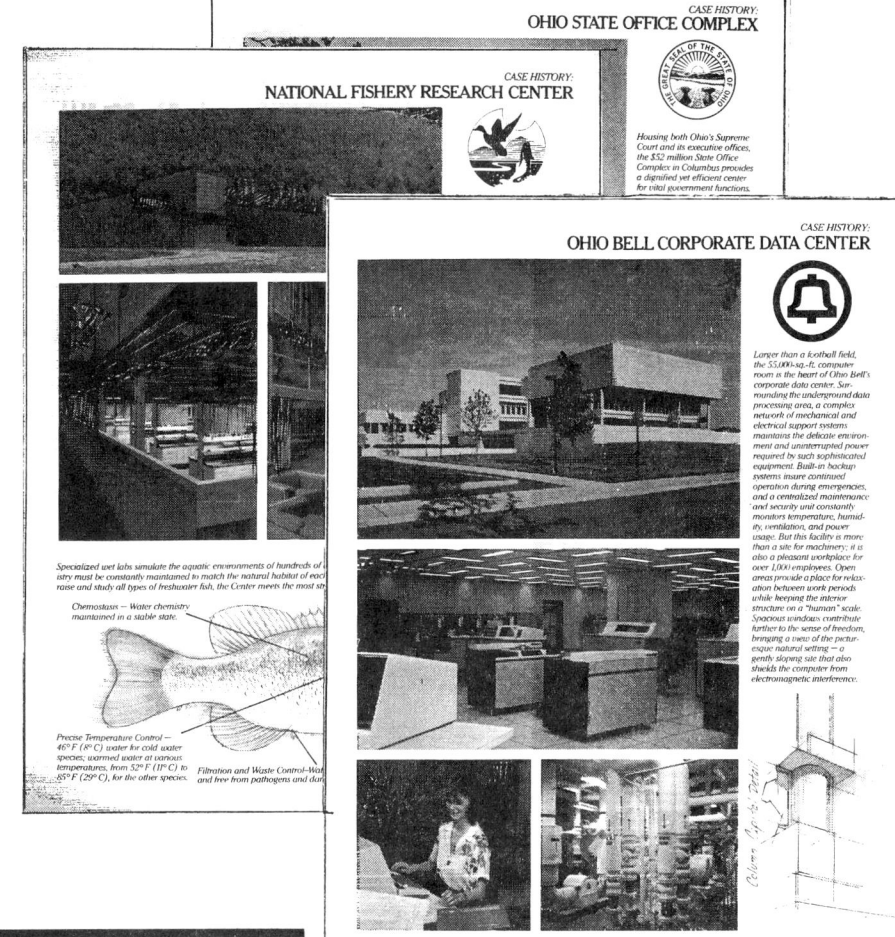

Anatomy Of Formats 43

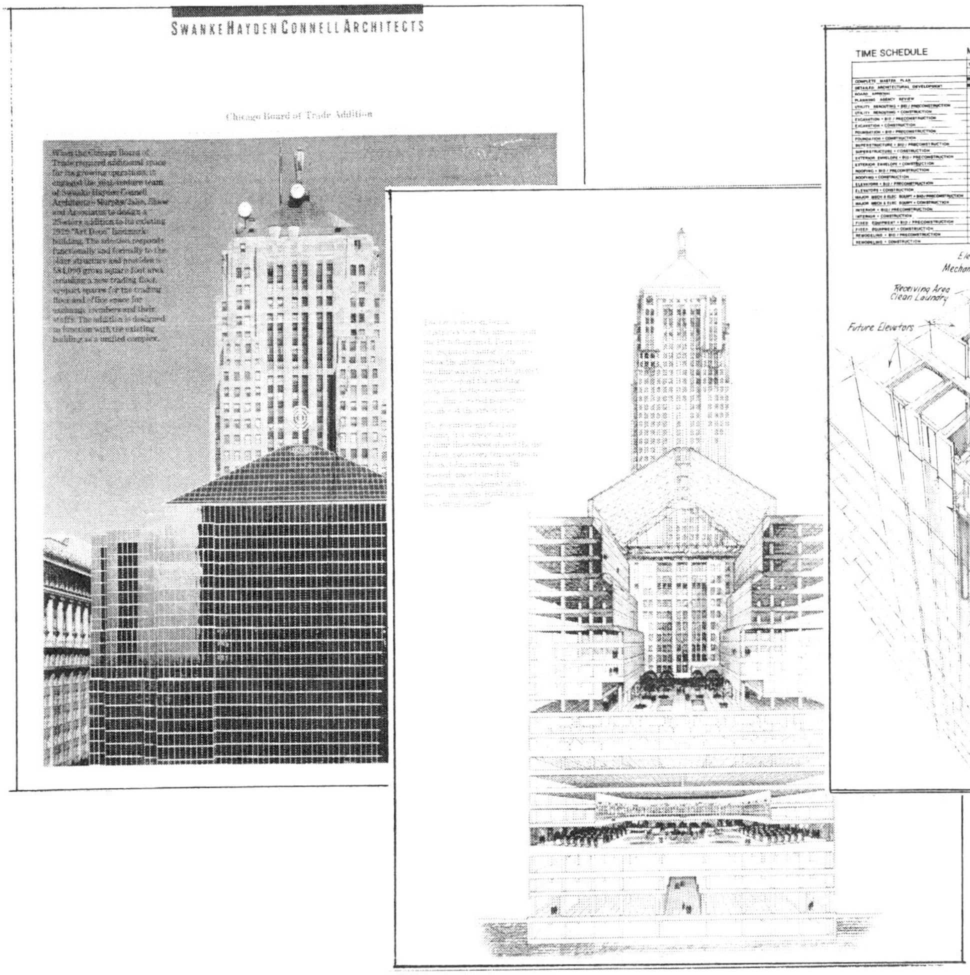

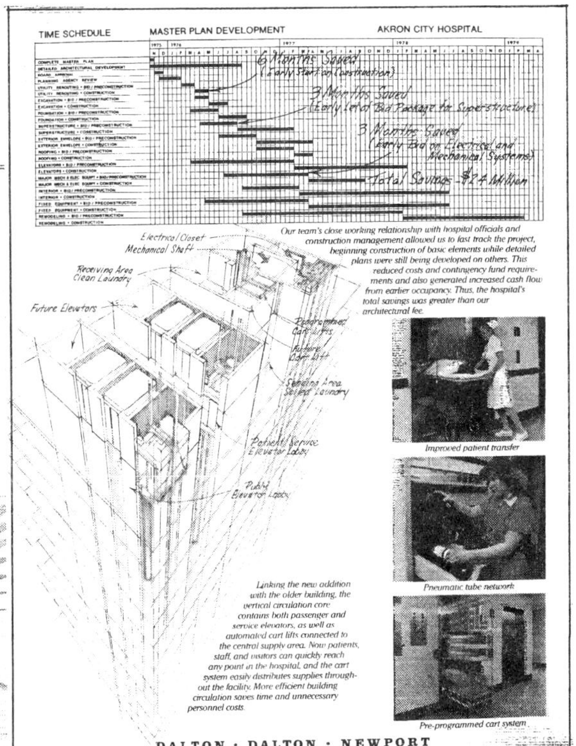

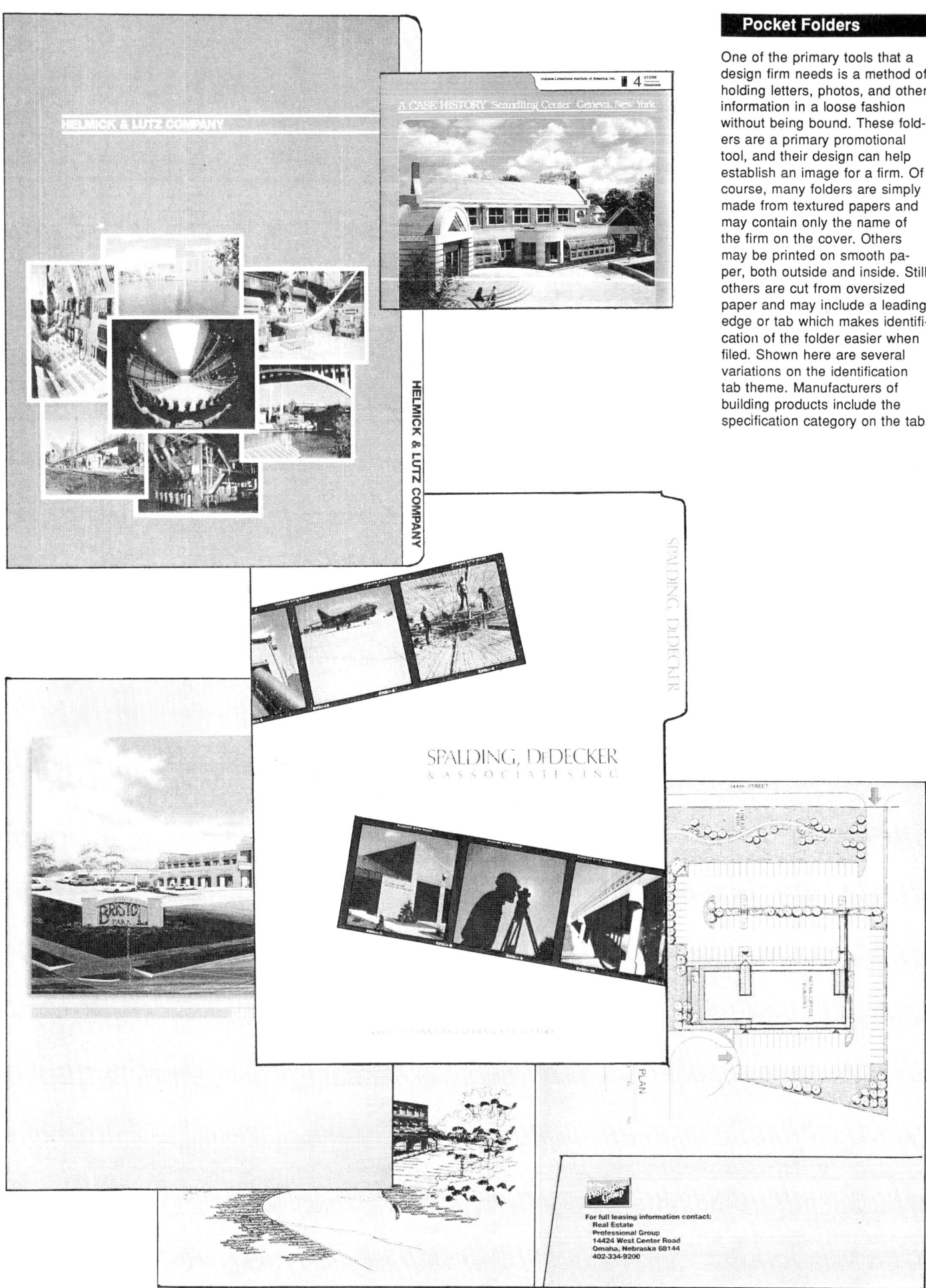

## Pocket Folders

One of the primary tools that a design firm needs is a method of holding letters, photos, and other information in a loose fashion without being bound. These folders are a primary promotional tool, and their design can help establish an image for a firm. Of course, many folders are simply made from textured papers and may contain only the name of the firm on the cover. Others may be printed on smooth paper, both outside and inside. Still others are cut from oversized paper and may include a leading edge or tab which makes identification of the folder easier when filed. Shown here are several variations on the identification tab theme. Manufacturers of building products include the specification category on the tab.

The inside pages of these folders can be printed with additional information, such as project plans or photos, or additional text. The inserted material is held in place by a pocket on the right hand page. This pocket flap is folded on the bottom and glued on the side. Shown here are variations on the design of these folders.

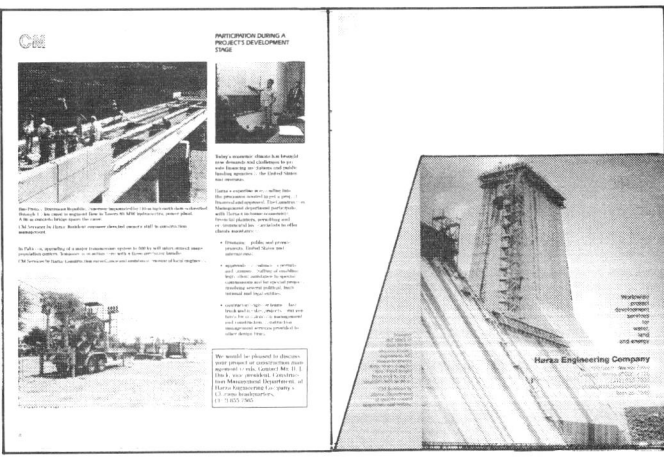

Opposite is shown the use of a die-cut on the cover, which reveals the material placed inside. When opened, a business card is visible. Most folders have four diagonally cut slots to hold a business card in the bottom flap.

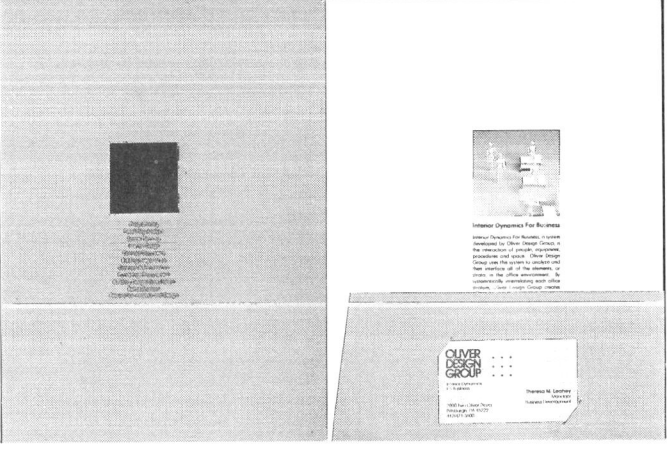

The three panel folder shown here used an illustration in color to depict some of the firm's services, beginning with the conceptual design, to construction-drawings of the project. The flap on the far right panel of the folder holds additional material.

This 12 page brochure carries a pocket on the last page to hold additional material. It has slots for a vertical format business card. The theme of the brochure relates to chosing a professional firm. The material carried in the pocket emphasizes this theme. It contains individual resumes of the key people which includes their photograph. It also holds an evaluation form and gives the client an opportunity to rate the firm's experience.

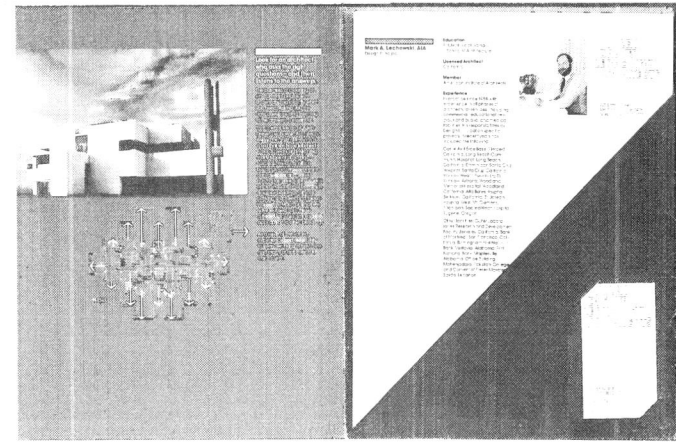

Anatomy Of Formats 45

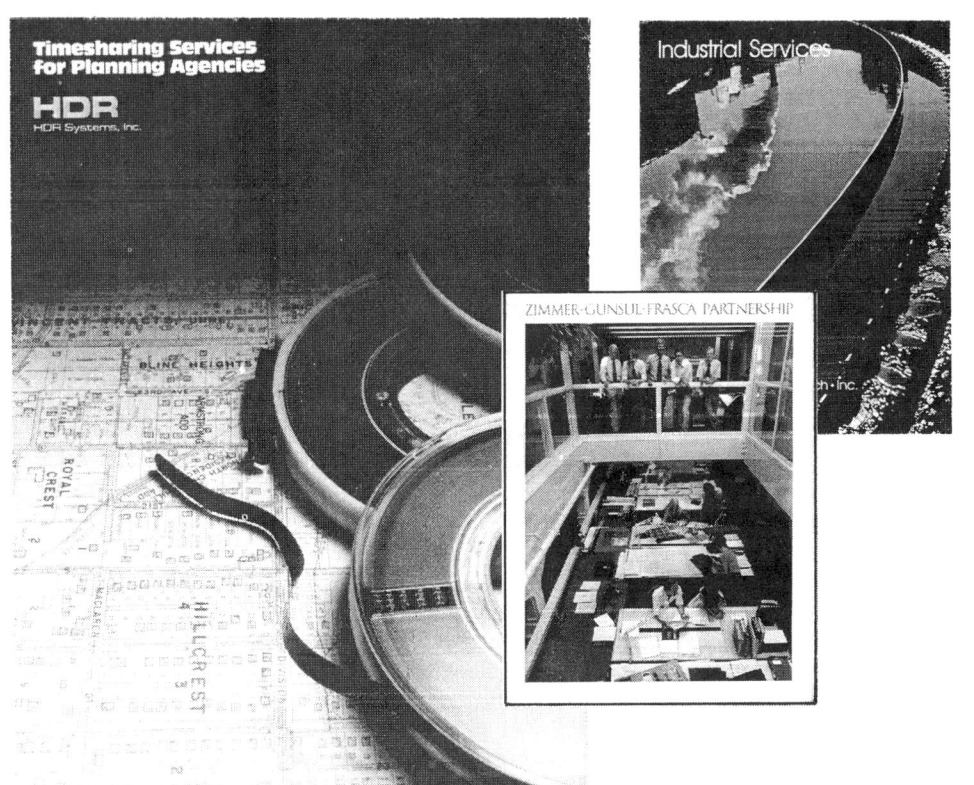

## Covers

Brochures for architectural and engineering firms are unusual in that they are selling a professional service rather than a product. The covers of these brochures relate to these services in a variety of ways. However, there are many brochures from related fields, such as manufacturers of building products and computer hardware and software programs, that can be adapted to these needs.

The series below shows several ways to incorporate people into the cover design. Most of them have a quotation accompanying the photograph. Others show multiple pictures of people, either in boxes or grid patterns. This helps personalize the marketing approach.

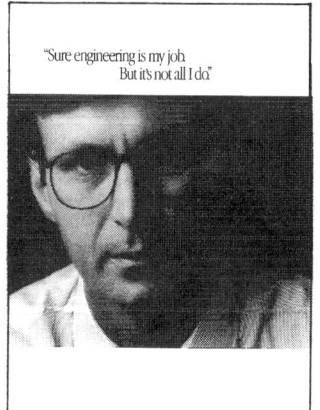

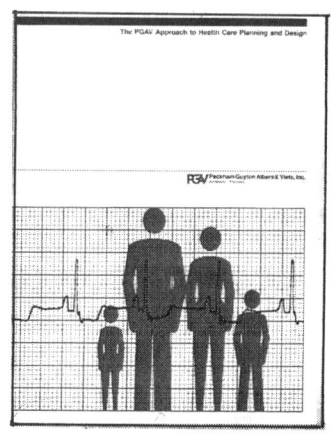

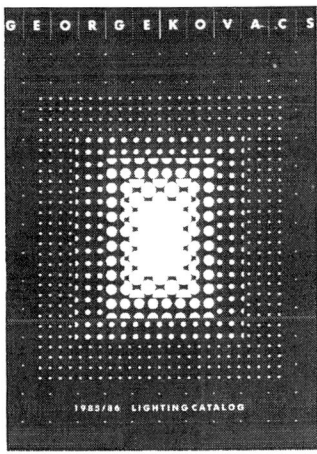

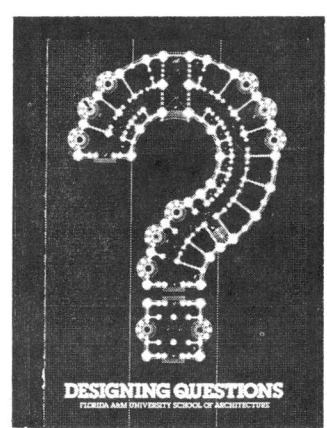

There is such a wide variety of solutions for covers that any display will simply open new doors to explore further. They can be grouped by any number of classifications depending on the predominant design device employed. Here, the elements are variations of a graphic approach, whether in design, by using numerals, in the case of the "Designing Questions" brochure, an imaginary floor plan bent into the shape of a question mark.

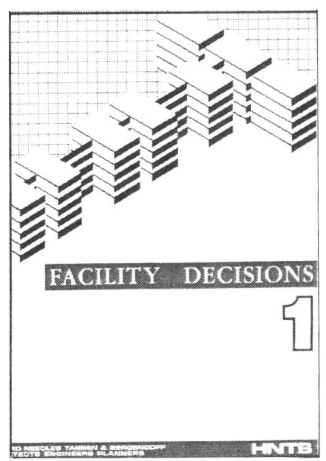
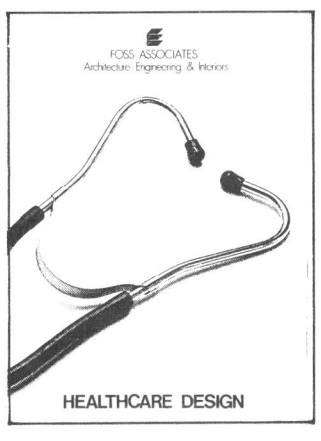
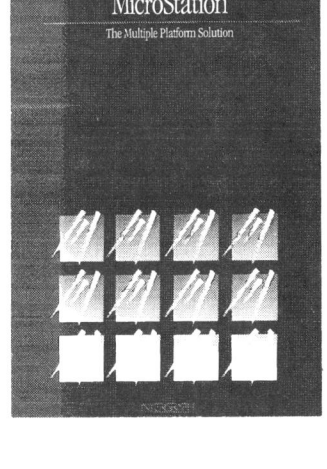

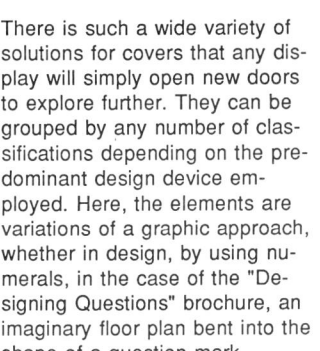

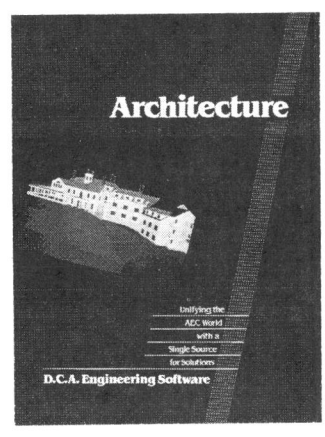

Anatomy Of Formats 47

## Covers

Perhaps one of the most commonly used items on covers of construction brochures is the hard hat. This not only signifies construction, but the hats usually carry the logo of the company. Next to the hard hat, actual construction scenes with reinforcing bars, beams and girders are the favorite images.

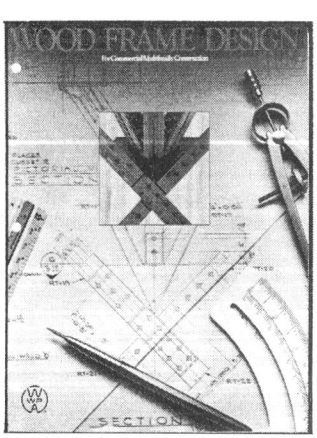
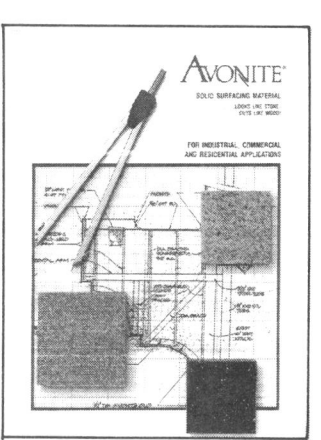

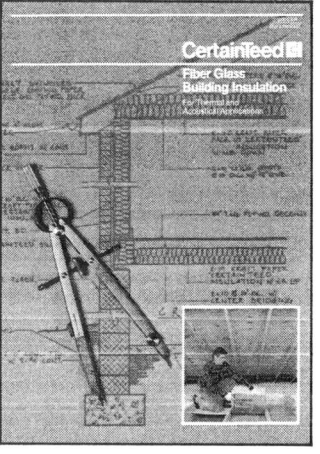

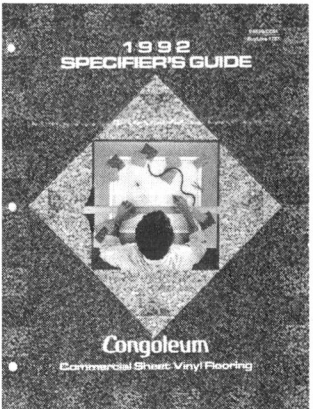

Compare the use of drawings and drawing instruments in this group of catalogs and brochures from building product manufacturers. They range from very precise to very loose, but they all make use of similar tools and similar images relating to drawing, design, drafting instruments, and pieces of buildings and materials.

In the design and construction industry today there is an emphasis on high-tech solutions to building problems. In addition, many computer companies, featuring either hardware or software, are approaching an architecture, engineering and construction audience through printed literature. The high-tech nature of their product is expressed in the design. One of the most common is the use of electronic images, grids vanishing in perspective and inserts showing digitizers or computer parts.

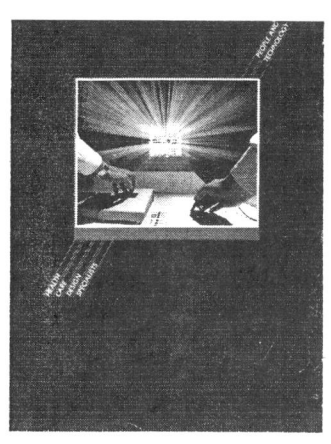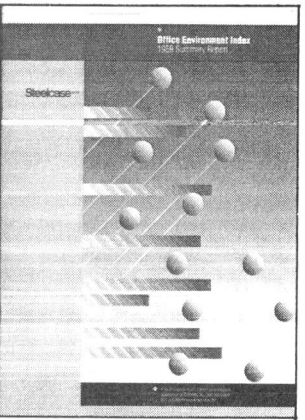

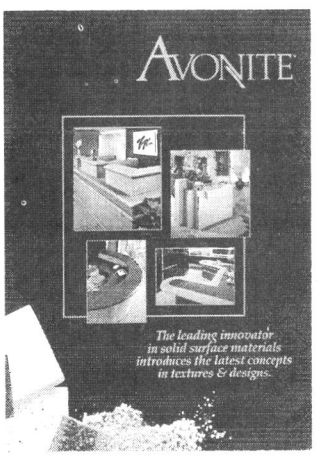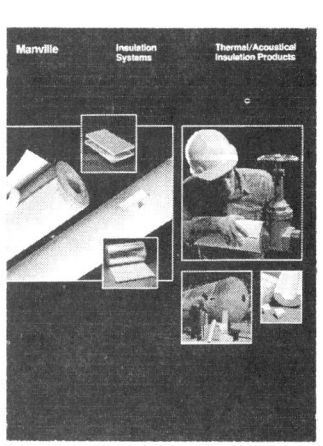

Manufacturers' brochures, or catalogs, run the gamut from very simple to complex. Generally the cover is either a project photograph set into a photo of the product or a full bleed photograph of an installation, particularly if it is an exterior material for a building.

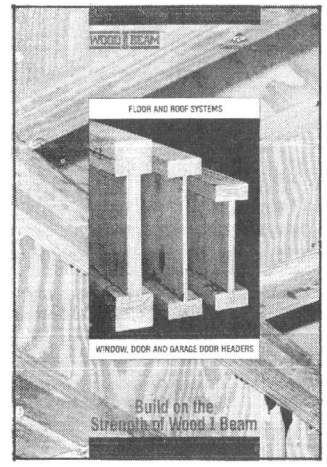

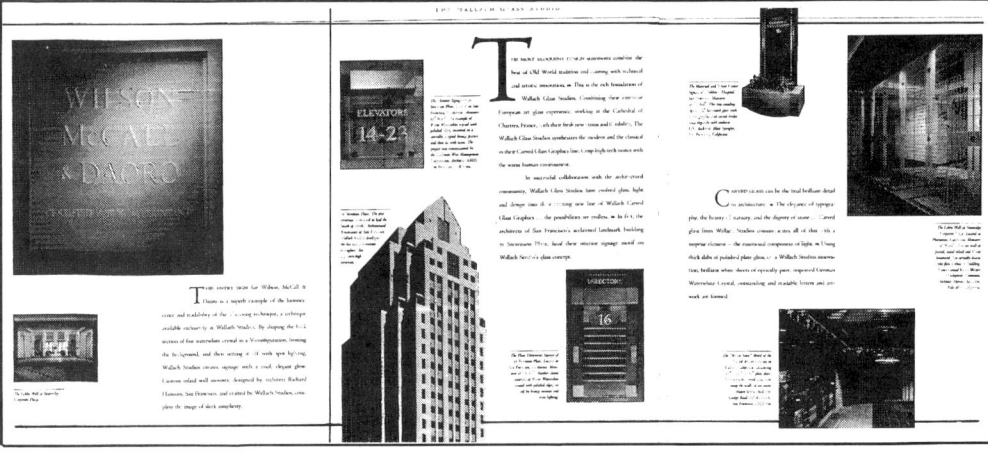

**Covers And Insides**

A stunning photograph of a frosted glass architectural capital is set into a black background on the cover of this brochure. Inside, the photographs and captions are surrounded by ample white space. The single column of text is headed by a very large letter, and the other columns are headed by a smaller version. This brochure folds out into three equal panels, so the generous white space flows from panel to panel across the spread.

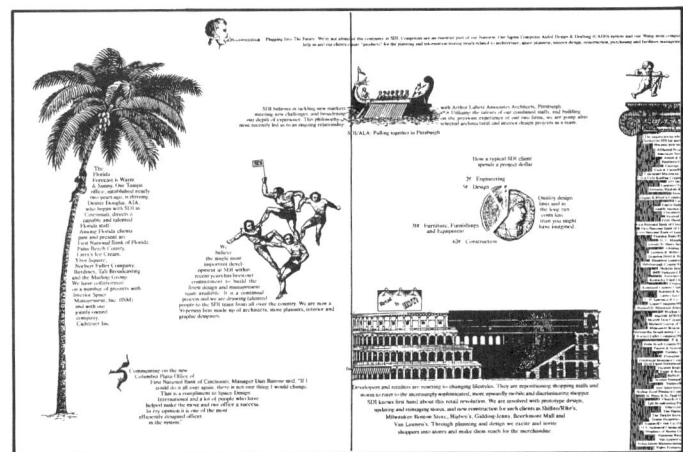

Unusual graphics and illustrations add interest to this four page folder. The type is used as an architectural element within the illustrations, making pediments, columns and pyramids. Shown above is the front cover, inside spread and back cover, all containing the unusual graphics.

It is clear to see how the cover design is reflected in the layout of all the pages in this four page folder. It is a case study approach, and the constant block of photographs is flanked by two equal columns of text. A color bar extends across the top of each page.

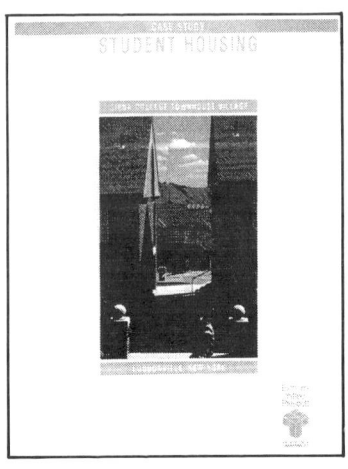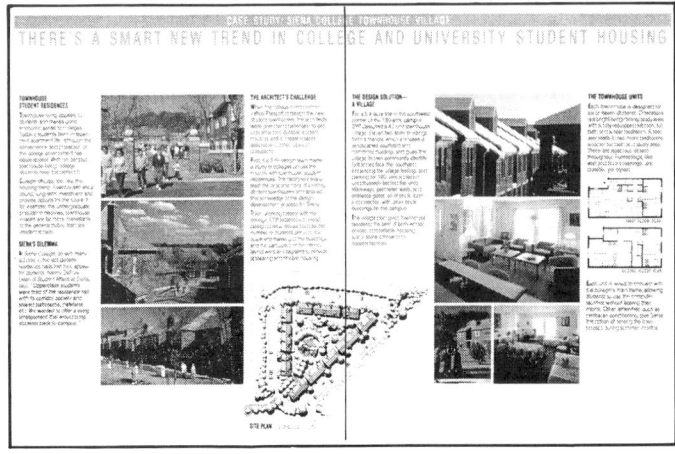

A framed portrait of an airplane appears on the front and back cover of this aviation consultant's folder. The cover features the front end of an airplane, and the back cover shows the plane from behind. The center spread has large photographs and a single column of type.

A highly visual approach is used in this four page folder combining photographs and colored illustrations. Text is minimal, and the message is carried in the one line of bold headline type, which runs across the top of the spread and back cover.

The simplicity of the cover design gives way to a busier inside spread that is hard to read. The reverse type is crowded, and although the photographs and color drawings are stunning, the overall effect is overdone. There is no relief on the back cover, as it is filled to capacity.

Anatomy Of Formats 51

## Overriding Graphics

The angular shapes on the cover of this brochure originate from the firm's logo, seen at the bottom of the cover. These shapes are repeated on the inside spread in outline form, surrounding the group of photos and columns of text.

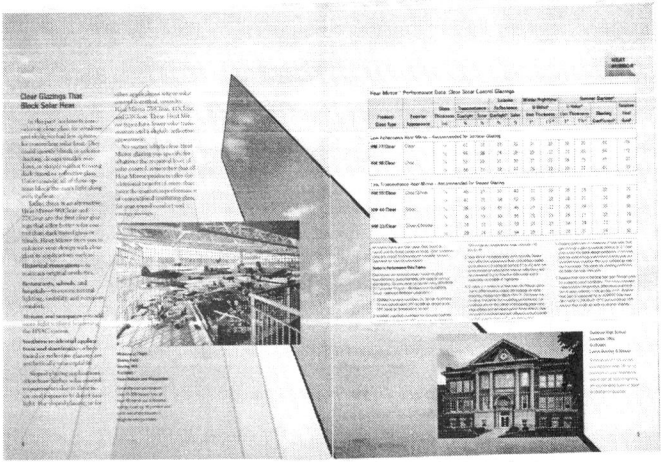

A striking cover photograph of overlapping transparent planes vanishing towards the sky represents a series of imaginary buildings. This same design is carried inside to provide a color background for the text and inserted photographs.

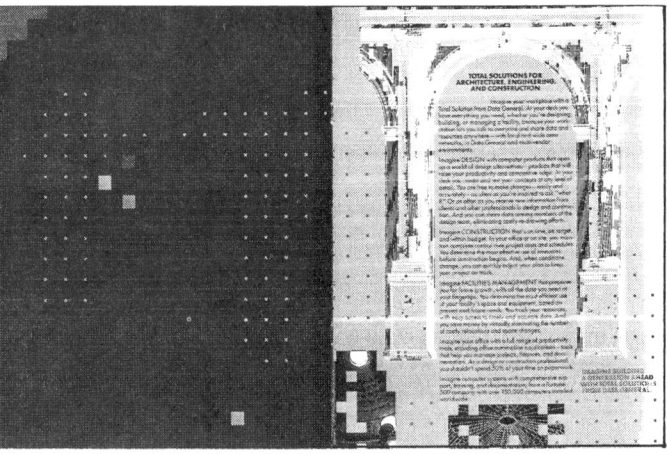

Although the diamond shaped die-cut works well on the cover, it is difficult to deal with on the inside of the cover. A cluttered look results from the inclusion of other text items and photographs. The inside of the die-cut opening would look better untouched.

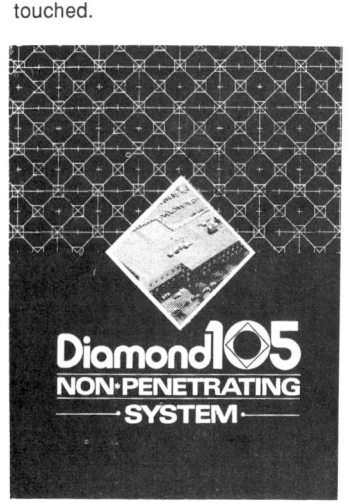

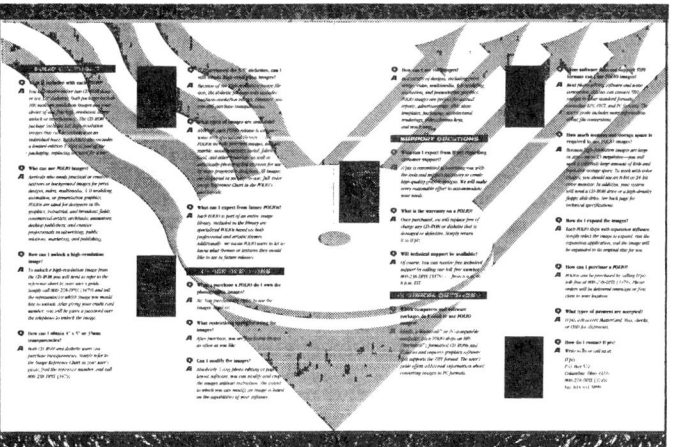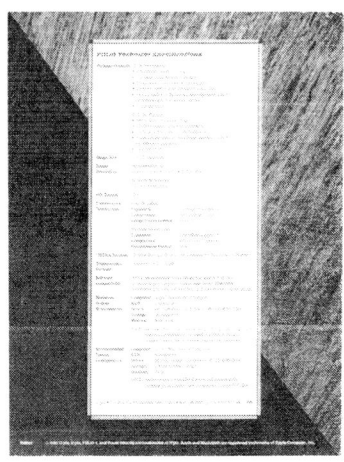

When the front cover, back cover and inside spread of this brochure are viewed together it has an organized look resulting from the angular graphic patterns. The outside cover echoes the inside graphics.

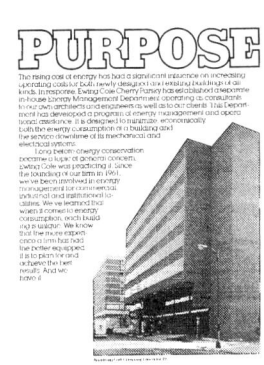

Text that is too large or clumsy in its design tends to overpower the page, and detract from the message. Type that is set solid against the margins or rules is harder to read. Crowded type or columns of information need blank space around it.

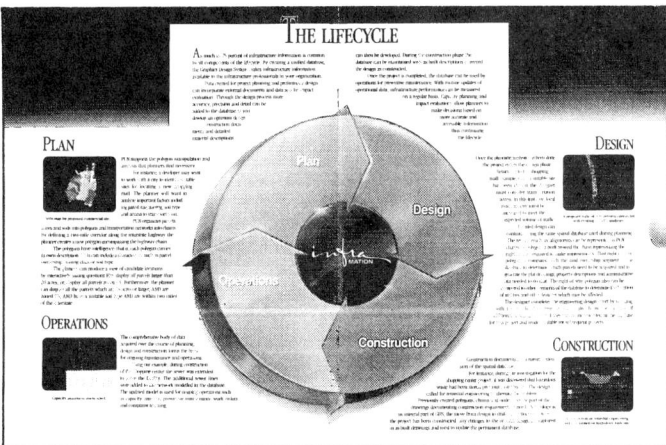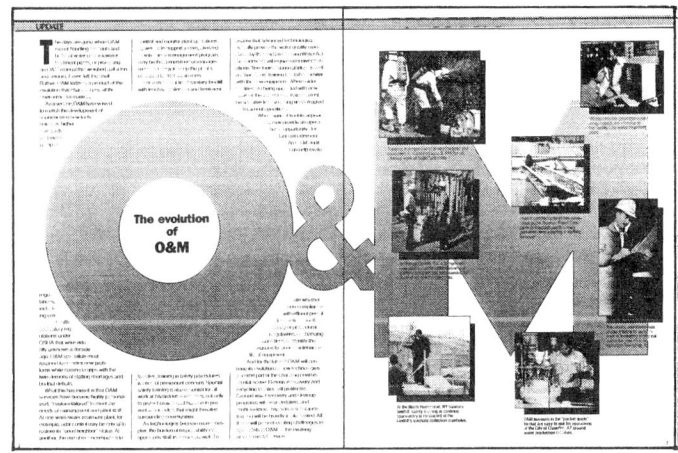

In this example a circular design dominates the center of the entire spread. The type is wrapped neatly around the shape and flanked in each of the four corners with an illustration.

The benefits of using a grid are obvious, but there can be too much of a good thing when the grid overpowers the material. Photos set within grid lines with drop-out borders and multiple rules are typical of the overuse of the grid.

The oversized initials "O&M" dominates the spread, with the type wrapped neatly around the circular shape. On the right-hand page a series of photographs is set over the "M." Each photo has a drop shadow on two sides, which adds to the visual clutter.

When a drop-out type is used over a photograph, it is important to consider the tonal values. Here is a comparison using the same type style over two similar exterior photographs. The type is far easier to read when it is placed on a dark background than one with a lot of detail in it.

Anatomy Of Formats 53

## Proposals

In a proposal, information must be presented clearly. Unfortunately, there is a lot of text involved for project descriptions, personnel resumes, and outlines of services. Sometimes this information is generic and can be pre-printed, commonly referred to as "boilerplate". However, this printed material may refer to people or projects that do not apply to the proposal at hand. Customizing and tailoring the proposal to fit the client's needs is the highest concern. Re-arranging, adding and deleting information is no longer the problem that it was with the "cut and paste" method. Most word processing programs make this an easy task. Items can be mirrored, stretched, shrunk into smaller columns, re-formatted in a different size and style typeface, all with a click of the mouse. Therefore, proposals can now become more of an extension of the firm's promotional material than something prepared by an outside print shop. Photo project pages and resume pages can be coordinated in both type style and design, and the proposal can have a more unified look from cover to cover.

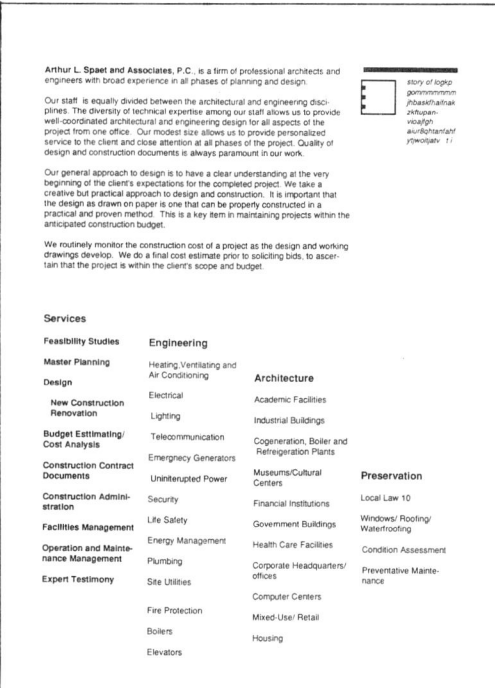

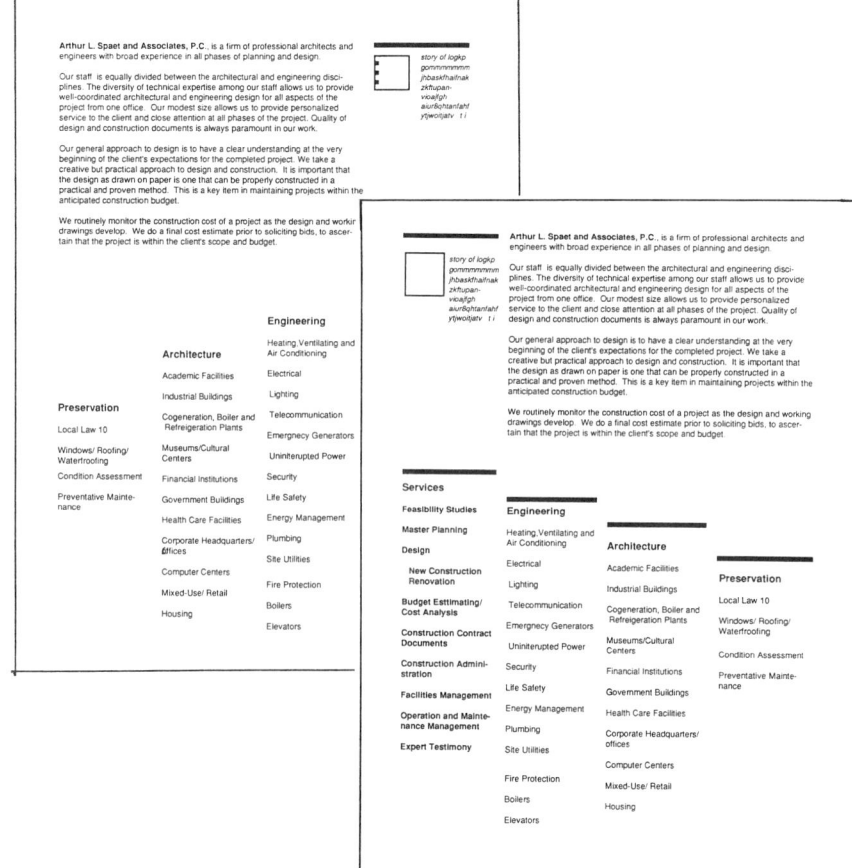

Anatomy Of Formats 55

### Process Chart

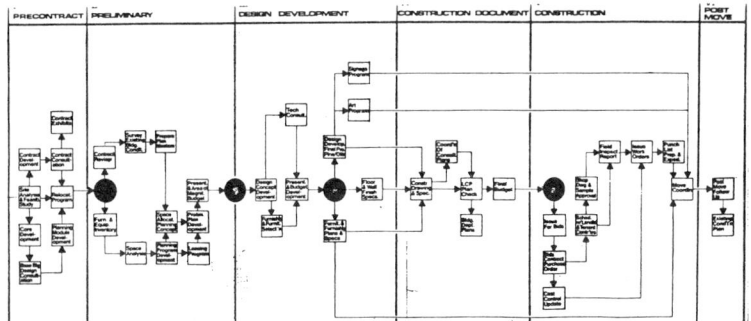

Most design process charts are hard to read and even harder to understand. Arrows point in one direction, then back again, until the viewer is totally confused. The development of this chart focused on the issue of clarity, continuity, and making it visually understandable.

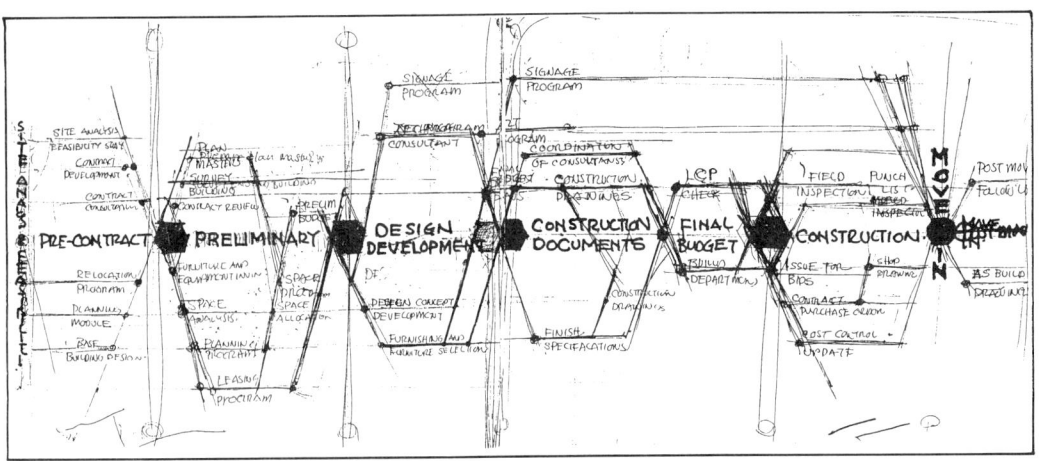

First, a rough sketch was made to outline the major elements that would go into the chart, from the pre-contract phase to the post-move-in. Then the items were divided into two main groups, standard services above, and additional services below the main line of the chart. As one proceeds across the chart, similar items can be traced horizontally through the different phases.

## PROJECT PHASING

**CLIENT INPUT**

**PRE-LEASE PHASE**
- ○ Relocation Program
- ○ Consult on Work Letter
- ○ Lease Consultation
- ① Preliminary Program
- ② Site Analysis
- ③ Feasibility Study
- ④ Contract Exhibits
- ⑤ Budget Consultation

**PRELIMINARY ANALYSIS**
- ○ Tenant Lease Program
- ○ Furniture Inventory
- ○ Base Building Survey
- ⑥ Final Planning Program
- ⑦ Furniture & Equipment
- ⑧ Space Allocation
- ⑨ Development of Plan
- ⑩ Preliminary Budget

**DESIGN DEVELOPMENT**
- ○ Art Program
- ○ Signage Program
- ○ Graphics Development
- ○ Technical Consultation
- ⑪ Design Concept
- ⑫ Furniture Selection
- ⑬ Final Preliminary Plan
- ⑭ Design Presentation
- ⑮ Budget Submission

Two different versions were made from this sketch. The first one has an angular framework suggesting movement across the chart. The other one is more static, with the items lining up in columns. The numbered items appear in the standard services. The sketch layout is then developed into a more accurate pencil layout with a preliminary spec for the size and style of the type. This should be tested and approved before any type is set or ordered.

This chart can be used in many ways. It can be enlarged and used as a board presentation to a client. It can be reduced to an overhead transparency, or photographed as a 35mm slide image. It can be printed in a proposal or as a foldout in a general brochure. Since it represents all the functions and steps in the interior design process, it can be a valuable tool for any design office to have in its files. If the design is kept generic, it can be used for many different situations.

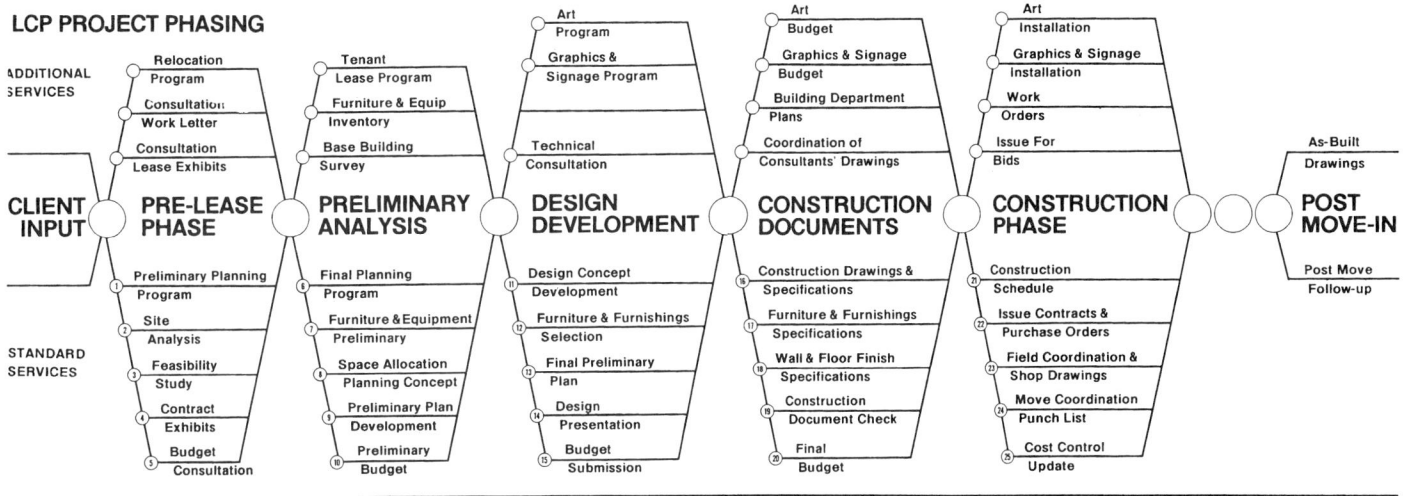

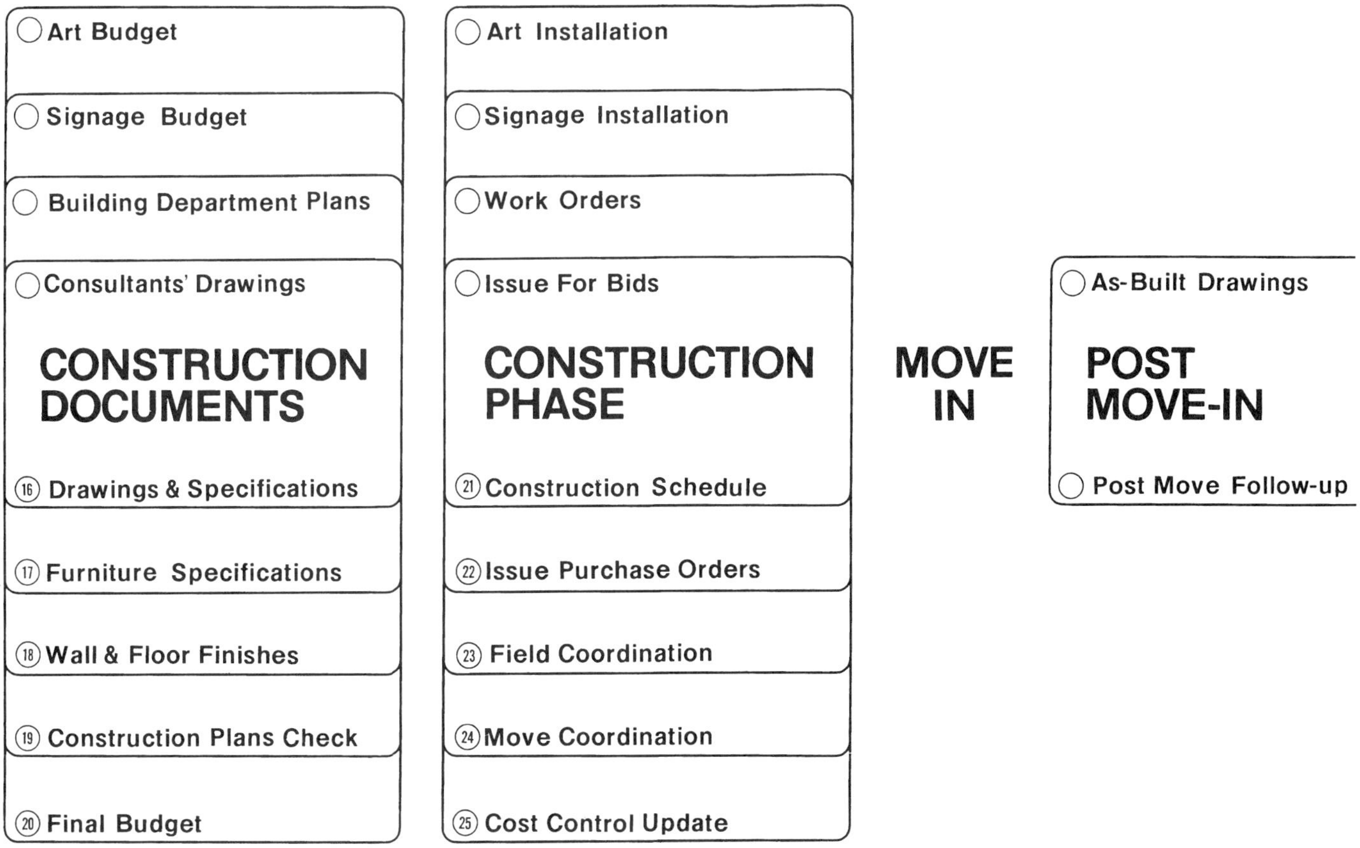

**LCP PROJECT PHASING**

○ Art Budget
○ Signage Budget
○ Building Department Plans
○ Consultants' Drawings

**CONSTRUCTION DOCUMENTS**

⑯ Drawings & Specifications
⑰ Furniture Specifications
⑱ Wall & Floor Finishes
⑲ Construction Plans Check
⑳ Final Budget

○ Art Installation
○ Signage Installation
○ Work Orders
○ Issue For Bids

**CONSTRUCTION PHASE**

㉑ Construction Schedule
㉒ Issue Purchase Orders
㉓ Field Coordination
㉔ Move Coordination
㉕ Cost Control Update

**MOVE IN**

○ As-Built Drawings

**POST MOVE-IN**

○ Post Move Follow-up

Anatomy Of Formats 57

## Project Workplan

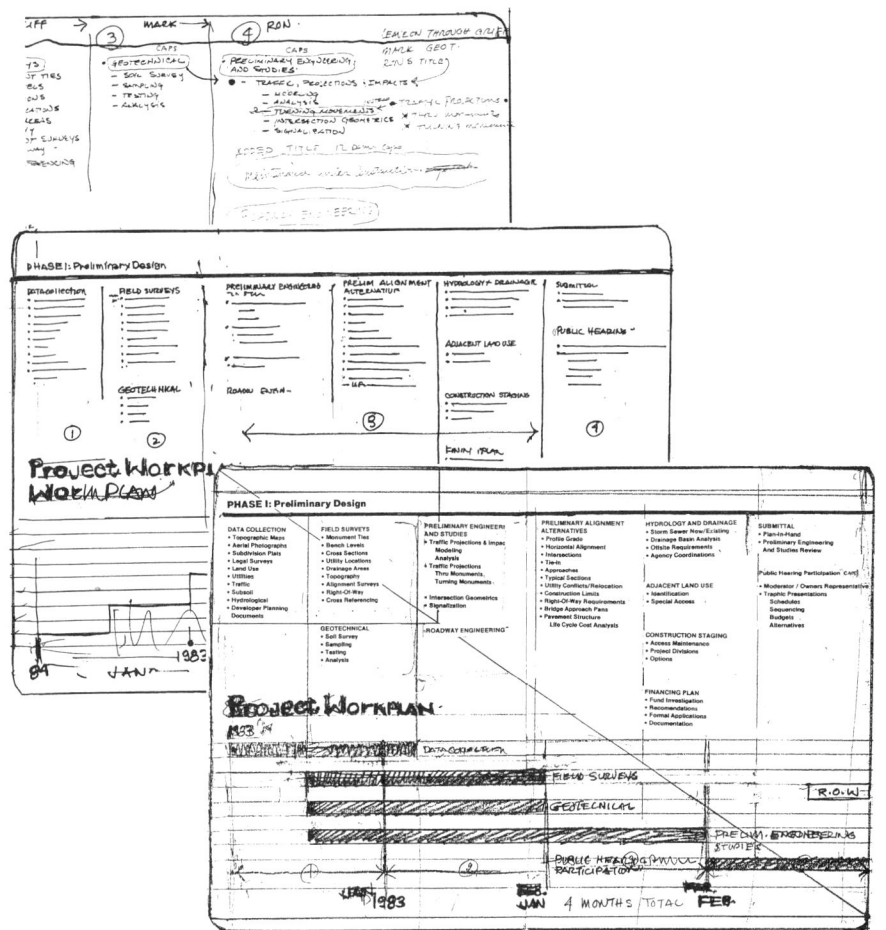

When making proposals for large scale projects a detailed outline of the sequence of work is a necessary item. The design firm needs one to calculate the size and scope of the job and to have specific items on which to base a cost estimate. The client needs one to determine if the design firm understands the scope. The workplan begins with rough notes prepared by the project manager in a very loose fashion. These notes are organized into a rough diagram outlining the various phases of the job. Individual items are set in type and arranged in columns that correspond to items requested in the proposal.

At the same time a manpower chart is developed and shown in relation to the proposed work items. In this proposal the chart was included at the bottom of the workplan. This was printed as a foldout in the proposal, and enlarged for a wall chart for the presentation. It was also made into two 35mm slides, one for the left chart, and one for the right one. These were projected onto two side-by-side screens for a slide presentation.

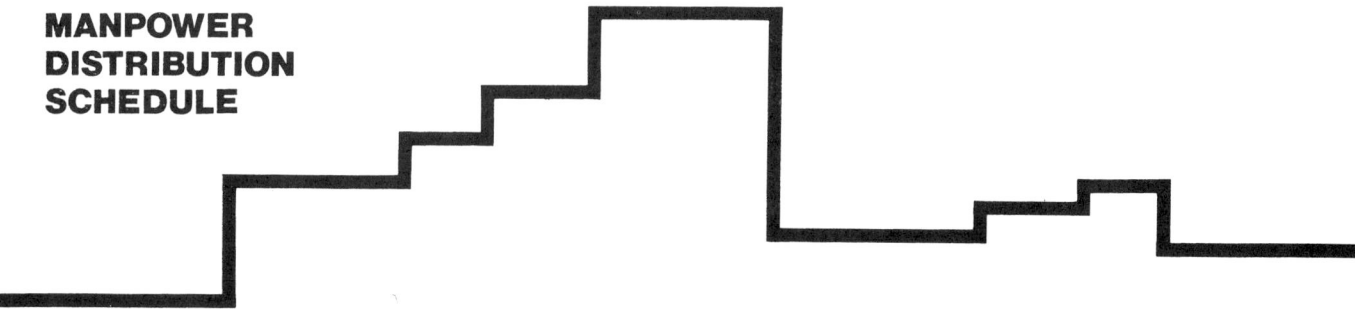

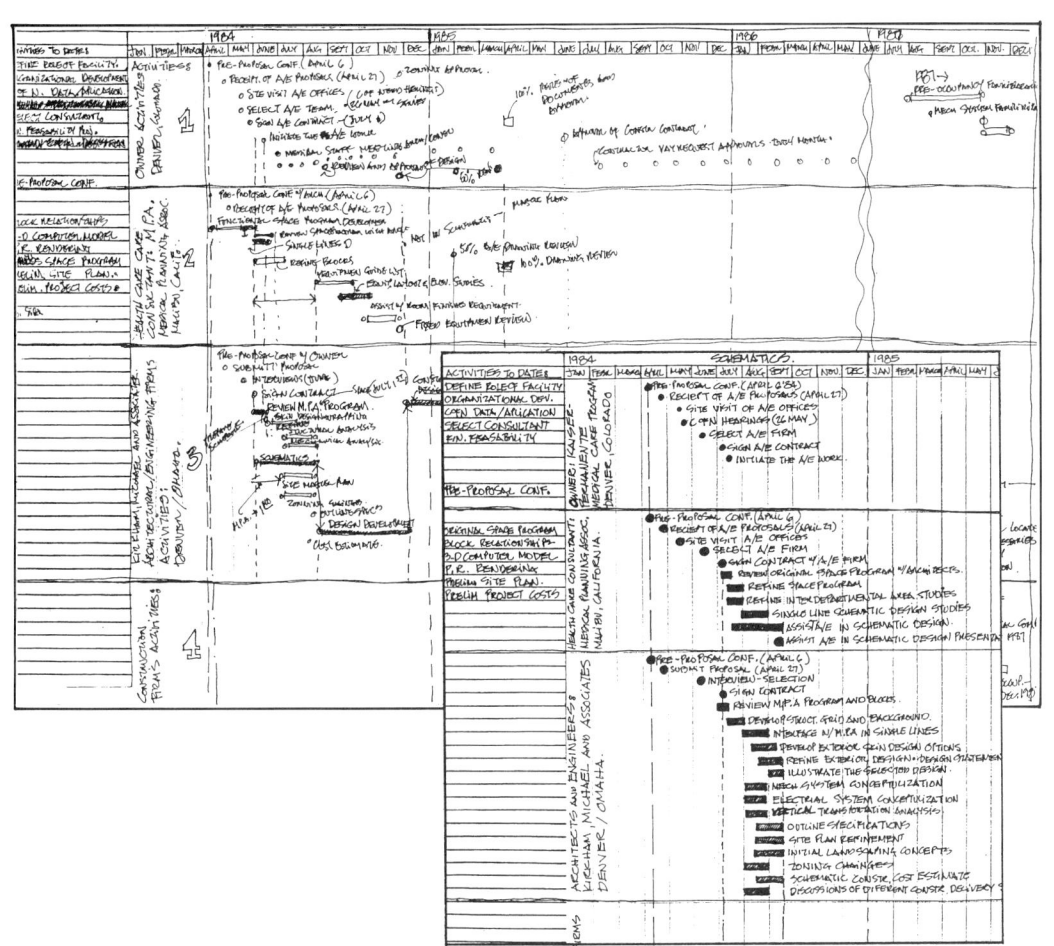

## Project Schedule

Project schedules are the mainstay of any proposal seeking to obtain a commission to design or build the future project. The accuracy of the presentation is paramount, as well as the readability in the proposal document. This schedule, designed for an addition to a major hospital addition, began as a rough outline prepared by the project manager of the proposing team. It was important to get every item in its proper location early on.

The next stage was still in the form of rough notes, dates and design items. When all the rough notes were complete, a sample was drawn up on paper, using a timeline base drawing that was drawn and printed out by a computer. Hand notes were then put onto this timeline.

Rather than use the computer to figure out the schedule within a pre-determined software program, the computer was used here to draw and produce the schedule accurately and to be visually attractive. The schedule could then be printed out at any scale to go as a foldout in a proposal or as a wall chart several feet long.

| JAN | FEB | MAR | APR | MAY | JUN | JUL | AUG | SEP | OCT | NOV | DEC |
|---|---|---|---|---|---|---|---|---|---|---|---|
| ● DESCRIBE ROLE OF FACILITY | | | ■ PRE-PROPOSAL CONFERENCE | | | | | | | | |
| ● ORGANIZATIONAL DEVELOPMENT | | | | ■ RECEIPT OF A/E PROPOSALS | | | | | | | |
| ● CERTIFICATE OF NEED | | | | | ■ SITE VISITS A/E OFFICES | | | | | | |
| DATA GATHERING AND APPLICATION | | | | | | ■ SELECT A/E FIRM | | | | | |
| | | | | | | ■ SIGN A/E CONTRACT | | | | | |
| | | | | | | | ■ INITIATE A/E WORK | | | | |
| | | | | | | | ■ | ■ | ■ | ■ | ■ |
| ● SELECT CONSULTANT | | | | | | | | | | | |
| ● FINANCIAL FEASIBILIY OF PROJECT | | | | | | | | | ◆REVIEW | | ◆REVIEW |
| ● SOLICITATION OF ARCHITECT | | | | | | | | | | | |

### OWNER

| | | | | | | | | | | | |
|---|---|---|---|---|---|---|---|---|---|---|---|
| ● DEVELOP BLOCK RELATIONSHIPS | | | ■ PRE-PROPOSAL CONFERENCE | | | | | | | | |
| ● 3D COMPUTER MODELING | | | | ■ RECEIPT OF A/E PROPOSALS | | | | | | | |
| ● PUBLIC RELATIONS RENDERING | | | | | | FUNCTIONAL SPACE PROGRAM DEVELOPMENT | | | | | |
| ● GROSS SQUARE FOOTAGE PROGRAM | | | | | | | | REVIEW PROGRAM WITH ARCHITECT | | | |
| ● PRELIMINARY SITE PLAN | | | | | | | | SINGLE LINE DRAWINGS | | | |
| ● PRELIMINARY PROJECT COSTS | | | | | | | | SCHEMATIC DESIGN | | | |
| | | | | | | | | ◆REVIEW | | | |
| | | | | | | | | | ● REVIEW SPACE PROGRAM | | |
| | | | | | | | | | ● REFINE SPACE PROGRAM | | |
| | | | | | | | | | ● INTERDEPARTMENTAL STUDIES | | |
| | | | | | | | | | ● ASSIST IN A/E'S SCHEMATICS | | |
| | | | | | | | | | | FIXED EQUIPMENT GUIDE | |
| | | | | | | | | | | | EQUIPMENT LA |
| | | | | | | | | | | | ASSI |

### MEDICAL PLANNER

| PRE-PROPOSAL CONF. | | | | SUBMIT PROPOSAL | | | | | | | |
|---|---|---|---|---|---|---|---|---|---|---|---|
| | | | | A/E INTERVIEW | | | SIGN CONTRACT | | | | |
| | | | | | | | REVIEW MPA SPACE PROGRAM | | | | |
| | | | | | | | | EXTERIOR SKIN DESIGN | | | |
| | | | | | | | ● DEVELOP STRUCTURAL GRID AND BACKGROUND | | ● BLOW UP TO L. SCALE | | |
| | | | | | | | ● INTERFACE WITH MPA | | ● LIFE CYCLE ANALYSIS | | |
| | | | | | | | SINGLE LINE DRAWINGS. | | ● FINALIZE MAJOR SYSTEMS | | |

### ARCHITECT ENGINEER

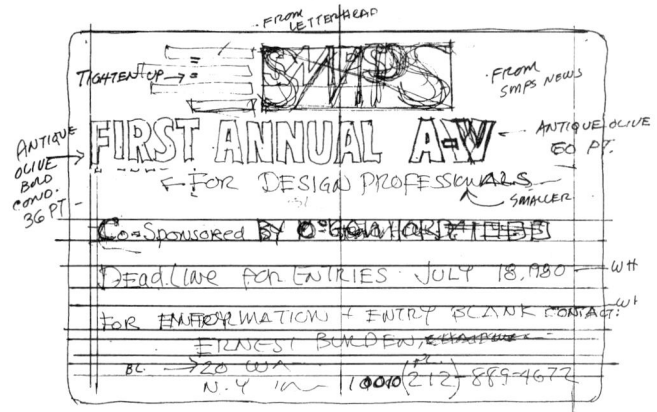

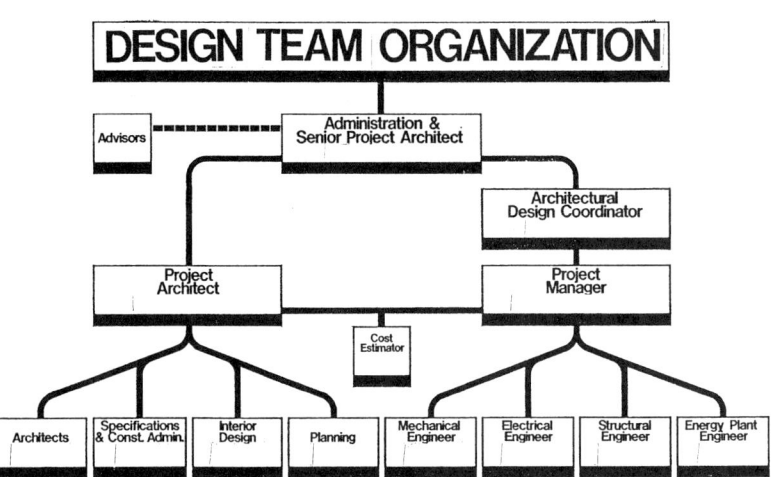

### Overhead Formats

The use of overhead transparencies has been a standard form of business presentations for many years. They were a popular format for presenting diagrams and charts, such as organization charts, time schedules, and other statistical information. The distinct advantages of using overheads is that they are easy to prepare, easy to use, and can be shown in a lighted room facing the audience. The disadvantage of the old projectors is that they were usually dull, consisting of small typed images which no once could read. While the most popular format for artwork is 8 1/2 x 11, the overhead transparency format shows an image of 7 1/2" x 10" divided into 1" increments with a finer grid of 1/4" squares. This is the dimension recommended for the smallest size of type within the frame.

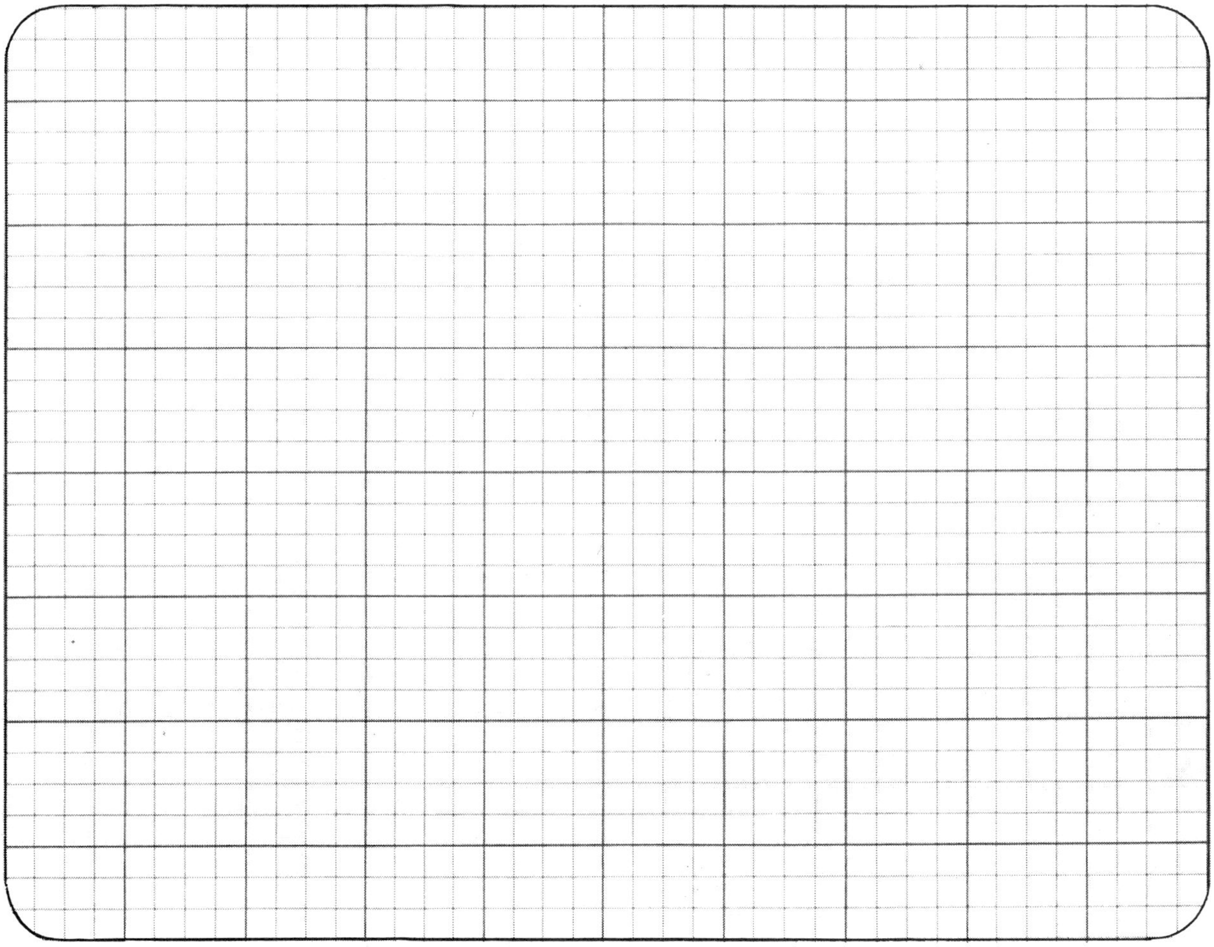

Today, overhead projectors have entered a new age of popularity among presenters. The old advantages are still there and there are new ones as well. Color images can be produced by the computer and projected and the projector itself can be hooked up to a computer for displaying live images. This combination can produce a wide range of possibilities on the production side with unlimited color and graphic capabilities. It can also contribute to a new range of presentation possibilities where preplanned images and live interactive information can be combined.

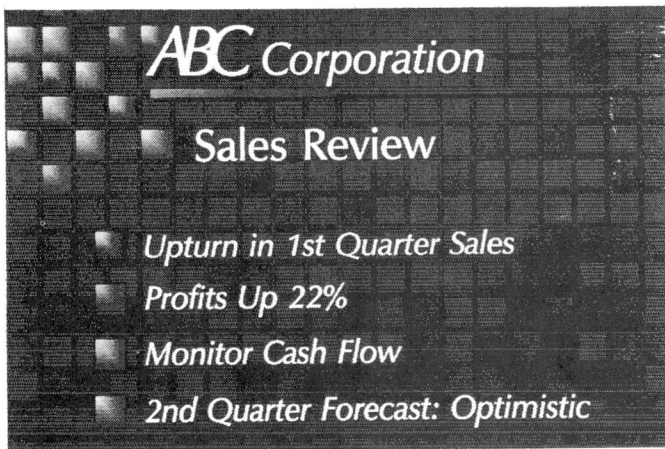

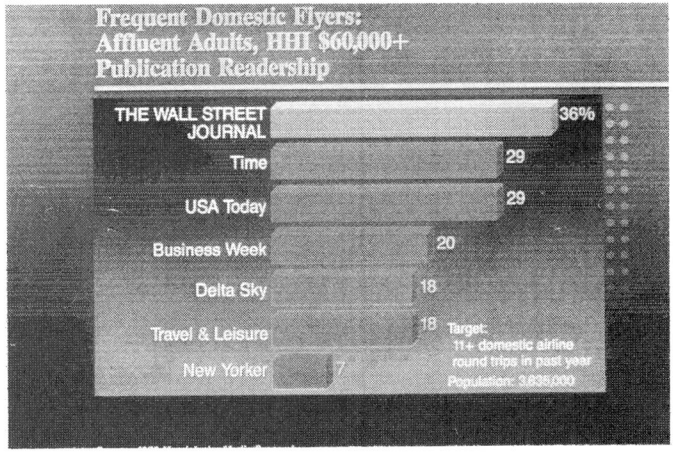

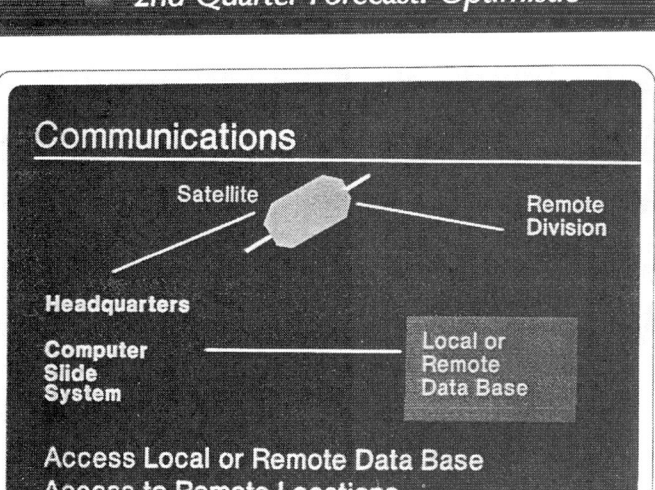

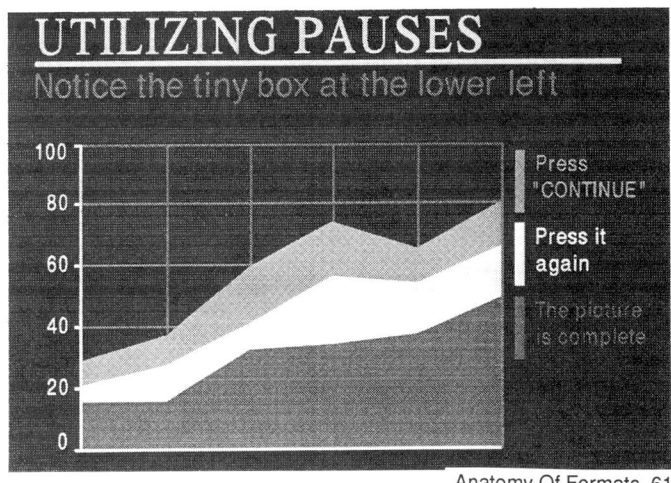

## Organization Charts

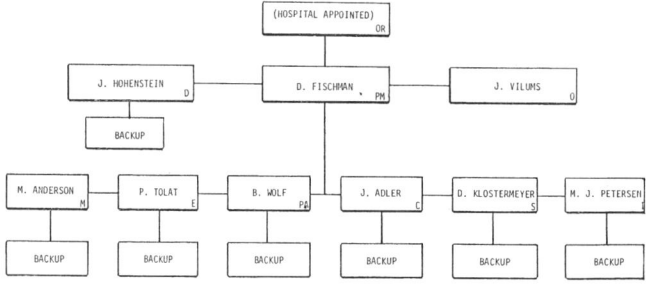

Most organization charts are difficult to read. This is because they are prepared to be read at a close distance in a proposal. Most organization charts are not interesting. This is because they are prepared as little more than names within boxes arranged in some form of hierarchical sequence. Most organization charts are impersonal, that is because there are only names and positions. The organization chart illustrated here attacks all three issues at once.

It begins with a conceptual sketch of the names and positions, just like any other chart, but the similarity ends here. In the following stages, the chart is organized into disciplines and boxes are added on top of the names. These boxes will eventually hold a photograph of the person, and we have the beginnings of a photo organization chart.

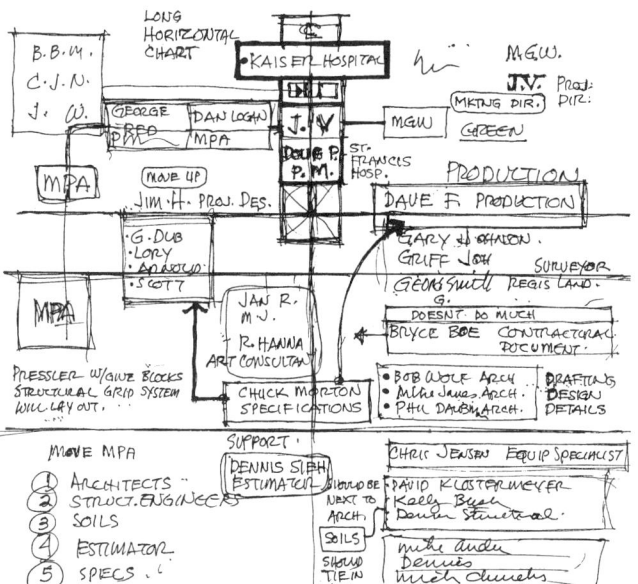

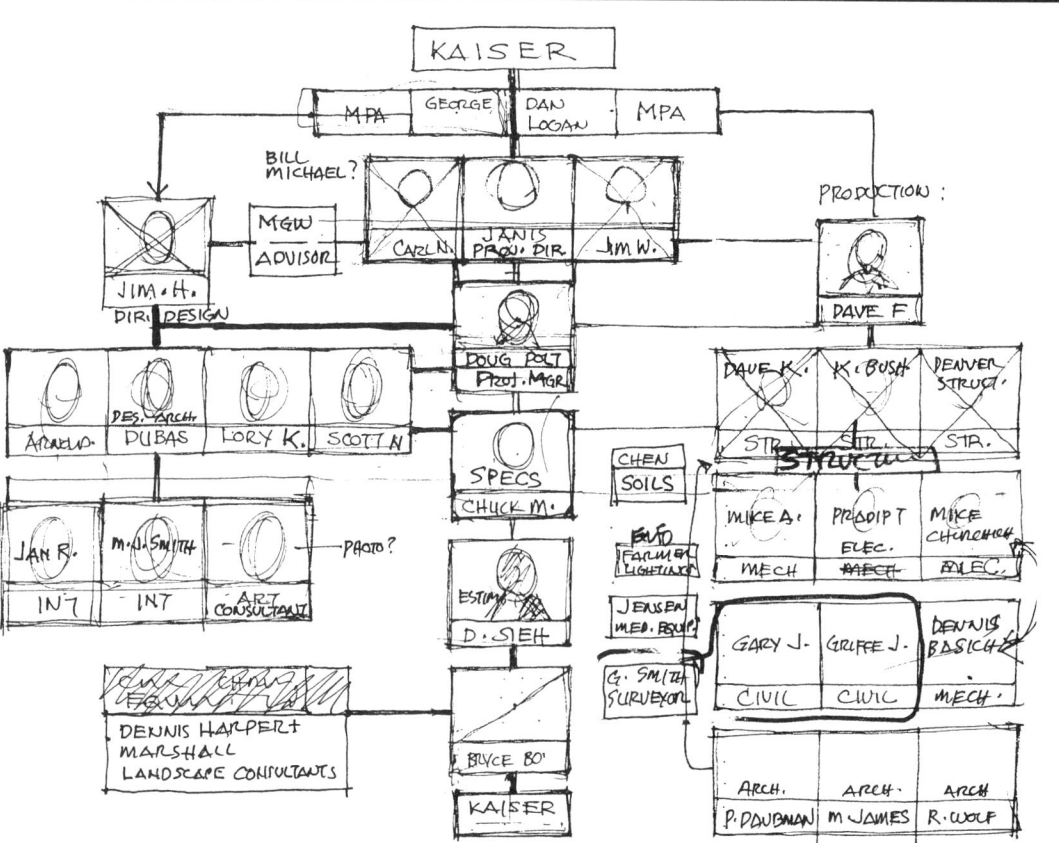

There are several ways to get it to the final stage. First, the pictures may be in slide form, so color prints can be made quickly and inexpensively with an instant slide to print copier. Once you have the image in print form you can gang the pictures together on a board and have screened, or "velox", prints made. They can be reduced to the size of the artwork at the same time.

The larger the size of the artwork, the more accuracy you will have in stripping in the photographs. The pictures will now have a dot pattern and can be further enlarged or reduced as line art, even by photocopy methods. The names of the individuals on the chart are set the same way, either in type or by manual means.

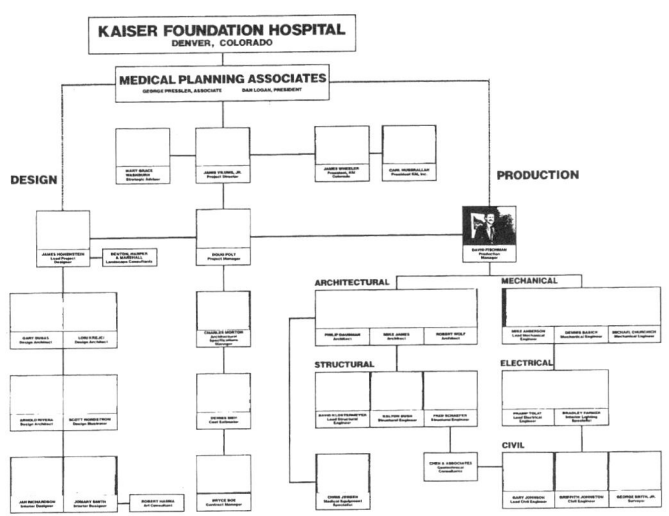

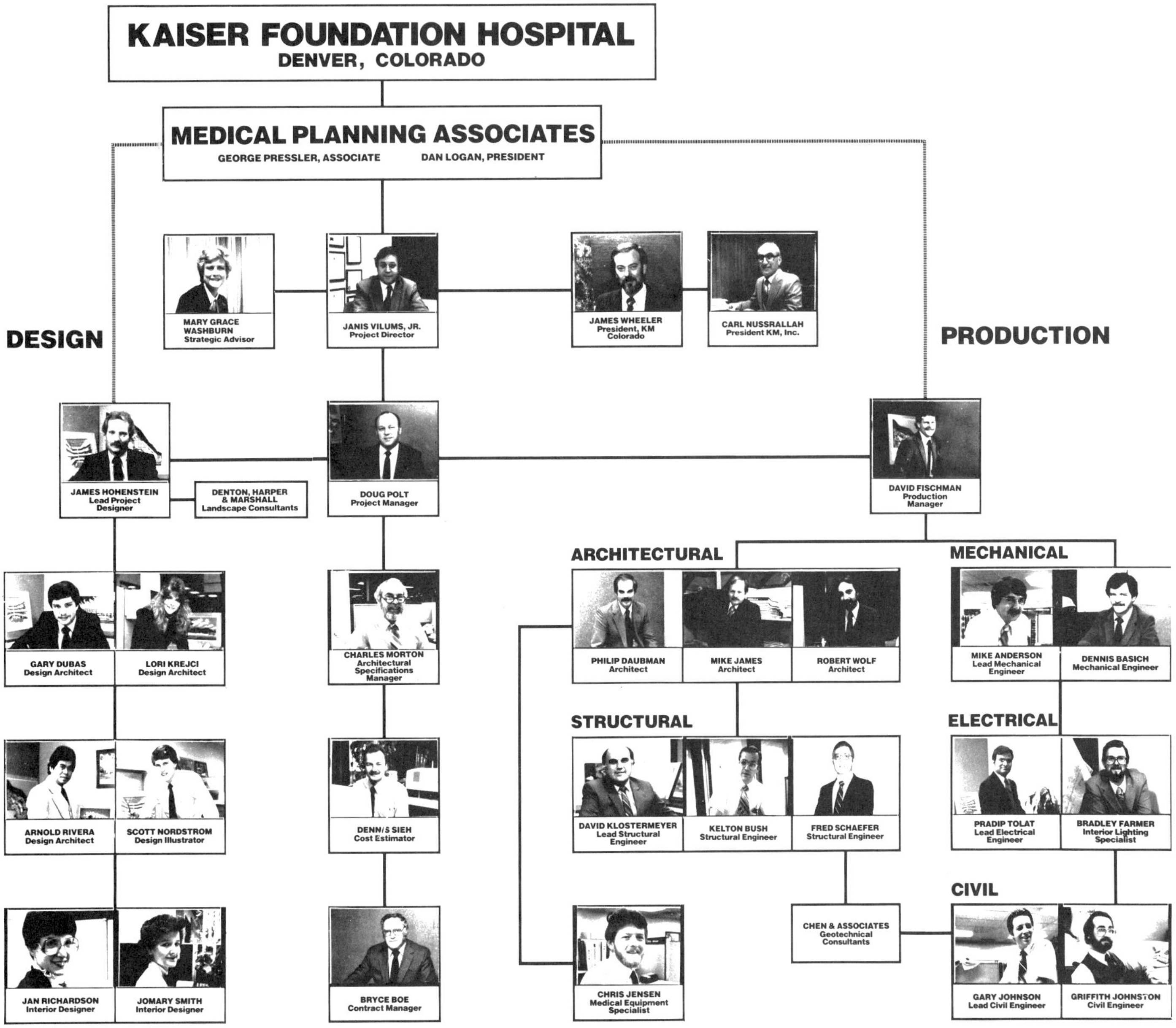

# KAISER FOUNDATION HOSPITAL
### DENVER, COLORADO

## MEDICAL PLANNING ASSOCIATES
**GEORGE PRESSLER, ASSOCIATE**     **DAN LOGAN, PRESIDENT**

**DESIGN**     **PRODUCTION**

- MARY GRACE WASHBURN — Strategic Advisor
- JANIS VILUMS, JR. — Project Director
- JAMES WHEELER — President, KM Colorado
- CARL NUSSRALLAH — President KM, Inc.

- JAMES HOHENSTEIN — Lead Project Designer
- DENTON, HARPER & MARSHALL — Landscape Consultants
- DOUG POLT — Project Manager
- DAVID FISCHMAN — Production Manager

**ARCHITECTURAL**
- PHILIP DAUBMAN — Architect
- MIKE JAMES — Architect
- ROBERT WOLF — Architect

**MECHANICAL**
- MIKE ANDERSON — Lead Mechanical Engineer
- DENNIS BASICH — Mechanical Engineer

- GARY DUBAS — Design Architect
- LORI KREJCI — Design Architect
- CHARLES MORTON — Architectural Specifications Manager

**STRUCTURAL**
- DAVID KLOSTERMEYER — Lead Structural Engineer
- KELTON BUSH — Structural Engineer
- FRED SCHAEFER — Structural Engineer

**ELECTRICAL**
- PRADIP TOLAT — Lead Electrical Engineer
- BRADLEY FARMER — Interior Lighting Specialist

- ARNOLD RIVERA — Design Architect
- SCOTT NORDSTROM — Design Illustrator
- DENN/S SIEH — Cost Estimator

**CIVIL**
- CHEN & ASSOCIATES — Geotechnical Consultants

- JAN RICHARDSON — Interior Designer
- JOMARY SMITH — Interior Designer
- BRYCE BOE — Contract Manager
- CHRIS JENSEN — Medical Equipment Specialist
- GARY JOHNSON — Lead Civil Engineer
- GRIFFITH JOHNSTON — Civil Engineer

Anatomy Of Formats 63

## Organization Charts

An organization chart has two main purposes: first, to visually show the structure of an entity, and second to list the individuals that make up the organization. In other words, it is a chart of people coordinated with their positions. What is missing from most charts is the recognition of the individuals. One can clearly see the difference that photos make, in the two sketch examples to the left. The first is impersonal compared to the one with a picture of the individual worked into the design.

Several organizations have made use of this technique in the development of their charts. To a stranger the faces do not mean any more than the names would, but to a client or person familiar with the firm, the personal touch adds a great deal. Clients ultimately hire design firms based on the quality and experience of their people, as much as they do on reputation of the firm. Therefore it is logical to express this personal concept in all the promotional materials.

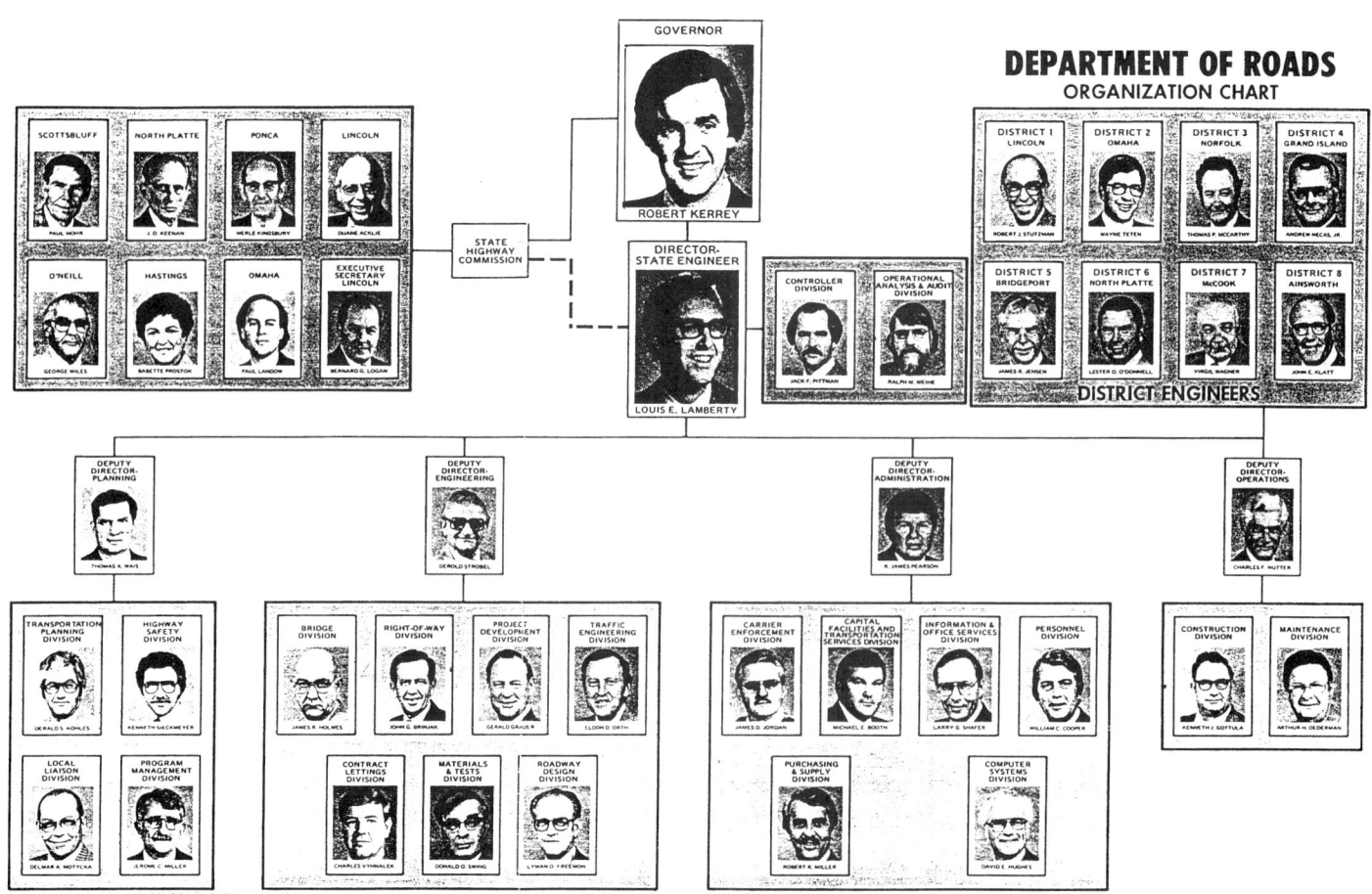

There are several ways to create a photo organization chart. The easiest and least costly is to prepare artwork, as described previously, and produce the chart as an oversized piece of artwork, then reproduce it to fit the application. There are digital photo-screen copiers, for example, that will faithfully reproduce a combination of line and photographic images.

Alternatively, the photos can be turned into a dot pattern or screened print, and then reproduced as a line image on any photocopier. In fact, it is no more trouble to produce one in color, using the technology of the color copiers. Color repros can be made of each individual, either from slides or prints, cut out and pasted on the chart to study the final positioning, as was done in the example below.

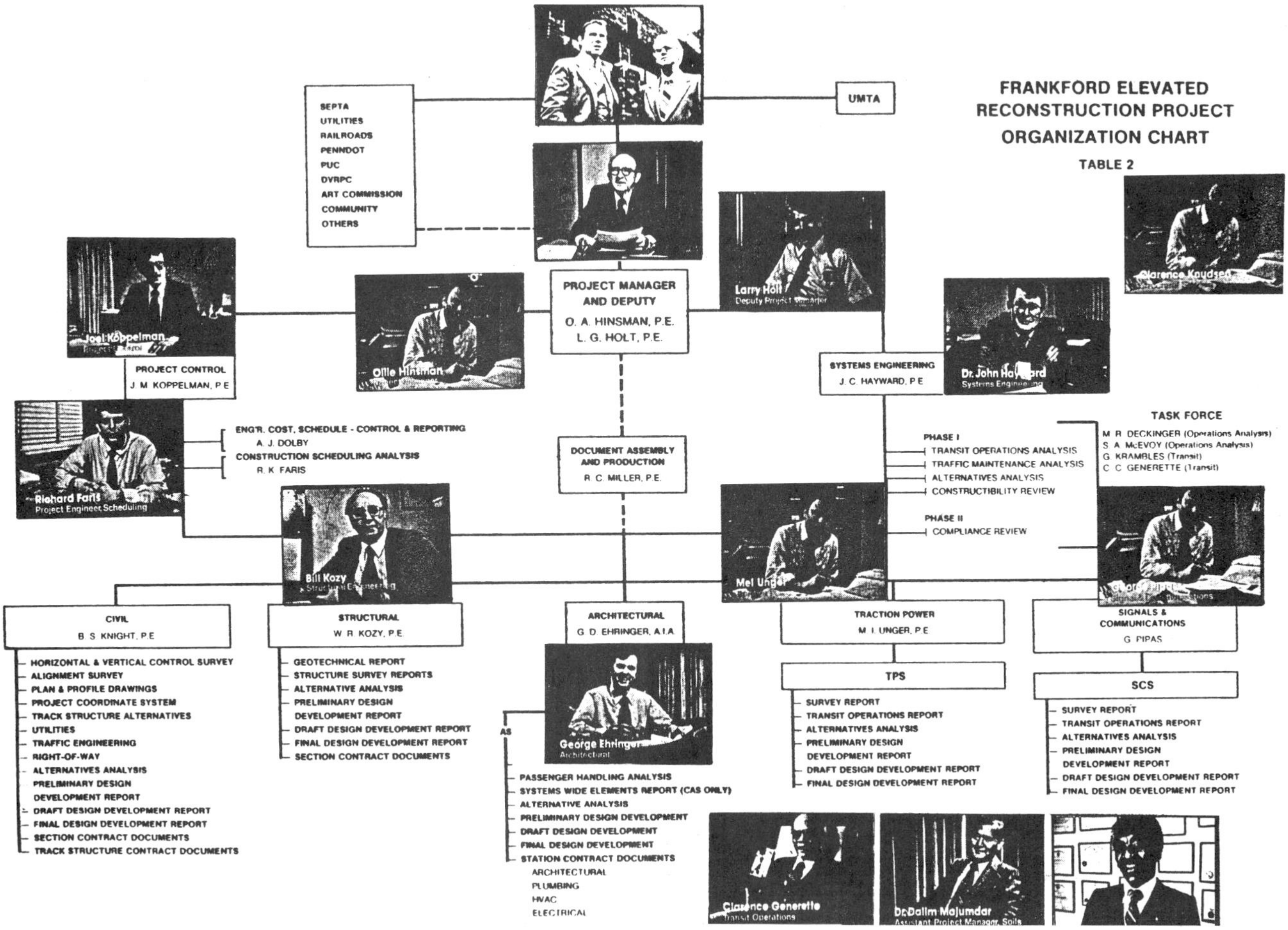

Anatomy Of Formats 65

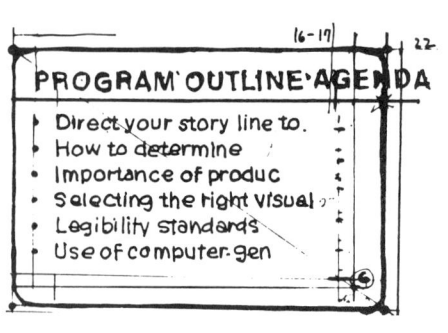
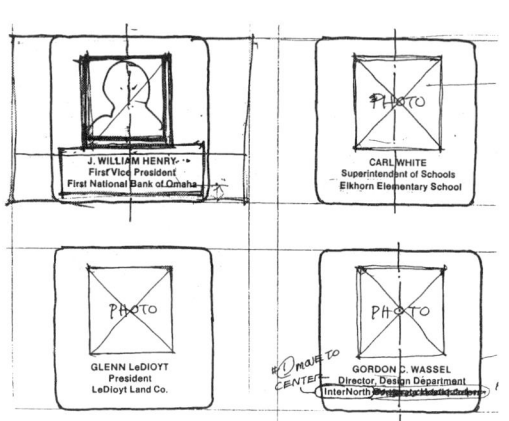

## Slide Formats

The production of 35 mm slides has changed dramatically as electronic imaging has replaced traditional production methods. The electronic formats have combined word processing programs with graphic capabilities to produce an infinite variety of colored type and graphic shapes. Charts, diagrams, and artwork can now be produced electronically at reasonable cost, and in less time than traditional methods. This does not mean that slides can be produced without first establishing a design format. This is usually done through rough sketches and accompanying notes or specs for the type, such as style and point size. These specs are most often related to a horizontal format of 2 x 3 proportion, although occasionally a vertical 3 x 2 or square 3 x 3 format can be used. There are also a wide variety of masks that fit within these formats for special effects or multi-image use, such as split screen images.

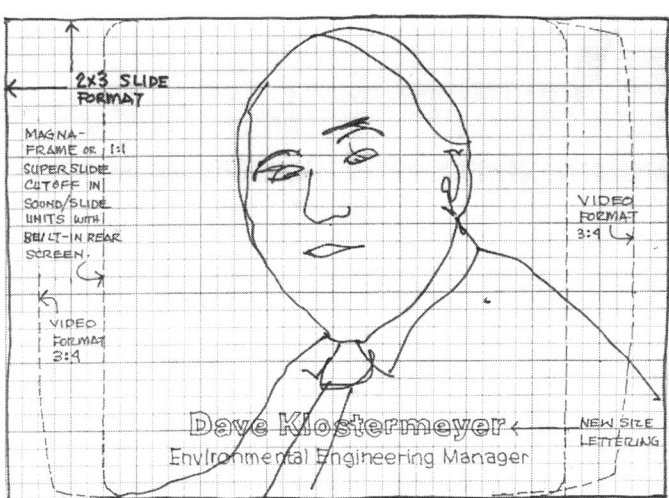

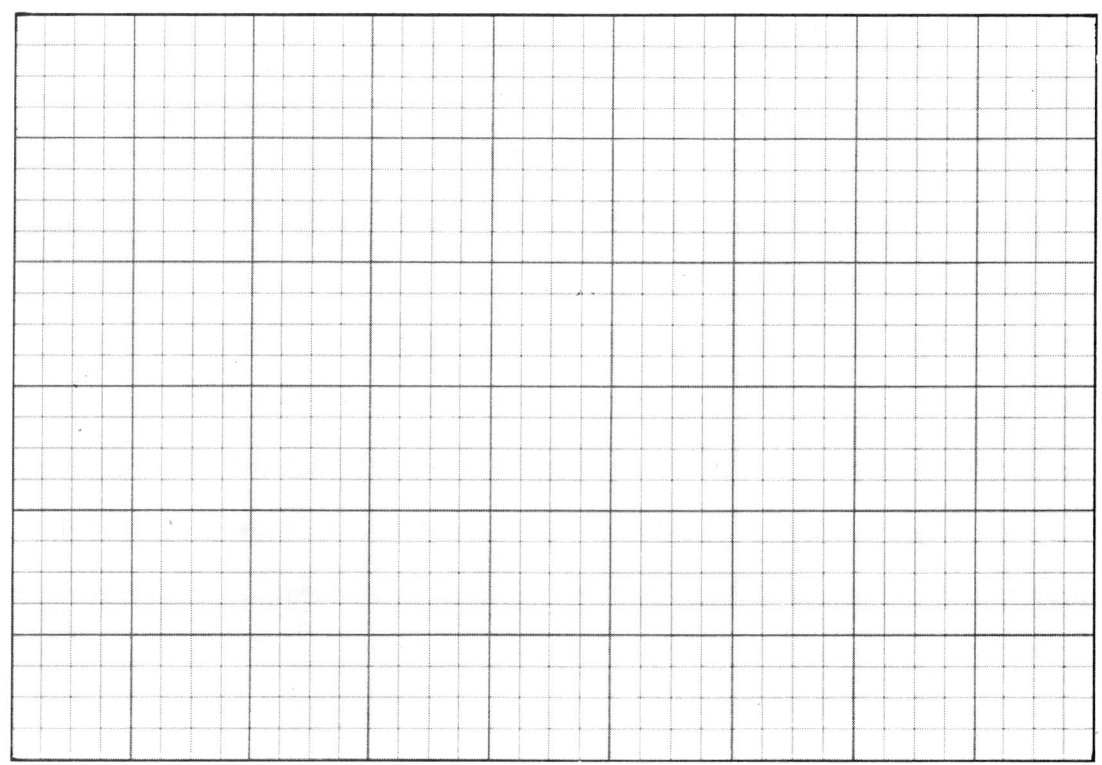

Once a format has been established, the program can be arranged on a storyboard. The storyboard is a series of miniature images representing the program in sequence. Ideas can be first sketched on a storyboard, then used throughout the production process for notes and future developments of the program. The final storyboard would include the script in relation to the visual images, as well as cue marks and other notations.

The primary concern in laying out the specs for slide images is legibility. This factor has less to do with how large the image is when projected than it does with the relative size of the type within the slide format.

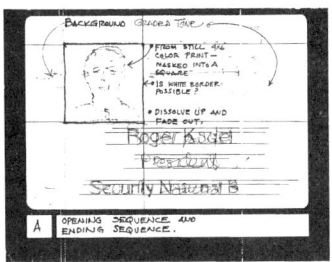

Title slides can also be produced in many different ways without electronic formats with the end result much the same. Titles can be produced on acetate overlays, which are then placed over a photograph. Positioning of the title is an easy matter. Another alternative is using transfer lettering placed directly on the photograph; however, the lettering must be accurately positioned and cannot be moved around. Shadow lettering is also available in transfer type and is effectively used when the background is too light for the letters to show. The electronic formats allow much more variety in "drop shadow" letters wherein variations can be created within the letters and their size, style, spacing and color.

Along the way laser prints can be made which show the elements that have been designed. Once the slides have been produced in an electronic format and all revisions have been made, they can be output on laser printers in color, plotted by thermal plotters for hard copy color reproductions, or fed into film recorders hooked to the computer for 8" x 10" transparencies, 35mm slides or video.

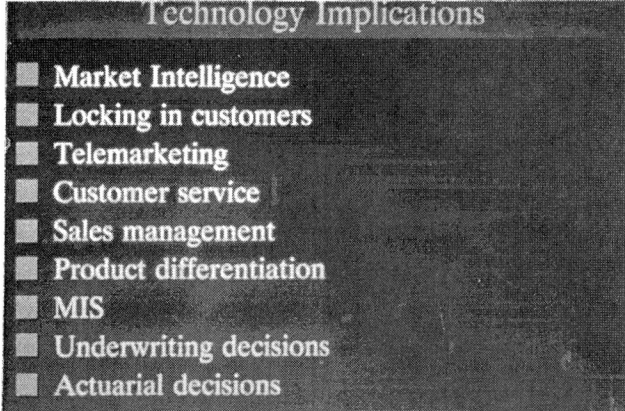
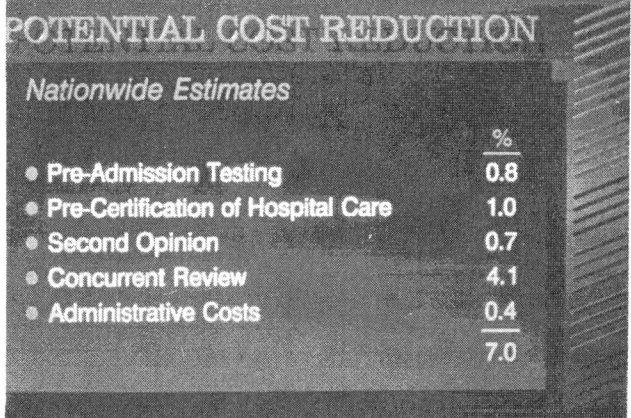
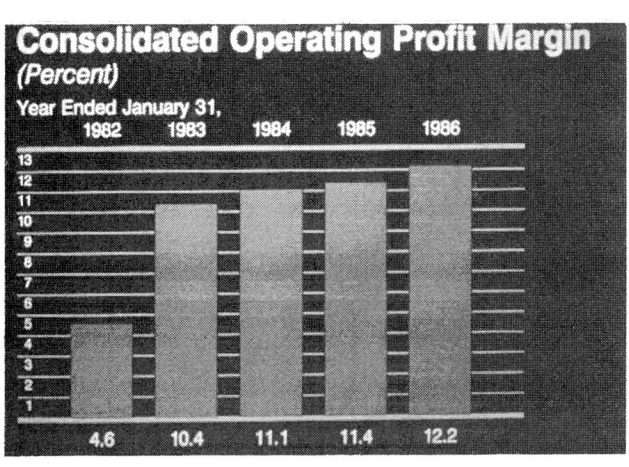
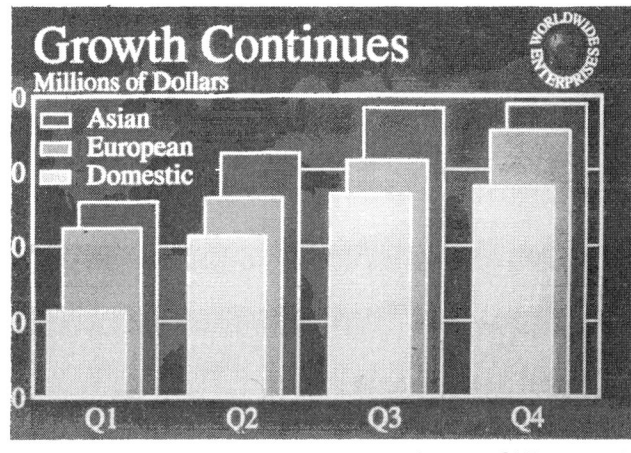

Anatomy Of Formats 67

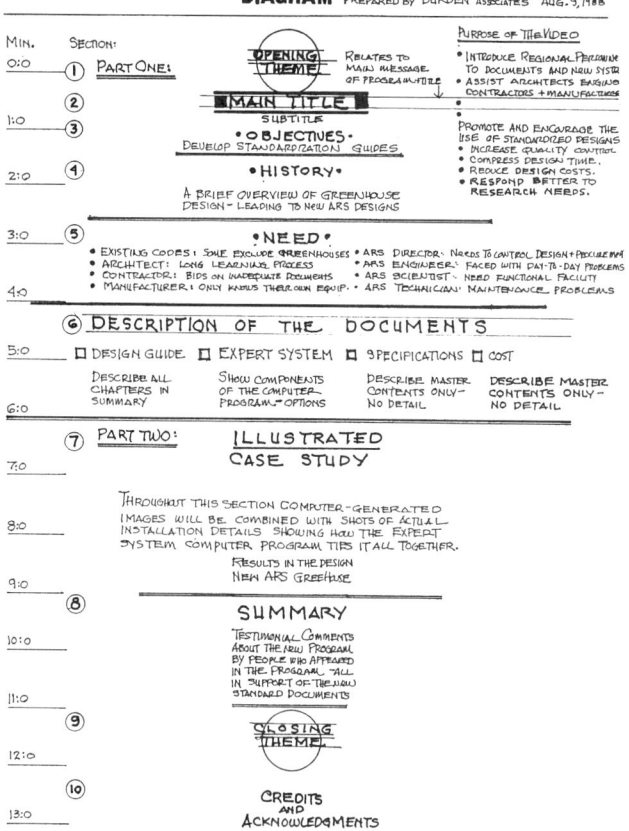

## Video Formats

In video production, "format" refers to the width of the tape, since all tapes show in the same aspect ratio of 3 vertical to 4 horizontal. There is 1" (broadcast), 3/4" (industrial), 1/2" (professional), 1/2 " (consumer), and 8 mm (consumer). The size of the tape has little to do with image, as horizontal resolution determines the quality. Some 8mm formats, for example, can deliver resolution equal to 1" broadcast.

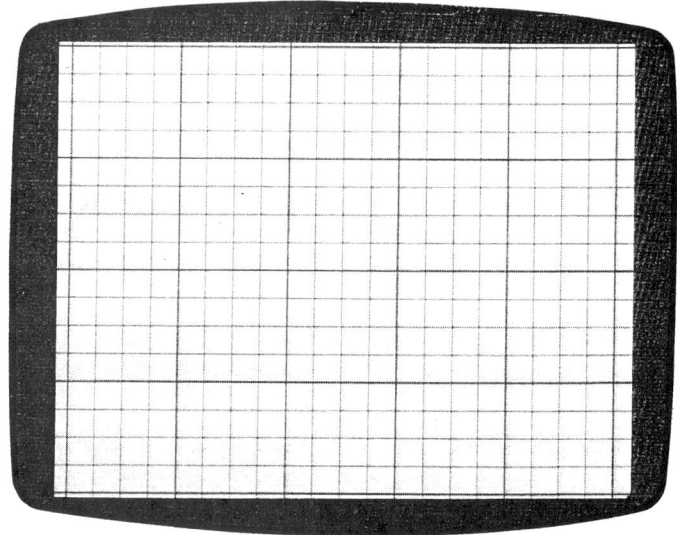

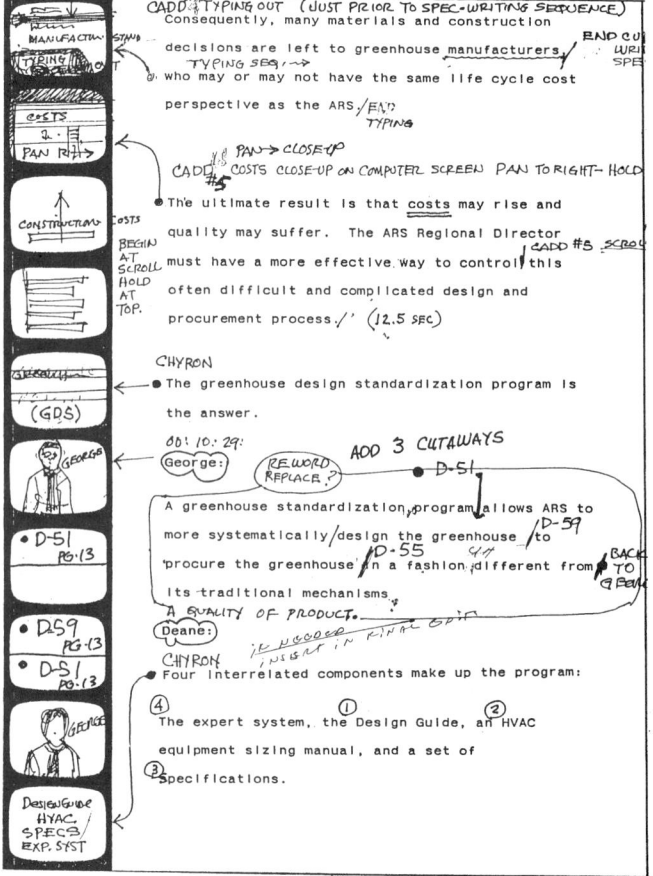

Therefore we need a monitor to view the images. These monitors are different in that they show different amounts of the picture. To overcome this, video studio monitors and editing equipment have a function that overscans the image to make sure that titles and other important information is not cut off in the final version of the program. There is another feature called the "safety zone" which is a grid that can be superimposed over the image to show that it will not be cut off by the monitor. Another grid can be superimposed over the image which shows the center line, plus horizontal and vertical lines. This grid can be used to align graphics and other diagrams.

The type used in video production is called "electronically generated." Type fonts are stored in a "character generator" (CG) and are used just like a typesetter in terms of size, spacing and type style. Also they can have special effects such as drop shadows, outlines and dropout characters.

The basis of any video program is the concept diagram and storyboard. The concept diagram outlines all the elements that will appear in the program in sequential outline form. The storyboard coordinates the script with the visuals so it is an important element throughout the entire planning and production process. It can be updated as the program develops. A storyboard format can also be used as a visual record of material that has been recorded on film or tape. Usually raw footage is taken to a lab and transferred to 1/2" stock and a visible "time code" is put on each reel. This shows up as running numbers on the video indicating hours, minutes, seconds and frames. (There are 30 frames for every second). These numbers, now referred to as the "address tracks" are used to edit the footage.

Everything can now be related to these numbers, such as the "in and out" points of a scene. The difference in the respective numbers indicates the length of a sequence. These edited sequences are jointed electronically with other sequences, and the program builds in this fashion. The only visible evidence of these maneuvers is the storyboard and script sheets.

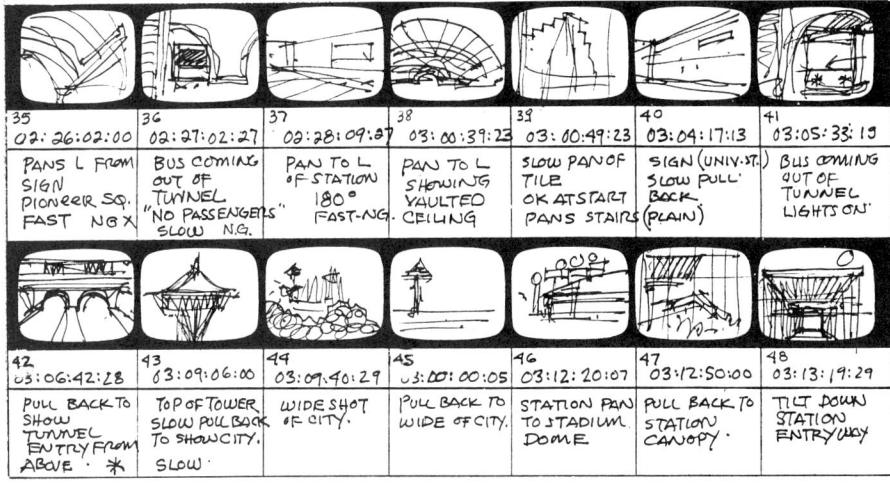

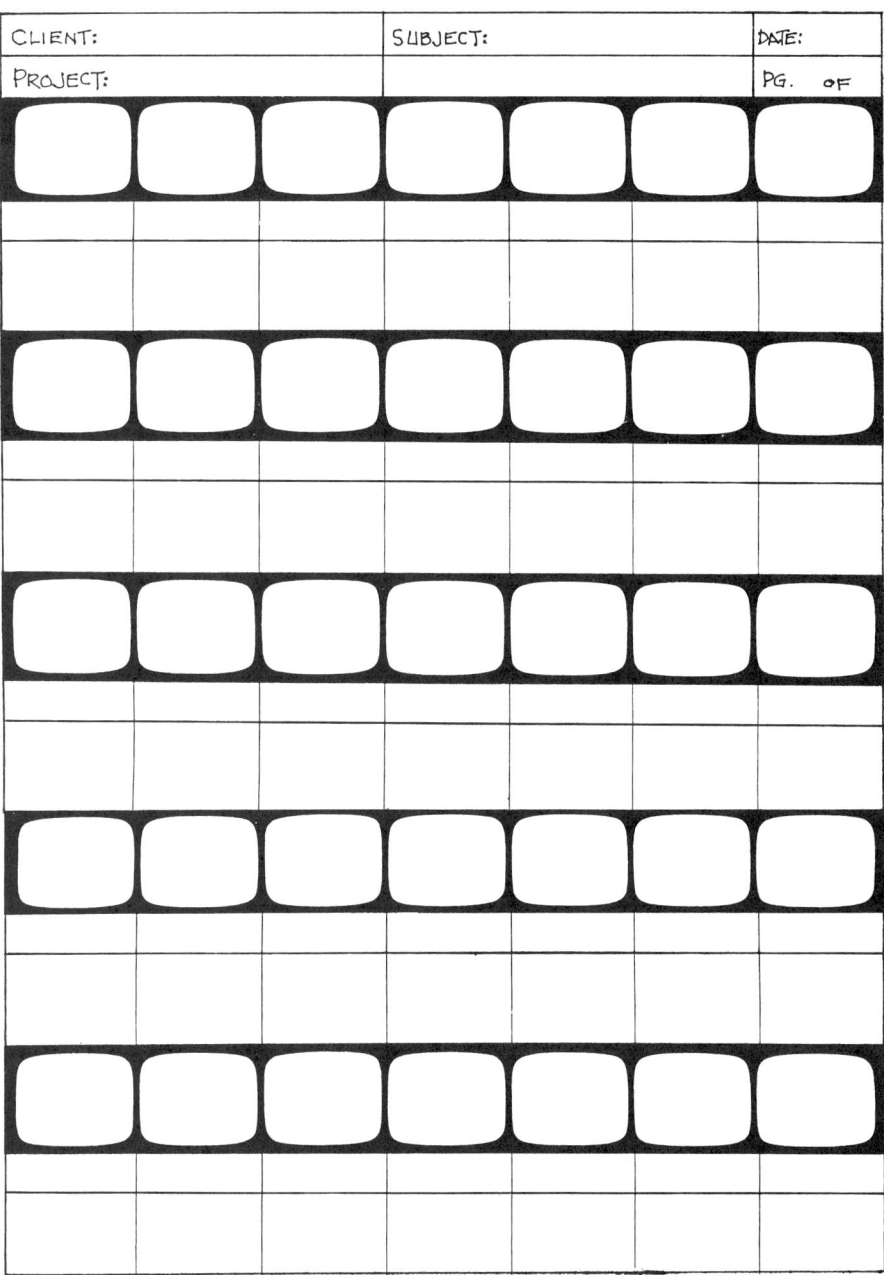

Anatomy Of Formats 69

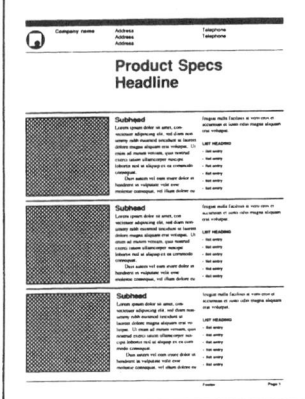
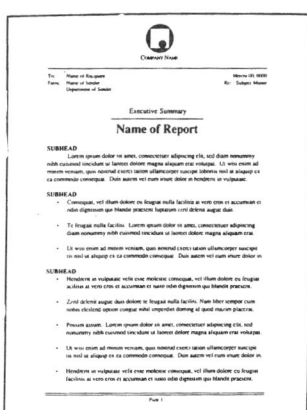
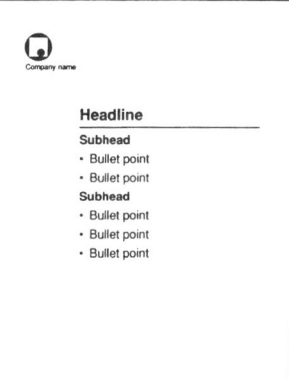
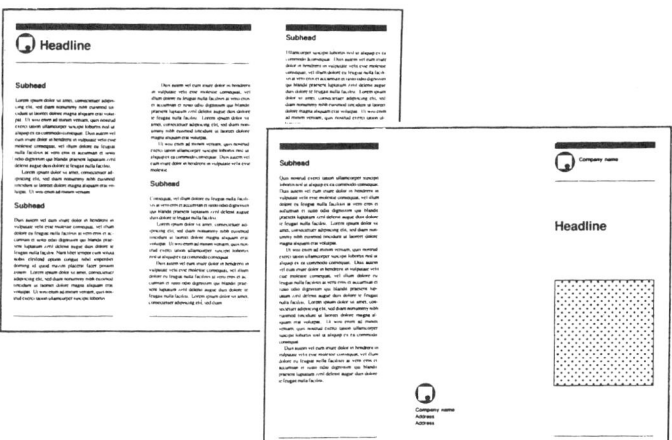

## DTP Templates

A template for a publication is a predetermined format that has been created previously. It can exist within the desktop publishing program or it can be created using that program. Many software programs come with certain built-in templates. They are, of necessity, very basic and very general to satisfy the most standard applications. The basis for a template is provided by the internal guides for margins, columns and rulers. These non-printing guides are dashed lines that help align text and graphics. They appear on the screen, but they do not appear in the printout.

Next, rules and column guides can be positioned using the rulers and "snap" commands. Once all the lines are in place, the grid created can be saved within the program, as a template file. Once it has been drawn in the DPT program it can be printed out as a master page. A handy complement to any electronic template is a manual that provides hard-copy examples of what the pages consist of, along with explanatory type specifications and general guidelines for

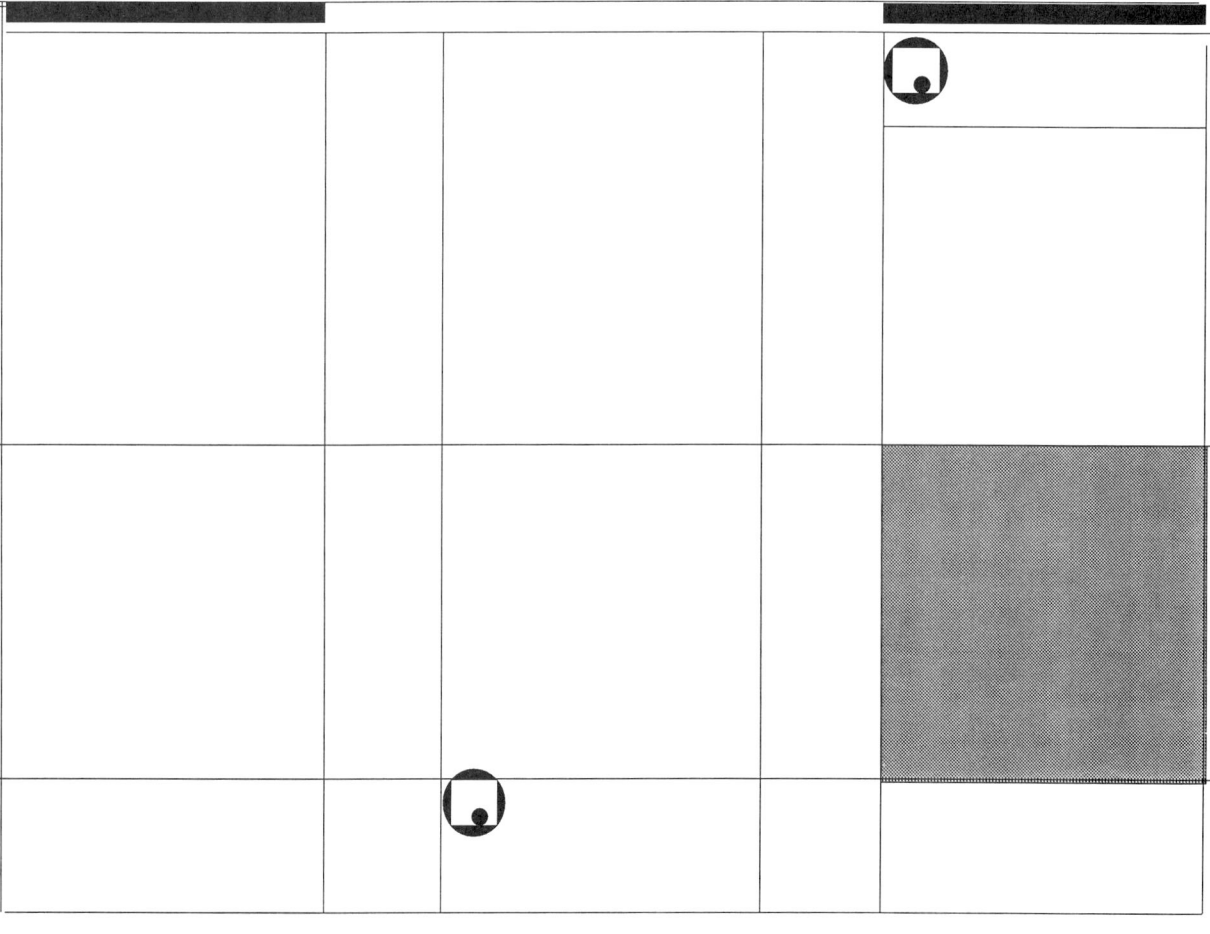

Templates are available in the software supplied by desktop publishing programs. Others can be ordered as add-on products. These provide fill-in-the-blank designs for all manner of promos and business forms, newsletters, magazines, fliers, brochures and catalogs. The dummy type used in the templates for headlines and body copy is replaced as one fills in the new material. These templates can be altered within the system to suit a particular use, then stored for future use. The grids in this book can be transferred to any desktop publishing format and made into templates. This can be done by either scanning them in and constructing them internally, or by constructing each element of the grid using the dimensions shown on the grid. They are shown in picas and points. Naturally, these grids can be altered to individual requirements.

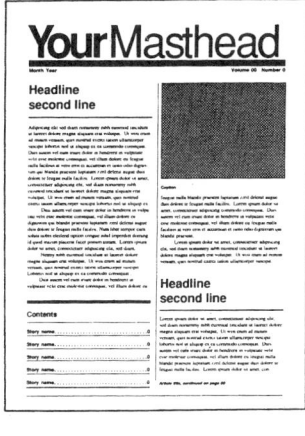 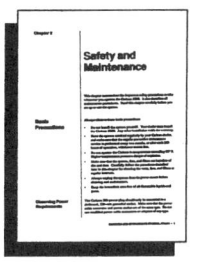 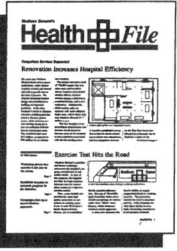 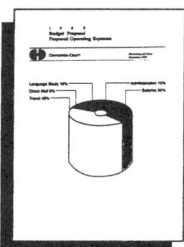

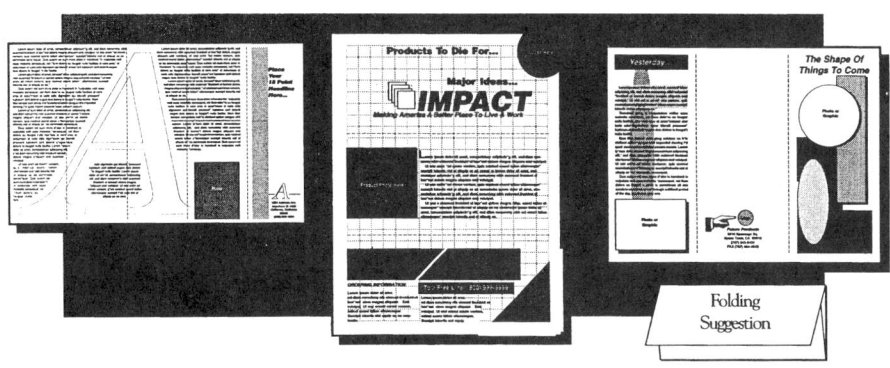
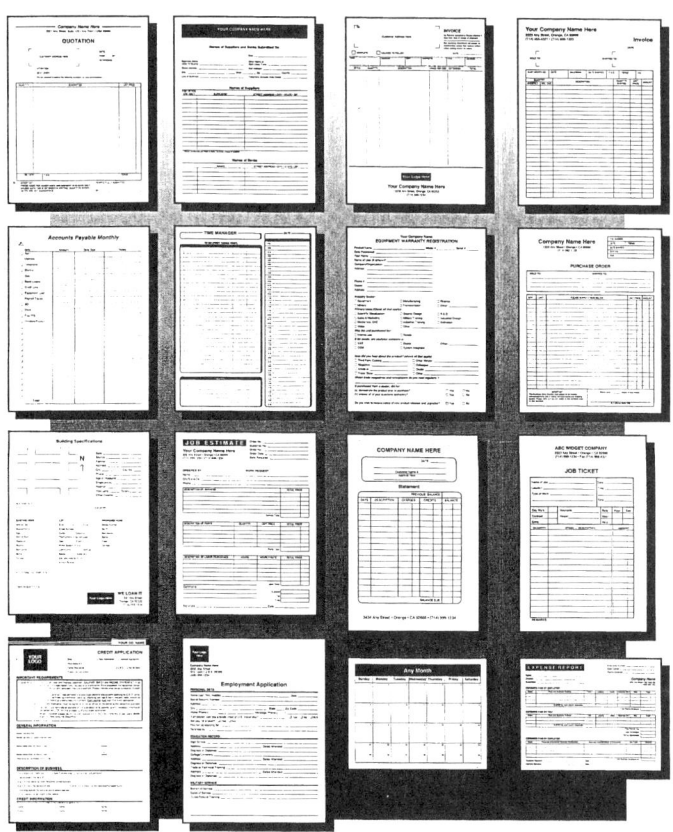 

## Electronic Formats

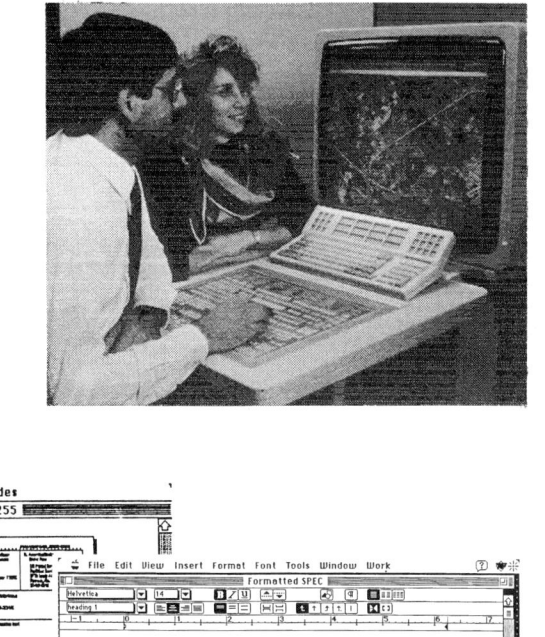

The computer revolution has changed every aspect of design and communication, not only in the way the material is produced but in the way it looks. There are electronic formats now for every task that we used to do manually, or even by calculator. This applies to all aspects of office practice, such as spreadsheet calculations and other financial management tasks. It applies to the design of publications and business forms, newsletters, proposals, and other periodicals. Electronic formats are used to study all forms of engineering data, from soil testing to beam and truss diagrams, and provide accurate calculations at the same time.

In the area of project management electronic formats can schedule projects, compute worker hours, provide cost estimates and display information that can be changed instantly, should conditions change. There are electronic formats for all forms of charts and diagrams, including organization diagrams. In the area of design, there are bubble diagrams programs, adjacency and stacking diagrams, and facilities planning and management programs. For production work there is a wide variety of programs for drafting, notation, and specifications. For advanced production techniques there are programs for advanced graphics and dimensioning in a two dimensional mode, and there are numerous drafting programs with standard details and specifications.

Cost Accounting

Project Management

Specifications

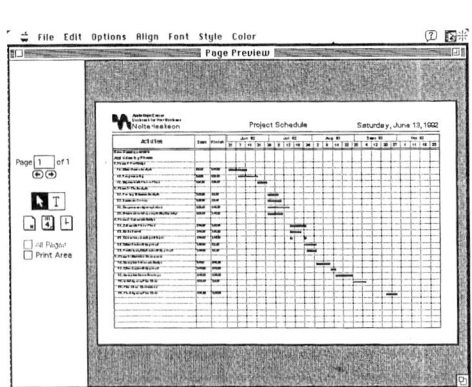

Project Scheduling

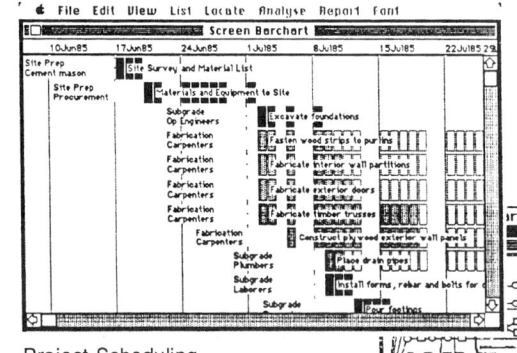

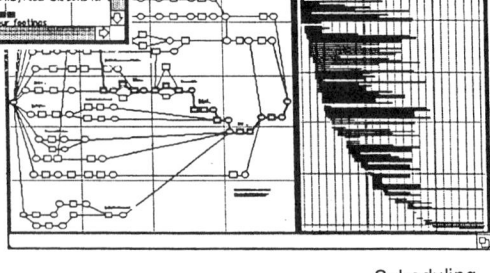

Scheduling

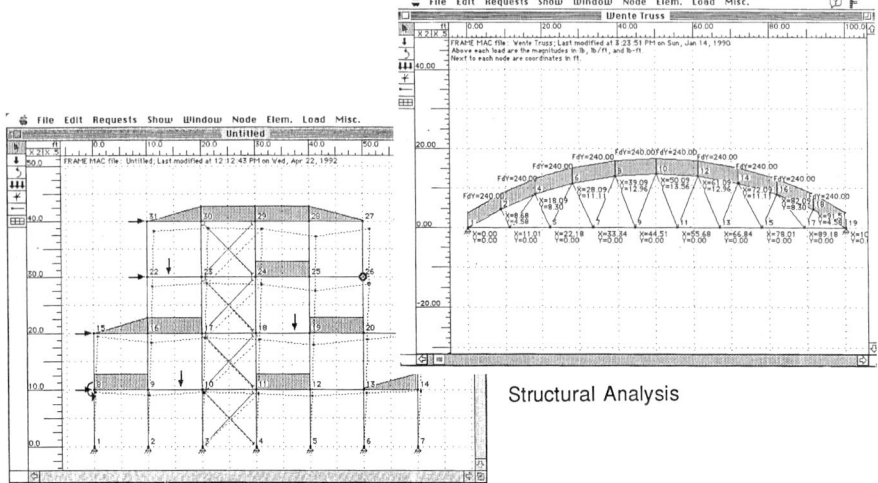

Structural Analysis

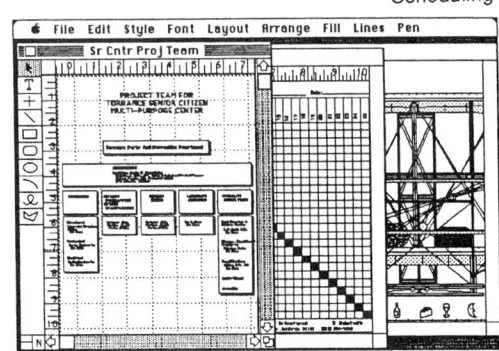

## CADD Formats

In the three dimensional mode there are a number of perspective programs whereby viewpoints of projects can be created that were previously impossible. These perspective viewpoints can be enhanced with other elements such as trees, cars and people to create a more realistic atmosphere. Materials, colors and textures can be added to these images to provide photorealism to any scene. Buildings can be created in one program, then imported into another when they can be superimposed over realistic views of the site. The combined formats of computer and video allow the recording of animated sequences through these spaces from virtually any angle and at any speed. The video image can now be controlled and edited by the computer. Sound can be added, and the entire presentation process can produce results never before imagined in a two-dimensional world. Multimedia is not only the format of the future, it is the format of the present.

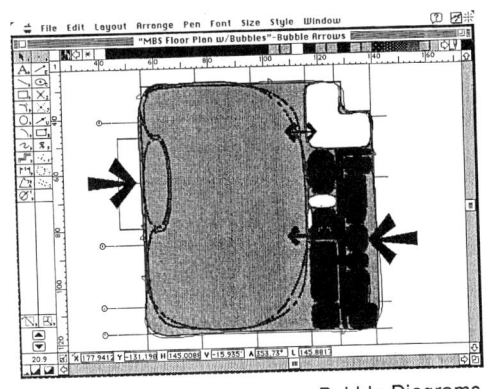
Bubble Diagrams

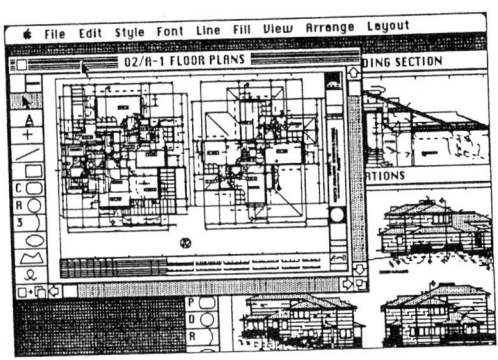
MacDraft

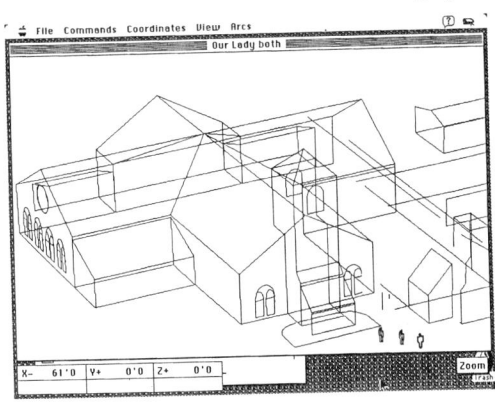
3D Orthographic

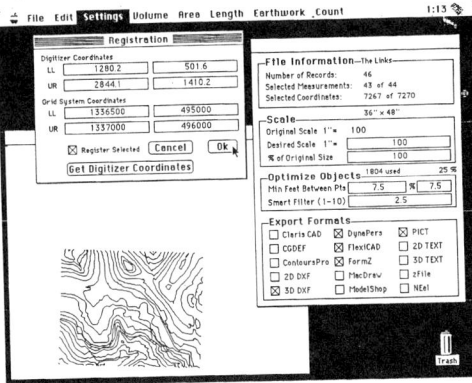
Topographic

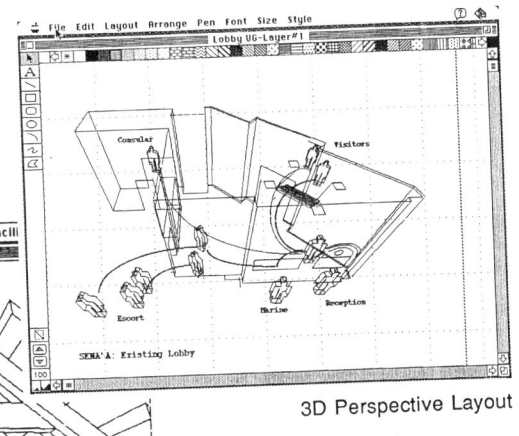

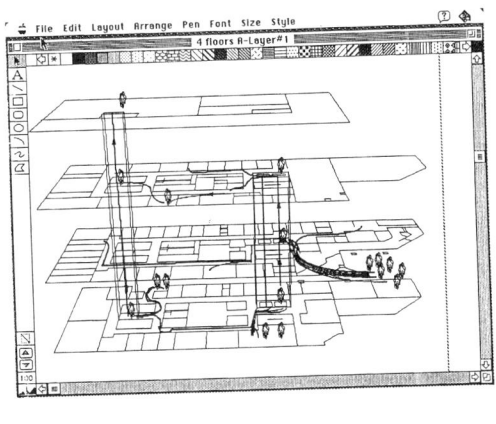
3D Perspective Layout

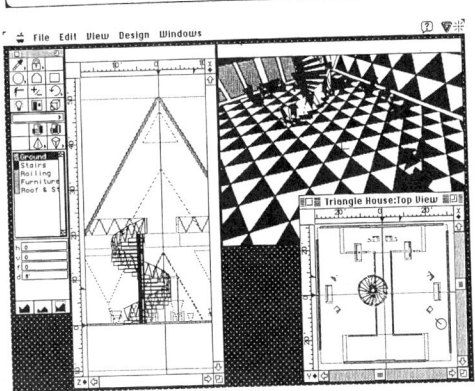

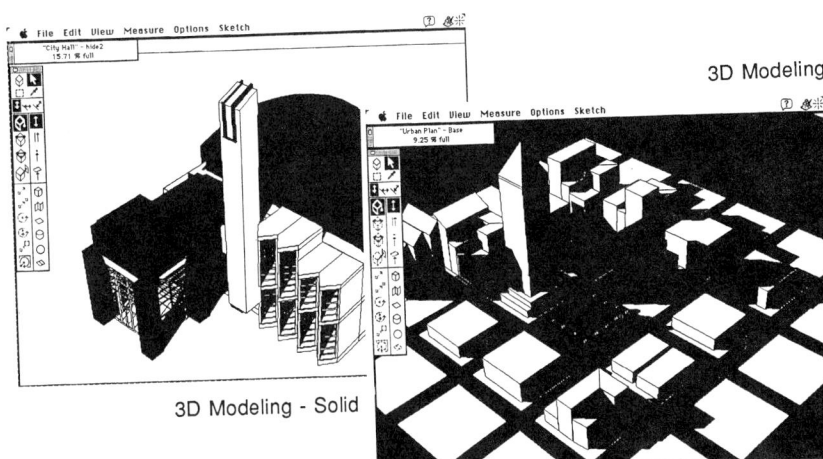
3D Modeling

3D Modeling - Solid

Anatomy Of Formats 73

# PART TWO

# 3 Generic Grids

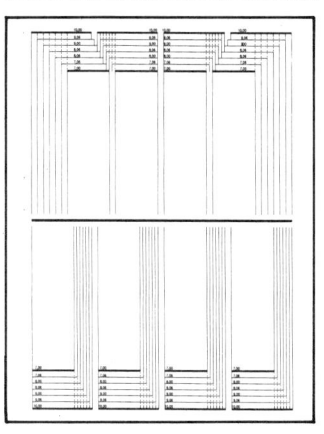

### Master Column Grids

Shown here is a page layout of all the grids in this section, from a single column grid to a twelve column grid. The width of the column is 44 picas, which is the maximum column width to work with on an 8 1/2"x11" page. This width is constant for all the grids in this section. If this width changed, everything would change in proportion to it. Each column is separated by a 1 pica margin, as indicated. The resultant column width, as expressed in picas and points, is also indicated in each column.

The smaller composite illustration on the left shows the relationship of similar modular columns. For example, the top left image contains the 1, 2, 4, and 8 column modules, all separated by the 1 pica margin. The one on the lower left shows the 3, 6, 9 and 12 column combinations. The one on the bottom right shows the 5 and 10, and the 7 and 11 column modules. Any of these can be combined in a page layout, provided certain transition measures are taken. Many of these combinations are demonstrated later on in the case study formats.

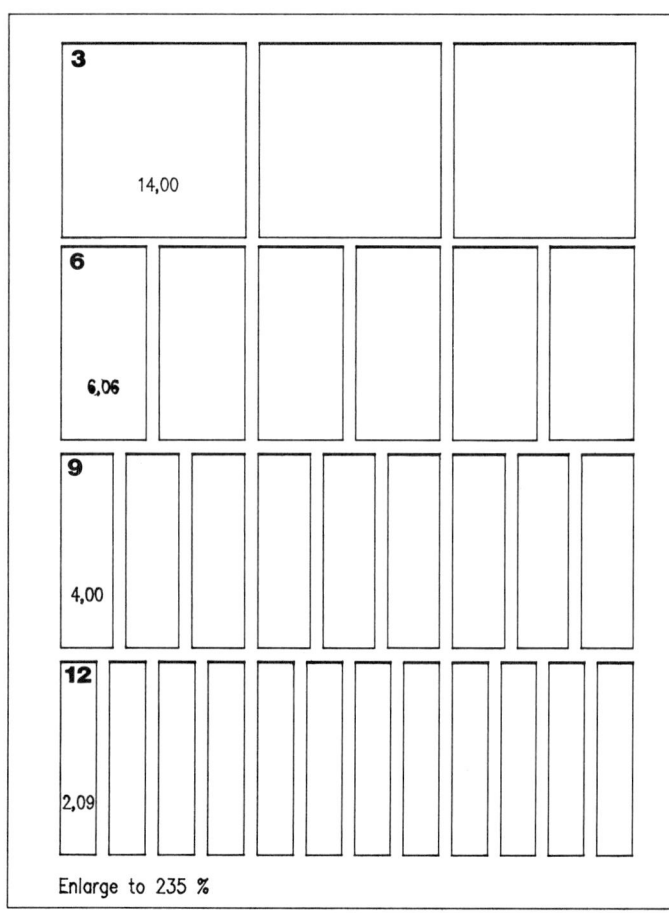

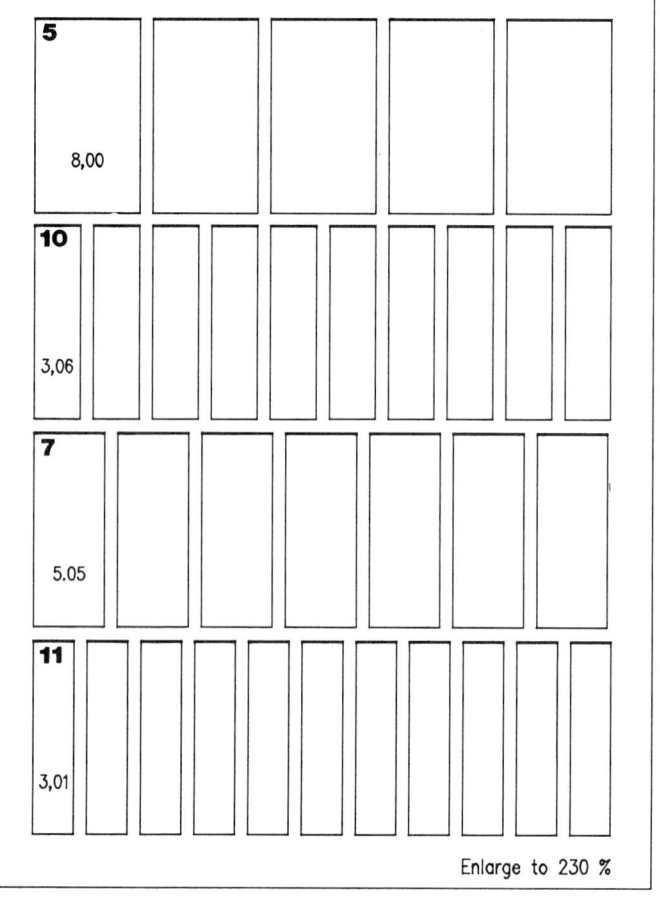

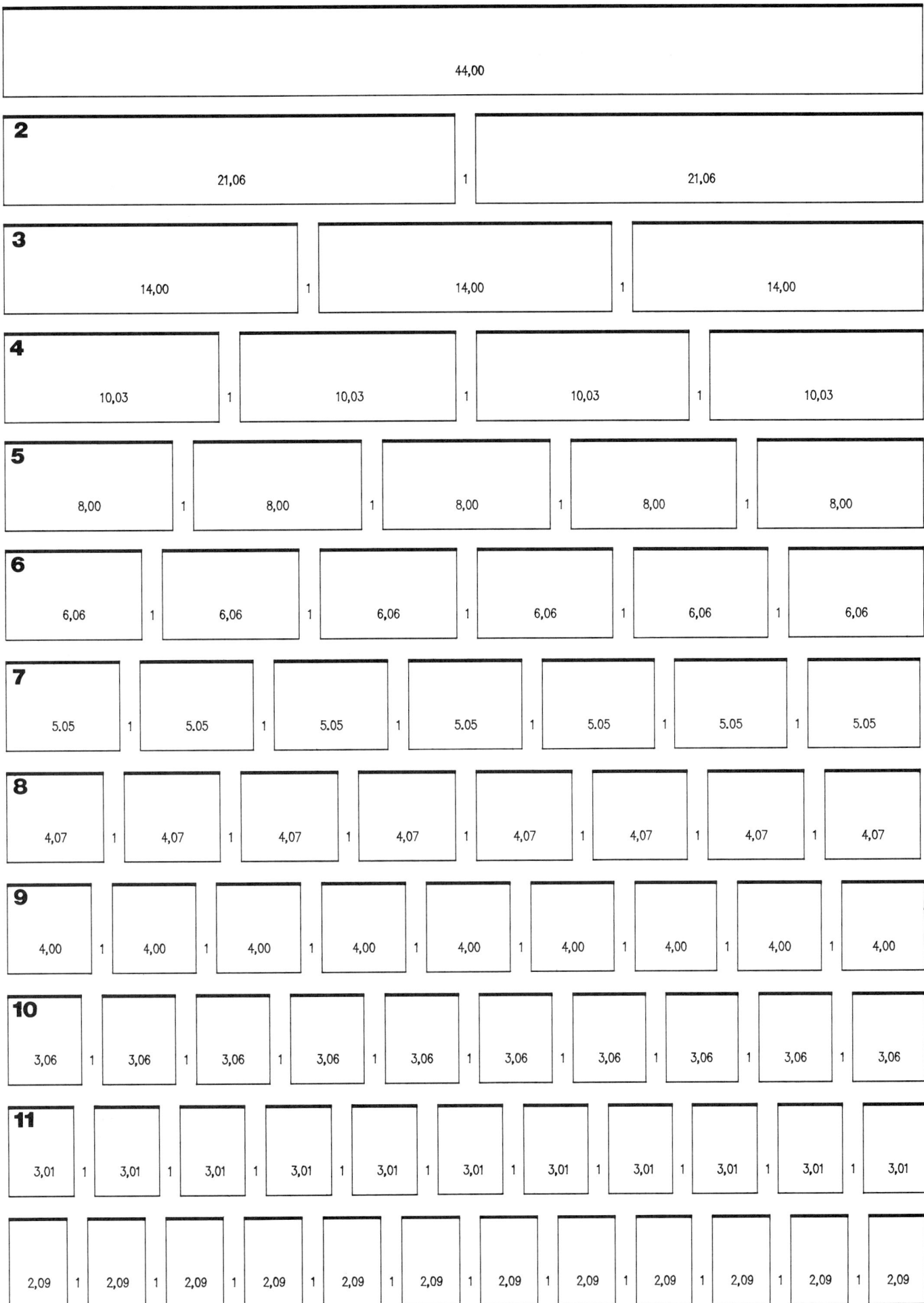

Generic Grids

## 1 Column Grid

The single column format shown here requires relatively little planning, and allows you to design the layout quickly. This format is very useful for business plans, reports, press releases, proposals, announcements, manuals and other internal communications. The symmetrical single column grid extends from the maximum width of 44 picas to a minimum of only 18 picas, in 2 pica increments, one from each side. The narrower columns can be positioned anywhere within the wide single column by simply following the dotted or dashed lines.

The symmetrical column can become dull and uninviting if not handled properly. One way to compensate for this is to offset it on a page. Therefore the margins on this grid vary in 1 pica increments on the left. This also yields columns of type that are 1/2 pica different than the one above. Naturally, the column can be positioned in any location on the page. The illustrations here show the various column widths in relation to the overall page. The thumbnail illustrations show the minimum and maximum for each set of margins.

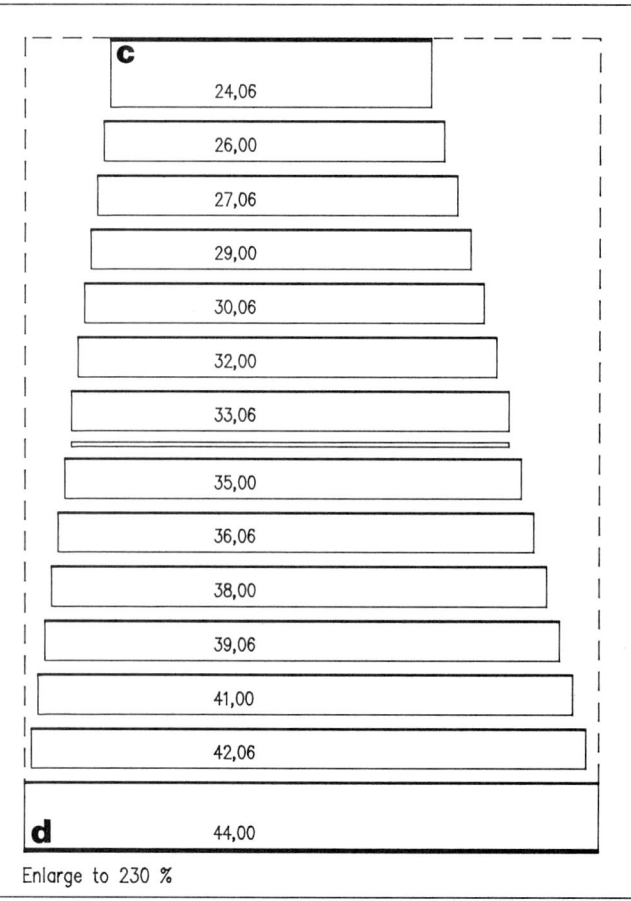

Enlarge to 230 %

Generic Grids

## 2 Column Grid

The grid with two equal columns is a very versatile and popular format. It can be used for a variety of publications such as newsletters, brochures, fact sheets, and catalogs. With two equal columns, the line length can vary from a narrow 11 picas to a width of 21 picas, depending on the width between the columns. The grids shown here make use of an expanding center margin, from 1 pica to 8 picas. At the same time the side columns reduce by 1 pica increments from each side.

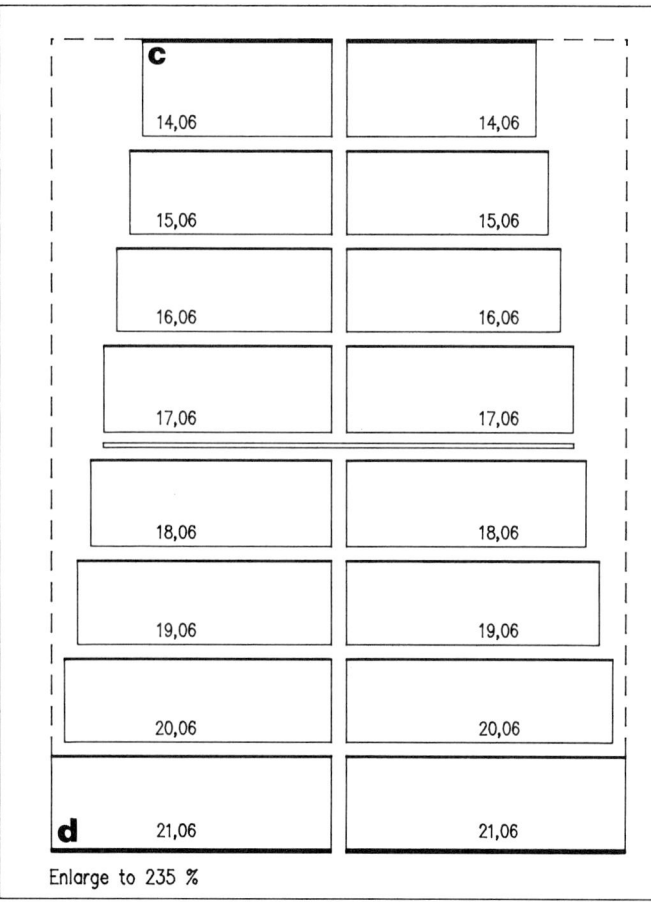
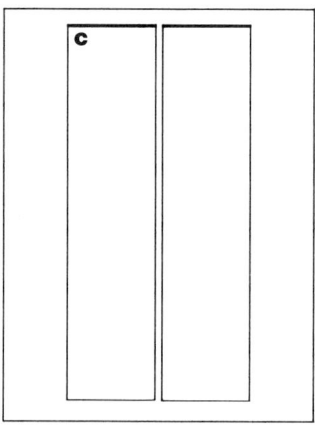
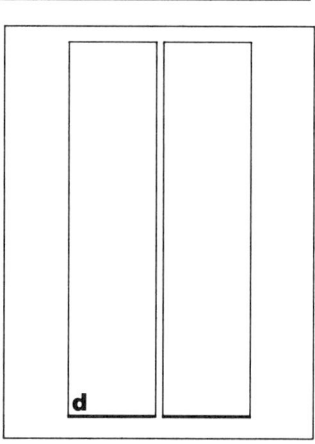

Enlarge to 235 %

The bottom half of this grid also shows two equal columns around a center margin of 1 pica which remains constant. With two columns there are more options in the use of both type and photographs. It is a good format for magazines, annual reports and newsletters, where there is a lot of text. The 21 pica width is the maximum line length for text. The illustrations show the grid above with the expanding center margin compared to the one with a constant center margin. The thumbnails show the maximum and minimum columns for each variation.

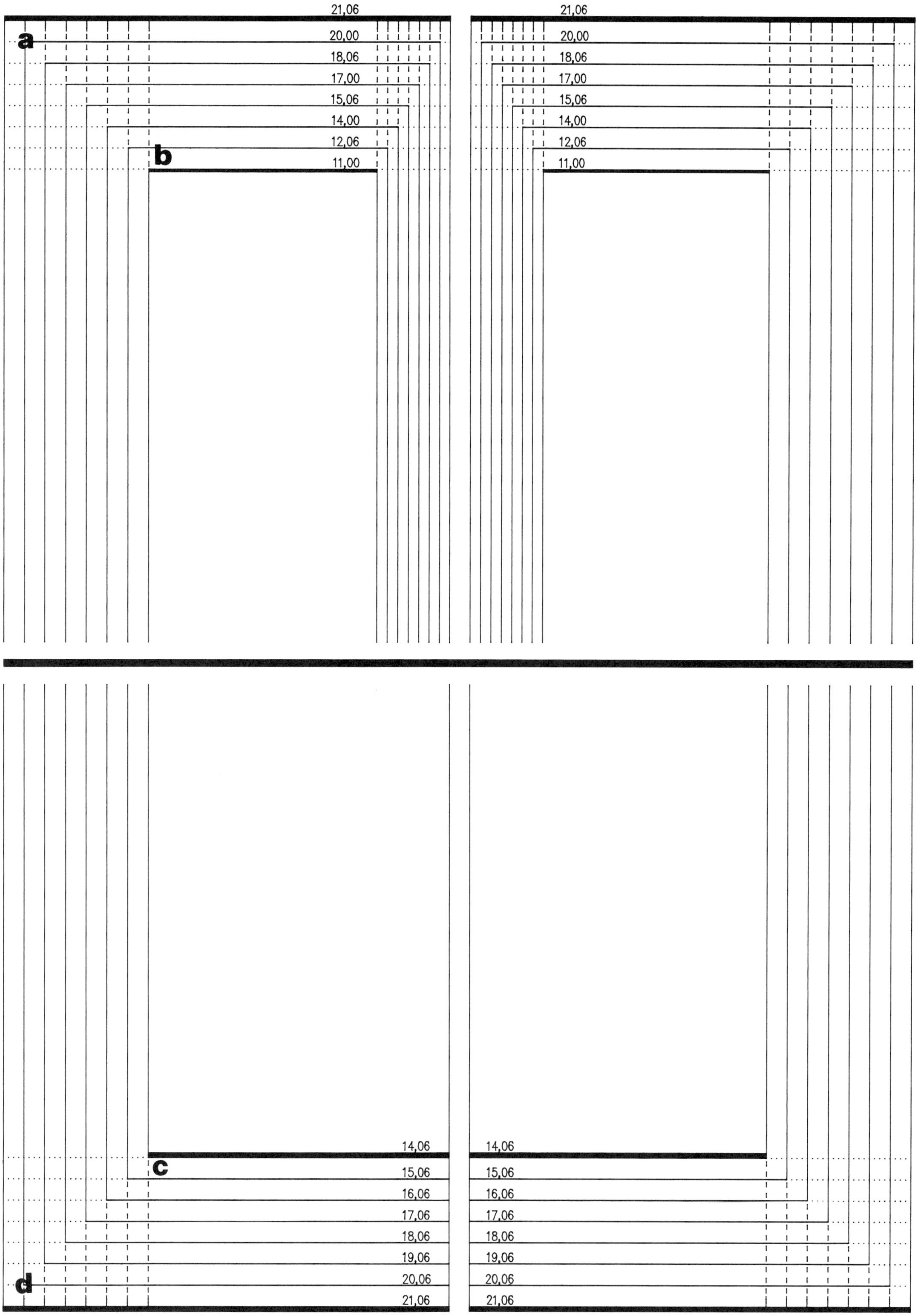

a

b

c

d

| 21,06 | 21,06 |
| 20,00 | 20,00 |
| 18,06 | 18,06 |
| 17,00 | 17,00 |
| 15,06 | 15,06 |
| 14,00 | 14,00 |
| 12,06 | 12,06 |
| 11,00 | 11,00 |

| 14,06 | 14,06 |
| 15,06 | 15,06 |
| 16,06 | 16,06 |
| 17,06 | 17,06 |
| 18,06 | 18,06 |
| 19,06 | 19,06 |
| 20,06 | 20,06 |
| 21,06 | 21,06 |

Generic Grids 81

## 2 Column Grid

This two column grid features a vertical rule in the center column, making it 2 picas wide with two unequal columns at the top. The left column is reduced at 1/2 pica increments, while the right column is reduced at 1 pica increments from the outer margins. The result is that in the final variation the two columns are equal in width. The two column format is very popular for layouts of magazines, newsletters, and annual reports, as well as brochures and catalogs.

The second variation on this two column grid begins at the bottom, where there are two equal columns separated by a vertical rule. The space between the columns is 2 picas. The left column is reduced on the outer edge by 1/2 pica units, and the column on the right is reduced by 1 pica. The resultant columns are unequal in width throughout all the variations.

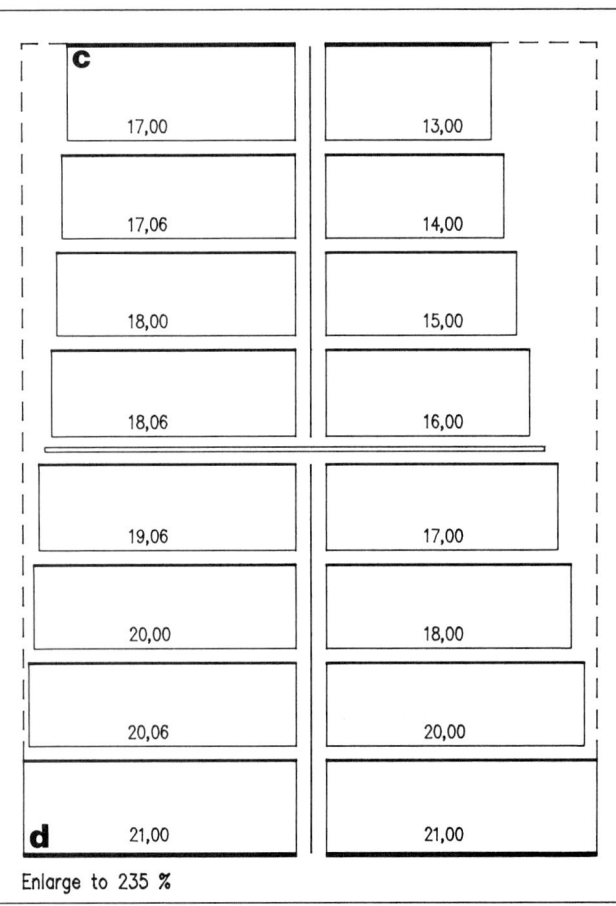

Enlarge to 235 %

| | | | |
|---|---|---|---|
| a | 19,00 | | 23,00 |
| | 18,06 | | 22,00 |
| | 18,00 | | 21,00 |
| | 17,06 | | 20,00 |
| | 17,00 | | 19,00 |
| | 16,06 | | 18,00 |
| | 16,00 | | 17,00 |
| | 15,06 | | 16,00 |
| b | 15,00 | | 15,00 |

| | | | |
|---|---|---|---|
| c | 17,00 | | 13,00 |
| | 17,06 | | 14,00 |
| | 18,00 | | 15,00 |
| | 18,06 | | 16,00 |
| | 19,00 | | 17,00 |
| | 19,06 | | 18,00 |
| | 20,00 | | 19,00 |
| | 20,06 | | 20,00 |
| d | 21,00 | | 21,00 |

Generic Grids 83

## 3 Column Grid

This is a slight variation on the three column grid where the center column decreases in 1 pica increments while the outside columns remain constant. The 1 pica margin is kept between the columns. This ends up with a format for three equal columns, with wider side margins resulting from this convergence. These formats are all adaptable to brochures, newsletters and annual reports. They also allow for a flexible combination of text and photographs.

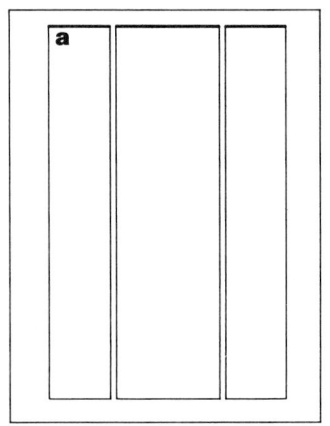
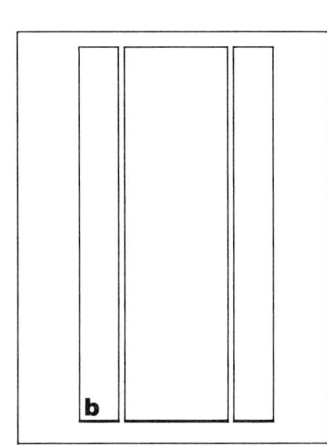
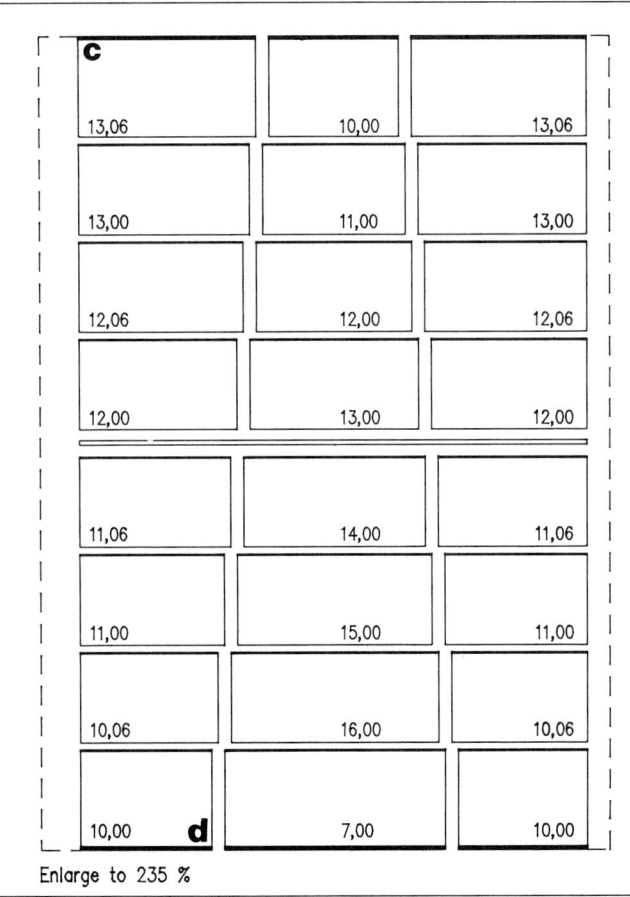
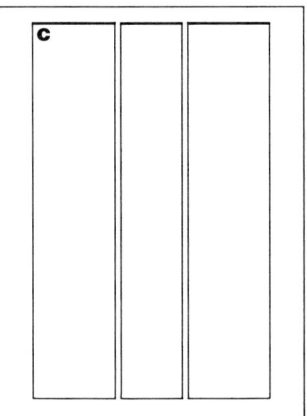
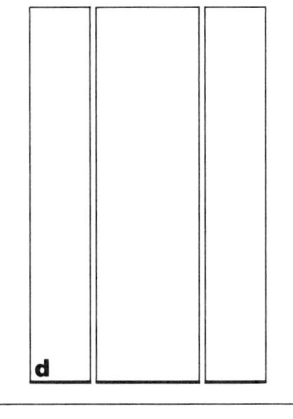

Enlarge to 235 %

This grid reverses the sequence of the one above in that it begins with three equal columns and changes into three unequal columns. Along the way the center column remains the same, and the 1 pica margin remains constant throughout. The two outer margins decrease in 1/2 pica increments until the grid format becomes a wide center column with two smaller outer columns. This leaves wide margins on both sides.

Generic Grids 85

## 3 Column Grid

The three column grid is the most common format in publishing. It is widely used in newsletters, magazines, brochures and annual reports because it is so versatile. The grid opposite represents a symmetrical arrangement using three unequal columns. There is a wide one in the center flanked by a narrower one on each side. The variations shown in this grid consist of the center 17 pica column remaining constant while the outer columns are reduced in 1/2 pica increments, resulting in a narrow 6 pica column each.

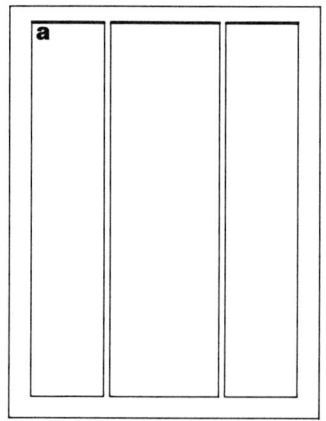
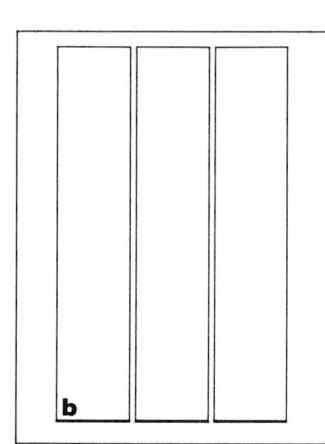

The grid opposite begins with the same dimensions on the outside as the one above. The difference is that the center column becomes smaller in 1 pica increments. This grid provides the option of using a wide center column with two small outer ones, three equal columns, or two wide outside columns with a narrow middle one. These are particularly useful in developing layouts for brochures and annual reports.

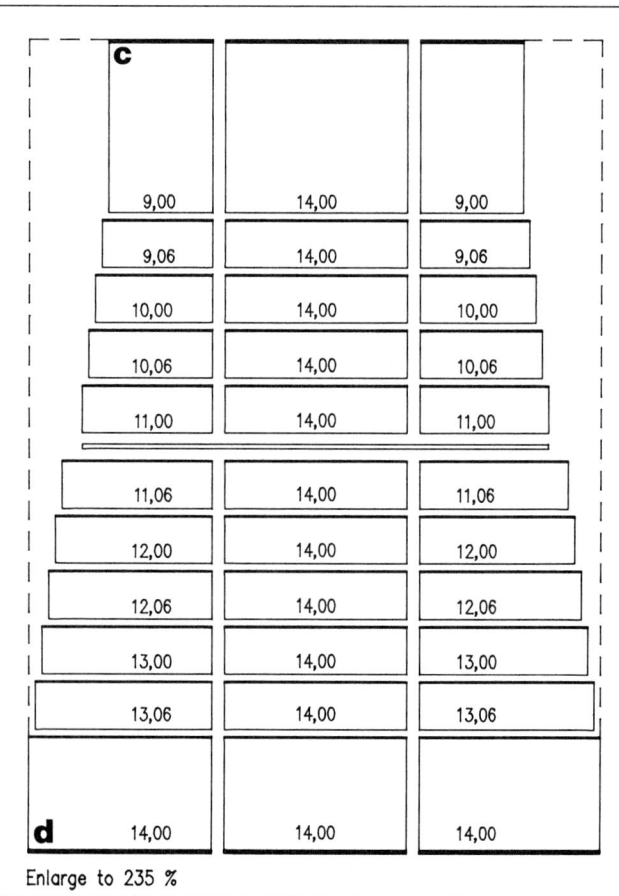

Enlarge to 235 %

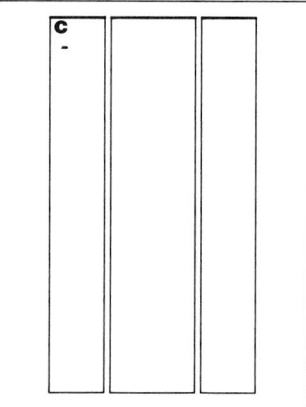
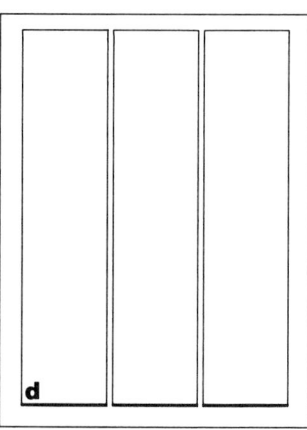

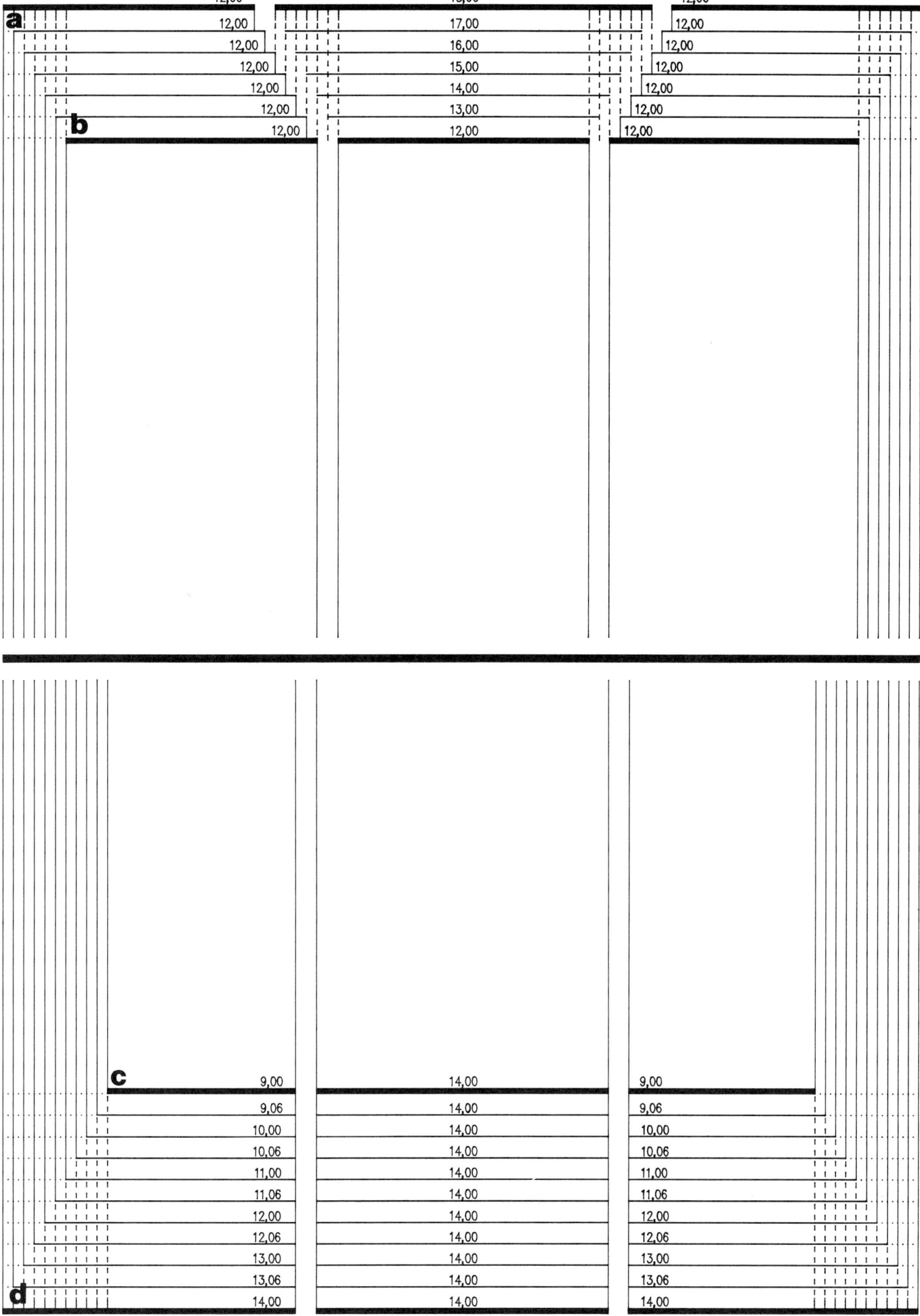

### 3 Column Grid

The grid in this case is a three column format of equal columns. This aspect is good for newsletters and other publications where standard columns of text are a major factor. From here the columns change by 1 1/2 picas for each of the outer columns and 1 pica for the center columns. This produces three columns of unequal widths. Another aspect of this grid is that the internal margins get increasingly wider, from 1 pica to 7 picas at the widest point.

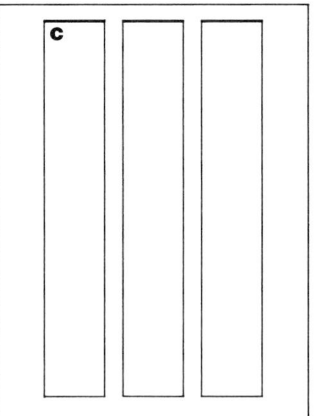

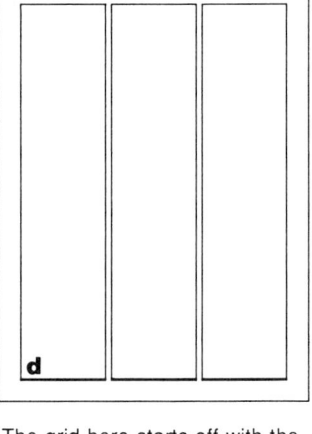

The grid here starts off with the same equal three columns as the one above, but keeps the three-equal-column format throughout the variations. Each of the outer columns are reduced at 1 pica each, and the center column loses 1/2 pica from each side, which keeps the three columns equal. Thus, each column loses less than the one above, and the margins between the columns are not as pronounced. This format is good for brochures, newsletters, and annual reports.

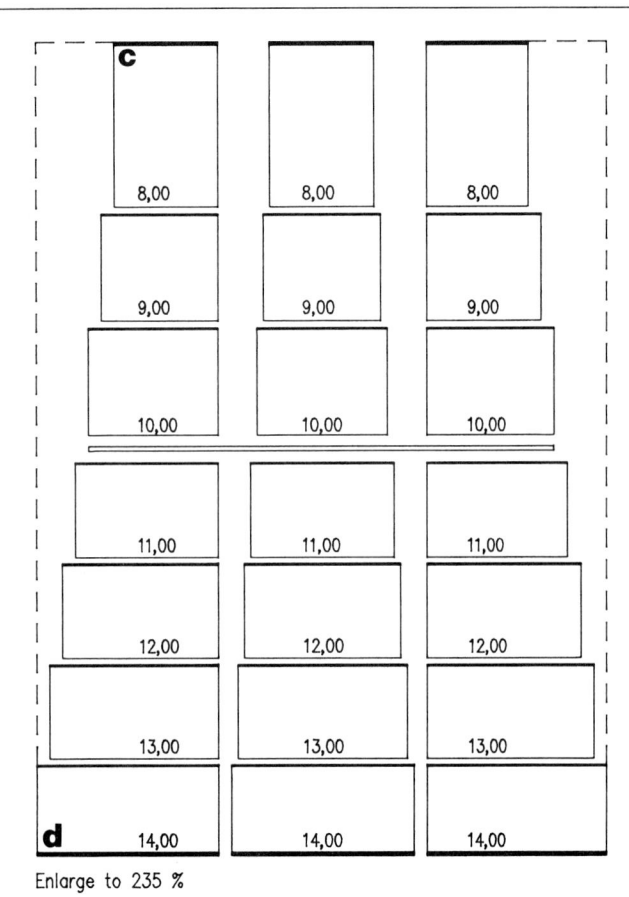

Enlarge to 235 %

Generic Grids

## 4 Column Grid

Four column grids are even more versatile than the three column formats and have a much wider application. They provide a basis for variety of page layouts, and are frequently used in magazines, brochures, newsletters and catalogs. This grid represents two equal and two unequal columns at the top. The two smaller outer columns remain the same as the center columns decrease in width by 1 pica each. At the last variation shown they are all equal in width.

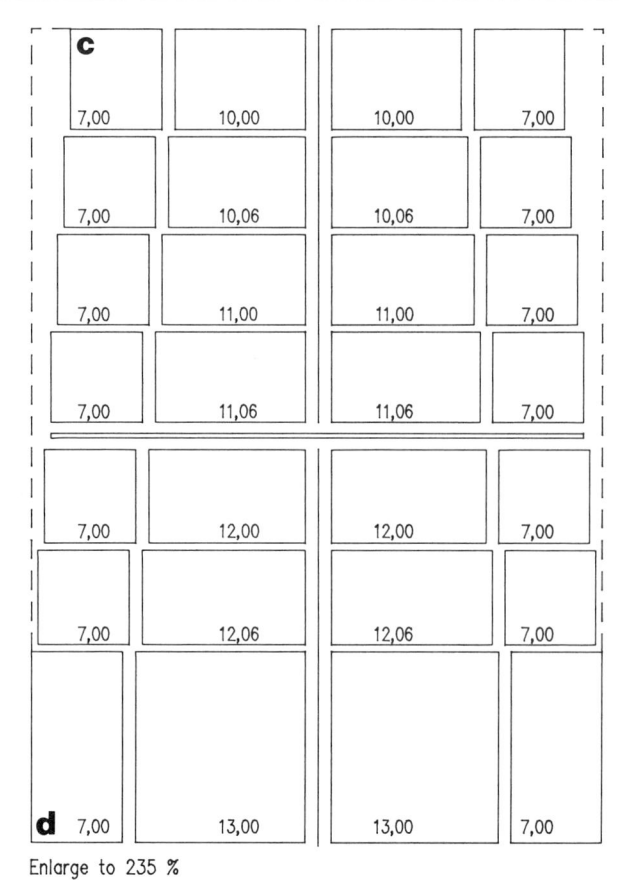

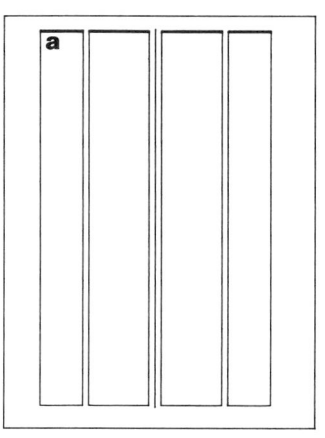
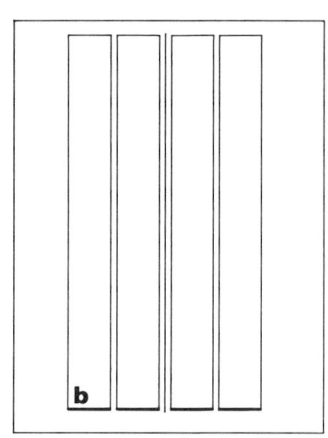
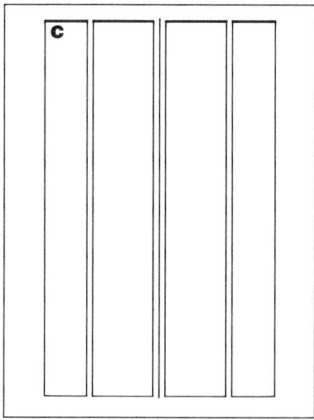
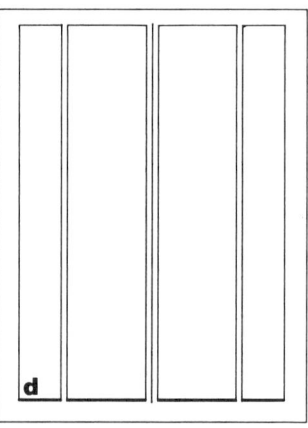

This grid is a modification of the one above in that the two outer columns remain the same width. The major difference is that the center two columns diminish by only 1/2 pica each, resulting in a wider center column. The page is divided by a rule, making the margin 2 picas wide. There are problems associated with working with narrow columns, such as requiring more decisions throughout the design and production process. These are overshadowed by the extreme versatility that it provides.

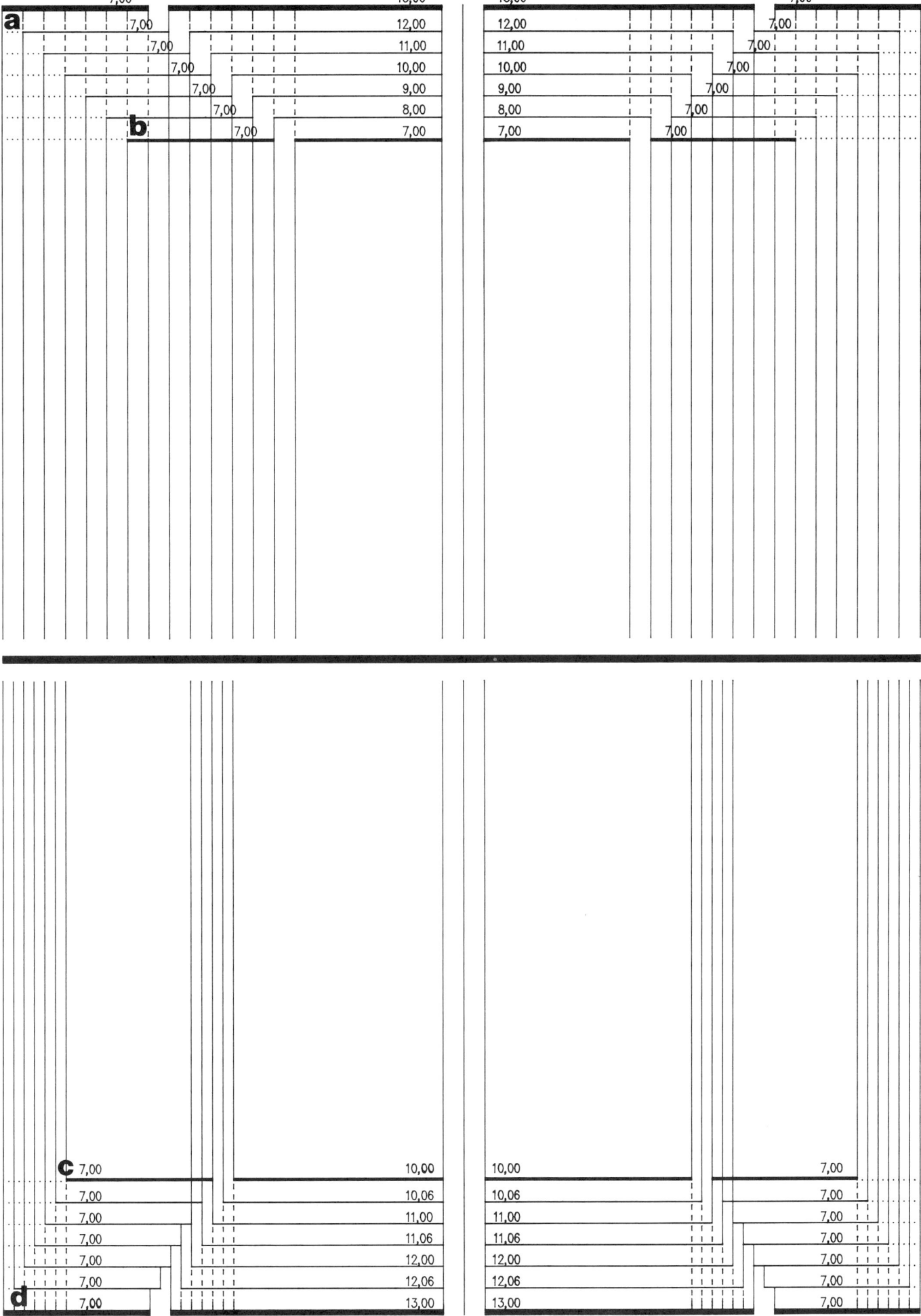

Generic Grids 91

### 4 Column Grid

Four column grids are widely used for newsletter design as well as for client lists and project identifications. This particular grid shows four equal columns at the top of the page with multiple variations to four smaller equal columns. By reducing the outer edge of the two outside columns by 1 pica each, while the center column reduces by 1/2 pica each, the four columns maintain an equal width throughout the variations. The final example is four columns of 7 picas each separated by a 1 pica margin.

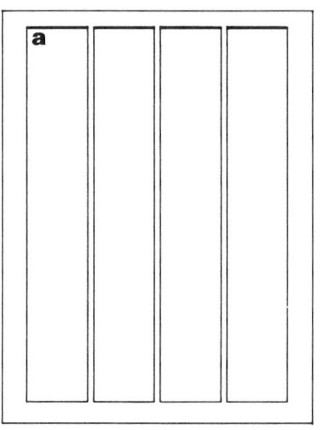
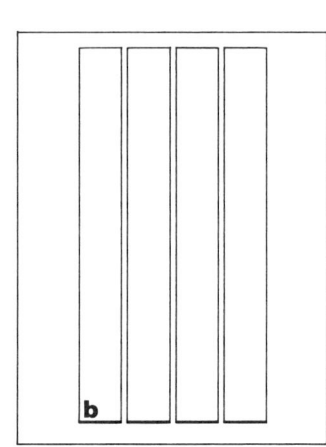

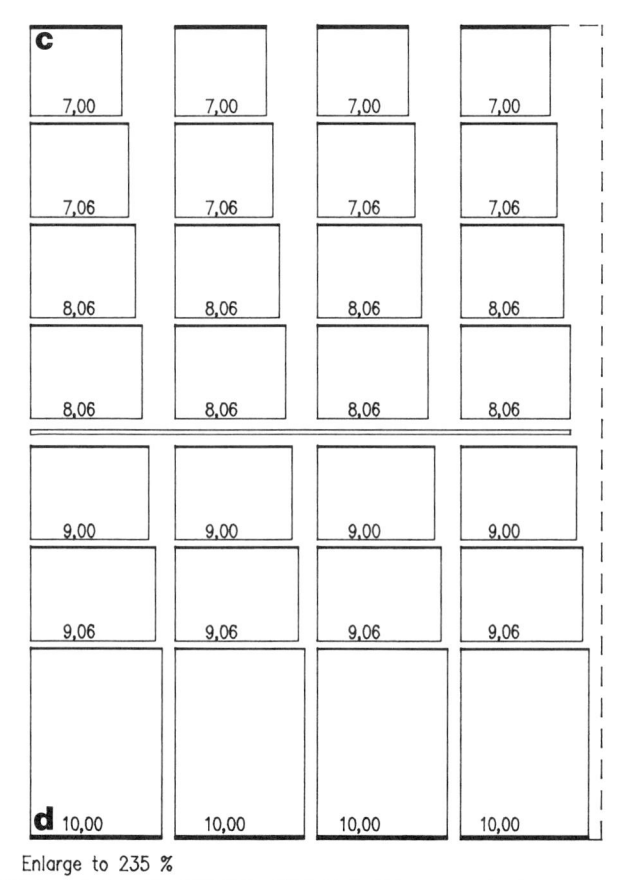

Enlarge to 235 %

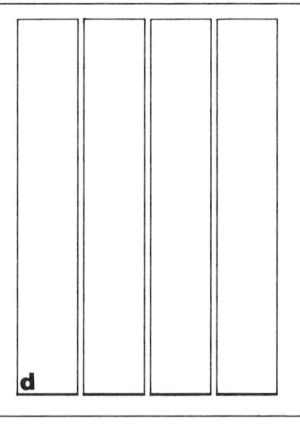

This example begins and ends with the identical width in the columns. The main difference is that the margins expand in width between the columns. They increase in size by 1/2 pica each, and conversely the columns decrease in size by 1 pica each. The left side of each column remains the same, while the right side decreases in 1 pica increments. The increased margins provide a clean, open look.

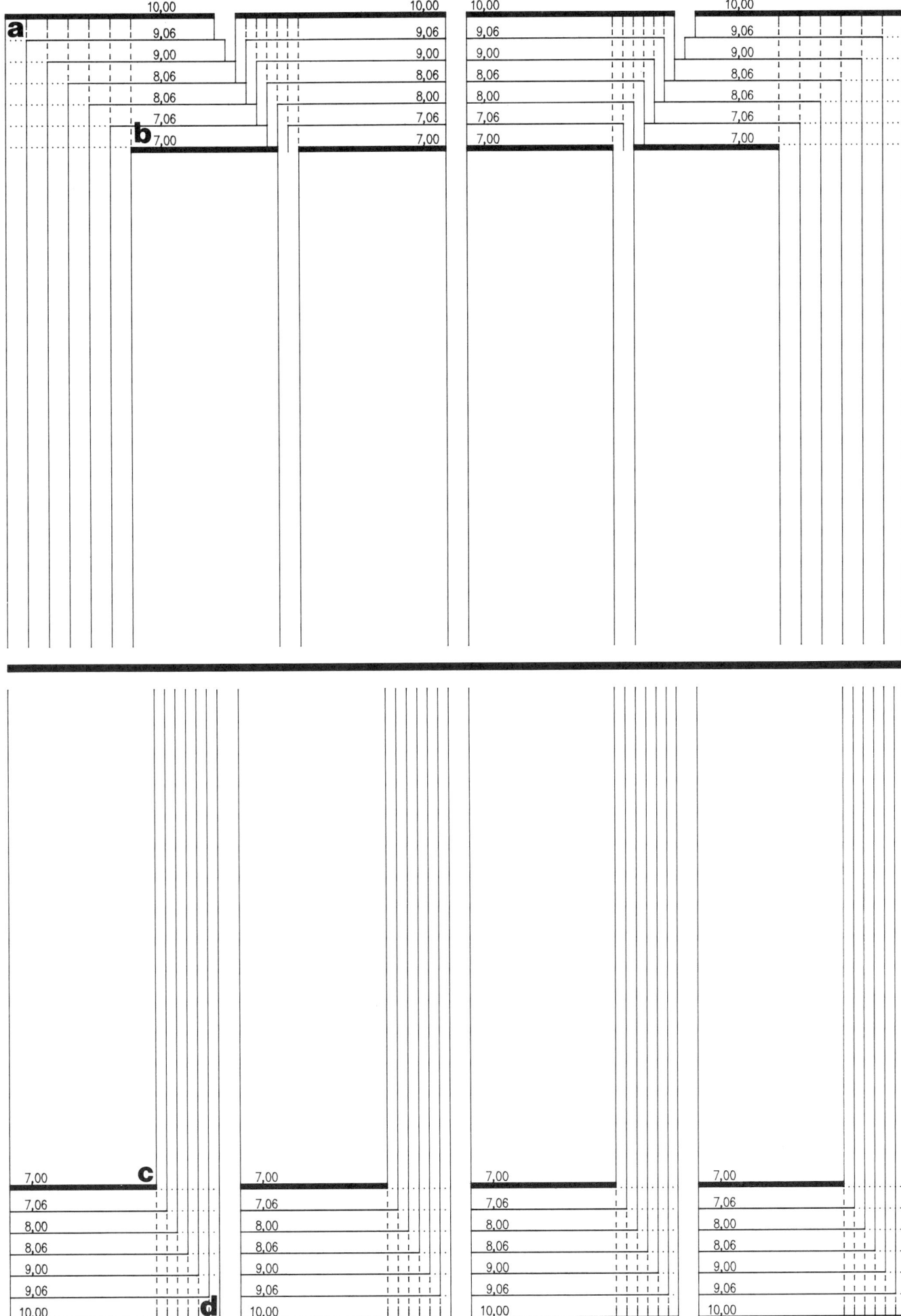

Generic Grids 93

## 5 Column Grid

Five column grids are not as common as the ones shown previously, but they offer a wide variety of layout possibilities. They are particularly good for layouts where a lot of photos are to be used since there are many ways to organize them. Here, an unequal arrangement of columns heads the page. The center column and the two outside columns remain the same in width. The other two columns decrease in size by 1/2 pica each, keeping the margins between them constant.

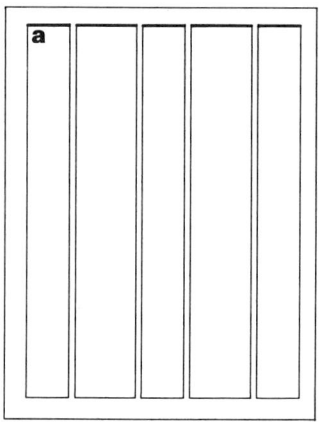
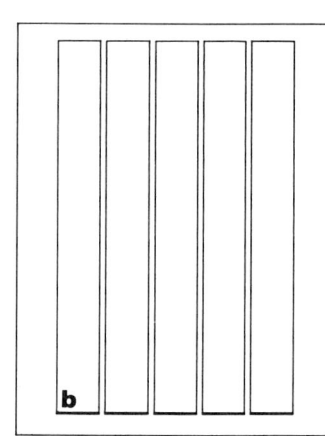

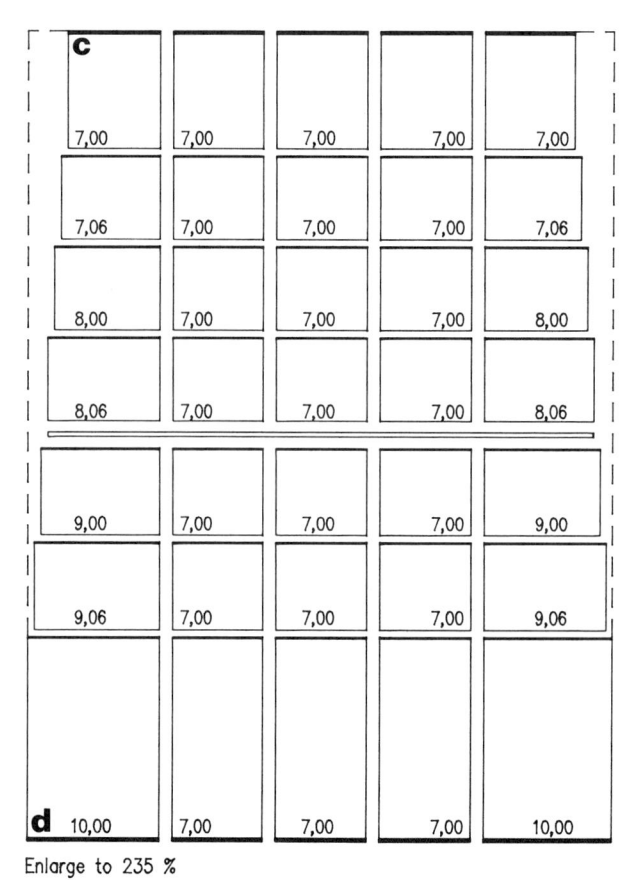

The example here shows the five columns with the three narrow columns in the center. They are flanked by the two wider columns. These wider columns reduce in size by 1/2 pica each from the outside edge of the columns. The margin space between the columns is maintained at 1 pica. These variations show only minor differences in the column widths, and any one of them could work equally well in providing a format for a creative layout.

Enlarge to 235 %

## 5 Column Grid

Five column grids have fewer applications than four or three column grids. Yet their usefulness for general listings and other specifications is readily apparent. Narrow columns require special care in the selection of type, and it should be relatively small in proportion to the column. However, in some applications the narrow columns can be used with larger type for emphasis. The narrow columns can become problematic when excessive hyphenation becomes necessary.

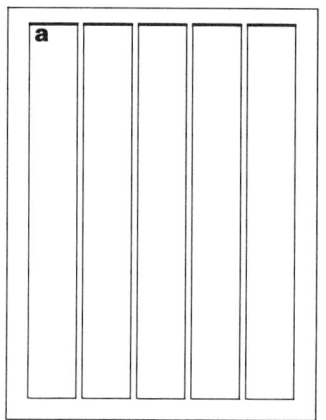

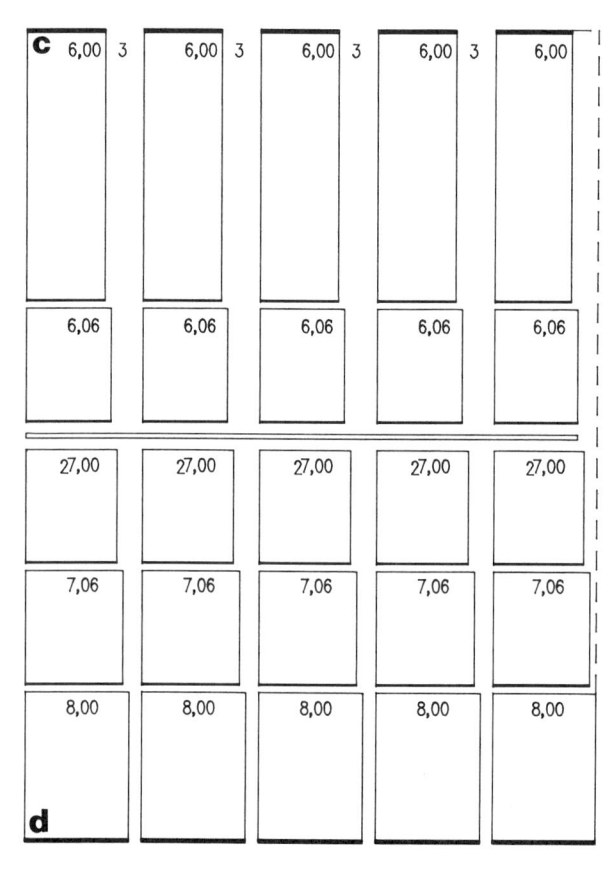

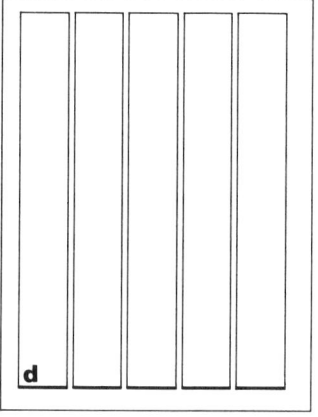

The five column formats shown here are all equal in width and vary by decreasing the width by 1/2 pica from the left edge of each column. This increases the margins between each column equally for all columns. On the bottom format the same variation is used except that the columns decrease by 1/2 pica from the right side of the column instead of the left. These formats will provide a base for a variety of page layouts.

Generic Grids

Enlarge to 235 %

## 6 Column Grid

The six column format yields a net column width of minimum dimensions for most applications. Therefore, the main advantage is that it is a borderline format. It can still be used for small columns of text, as seen in a few of the case study grids. In addition, it is a good module for sizing photographs. Finally, it becomes extremely useful as a subdivision of wider columns such as the three column and even the two column format. Some of the combinations are displayed on the opposite page.

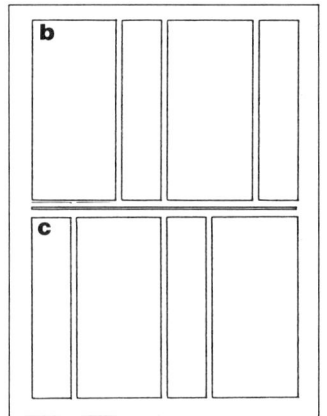
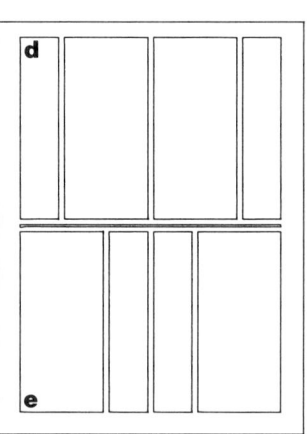
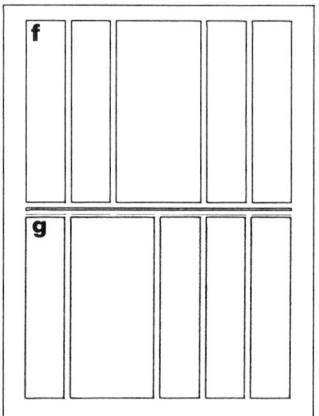
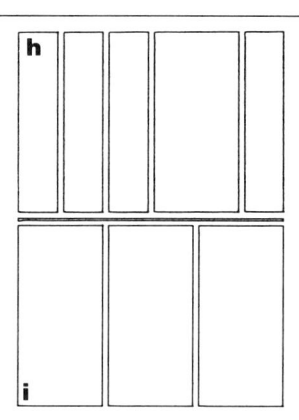
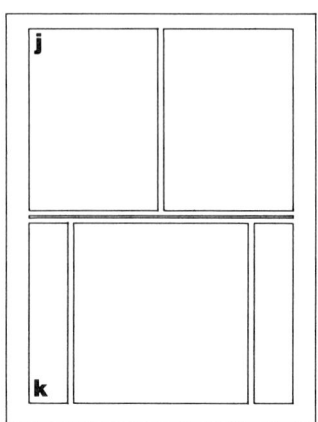
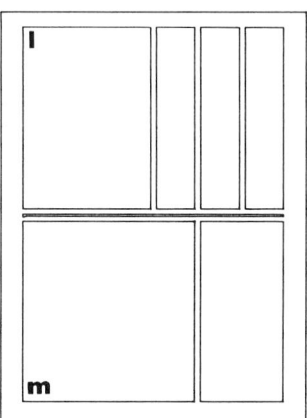
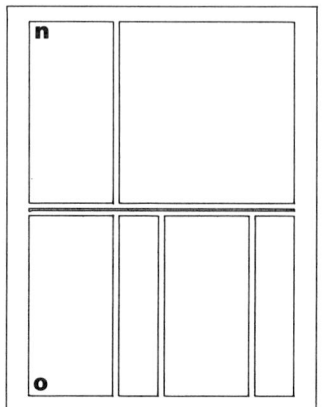

| | | | | | |
|---|---|---|---|---|---|
| **a** 6,06 | 6,06 | 6,06 | 6,06 | 6,06 | 6,06 |

| | | | |
|---|---|---|---|
| **b** 14,00 | 6,06 | 14,00 | 6,06 |

| | | | |
|---|---|---|---|
| **c** 6,06 | 14,00 | 6,06 | 14,00 |

| | | | |
|---|---|---|---|
| **d** 6,06 | 14,00 | 14,00 | 6,06 |

| | | | |
|---|---|---|---|
| **e** 14,00 | 6,06 | 6,06 | 14,00 |

| | | | | |
|---|---|---|---|---|
| **f** 6,06 | 6,06 | 4,00 | 6,06 | 6,06 |

| | | | | |
|---|---|---|---|---|
| **g** 6,06 | 14,00 | 6,06 | 6,06 | 6,06 |

| | | | | |
|---|---|---|---|---|
| **h** 6,06 | 6,06 | 6,06 | 14,00 | 6,06 |

| | | |
|---|---|---|
| **i** 14,00 | 14,00 | 14,00 |

| | |
|---|---|
| **j** 21,06 | 21,06 |

| | | |
|---|---|---|
| **k** 6,06 | 29,00 | 6,06 |

| | | | |
|---|---|---|---|
| **l** 21,06 | 6,06 | 6,06 | 6,06 |

| | |
|---|---|
| **m** 29,00 | 14,00 |

| | |
|---|---|
| **n** 14,00 | 29,00 |

| | | | |
|---|---|---|---|
| **o** 14,00 | 6,06 | 14,00 | 6,06 |

| | |
|---|---|
| **p** 36,06 | 6,06 |

| | |
|---|---|
| **q** 6,06 | 36,06 |

Generic Grids

## 7 Column Grid

The seven column grid is best used for asymmetrical layouts. This is particularly useful when the center column remains the same and the outer columns are combined in various ways. You can produce at a minimum a three column format with a narrow center column, or a five column format with a small center column and two wider outside columns. The odd number of columns provides many opportunities for asymmetrical layouts of many varying sizes.

Enlarge to 235 %

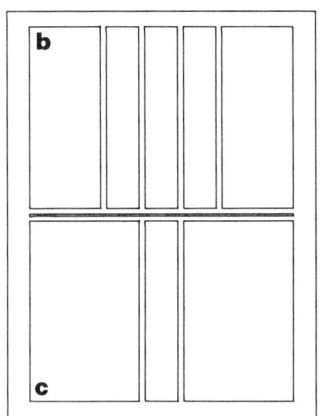 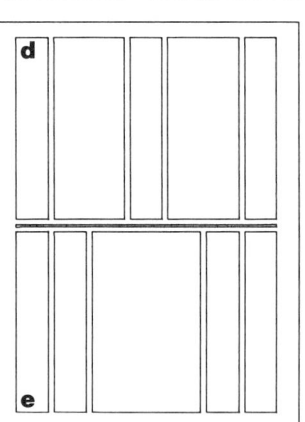 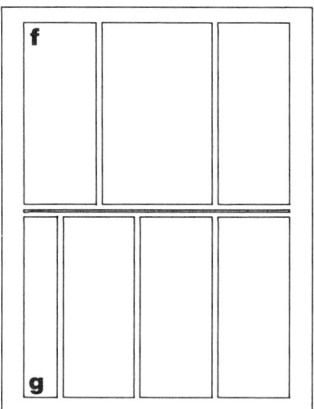 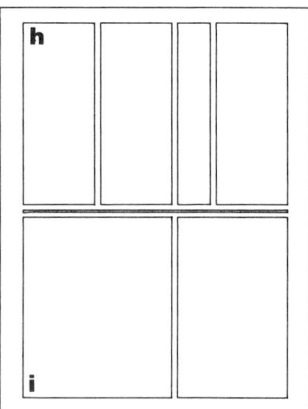

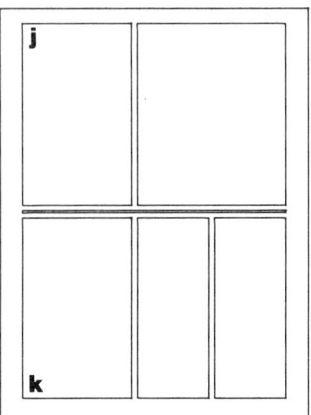 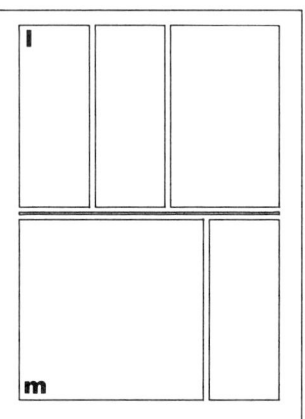 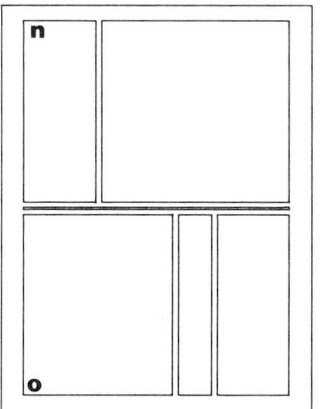 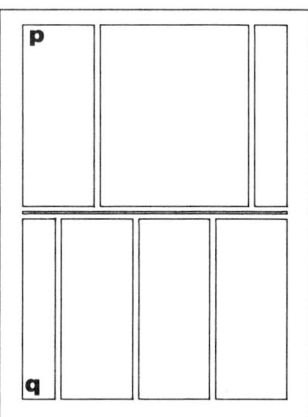

| | | | | | | |
|---|---|---|---|---|---|---|
| **a** 5,05 | 5,05 | 5,05 | 5,05 | 5,05 | 5,05 | 5,05 |

| | | | | |
|---|---|---|---|---|
| **b** 11,10 | 5,05 | 5,05 | 5,05 | 11,10 |

| | | |
|---|---|---|
| **c** 18,02 | 5,05 | 18,02 |

| | | | | |
|---|---|---|---|---|
| **d** 5,05 | 11,10 | 5,05 | 11,10 | 5,05 |

| | | | | |
|---|---|---|---|---|
| **e** 5,05 | 5,05 | 18,02 | 5,05 | 5,05 |

| | | |
|---|---|---|
| **f** 11,10 | 18,02 | 11,10 |

| | | | |
|---|---|---|---|
| **g** 5,05 | 11,10 | 11,10 | 11,10 |

| | | | |
|---|---|---|---|
| **h** 11,10 | 11,10 | 5,05 | 11,10 |

| | |
|---|---|
| **i** 24,07 | 18,08 |

| | |
|---|---|
| **j** 18,03 | 24,07 |

| | | |
|---|---|---|
| **k** 18,03 | 11,10 | 11,10 |

| | | |
|---|---|---|
| **l** 11,10 | 11,10 | 18,03 |

| | |
|---|---|
| **m** 31,00 | 11,10 |

| | |
|---|---|
| **n** 11,10 | 31,00 |

| | | |
|---|---|---|
| **o** 24,07 | 5,05 | 11,10 |

| | | |
|---|---|---|
| **p** 11,10 | 24,07 | 5,05 |

| | | | |
|---|---|---|---|
| **q** 5,05 | 11,10 | 11,10 | 11,10 |

Generic Grids

## 8 Column Grid

An eight column grid is one that can be combined in many different ways to provide symmetrical formats as well as asymmetrical ones. The most common combinations are two and four column formats of equal column widths. Beyond these are formats of unequal columns, which expand the number of possibilities. Within the four columns of unequal formats, for example, are several more combinations. There is even a variety of odd numbers of columns within this evenly divided master.

Enlarge to 235 %

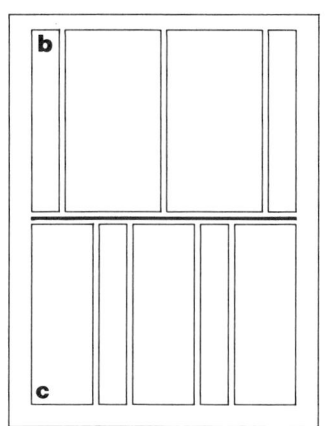

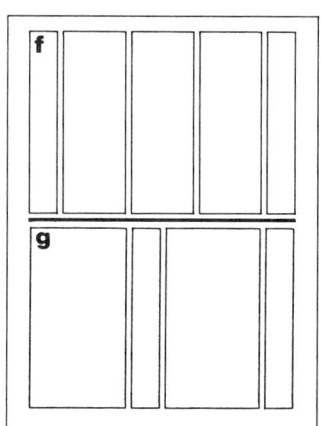
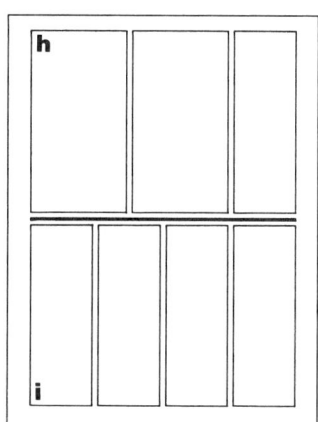
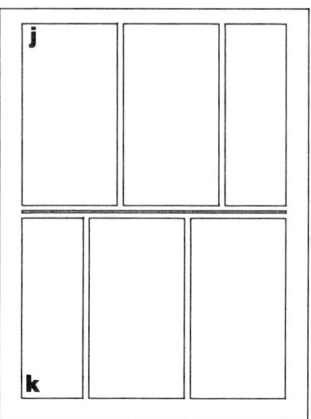

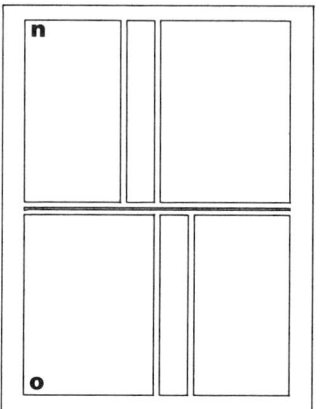
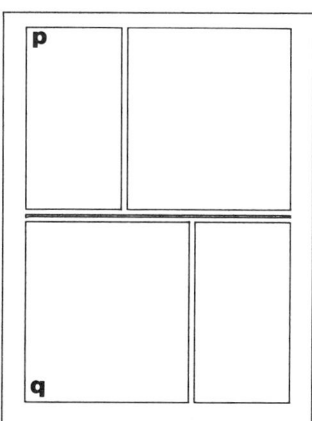

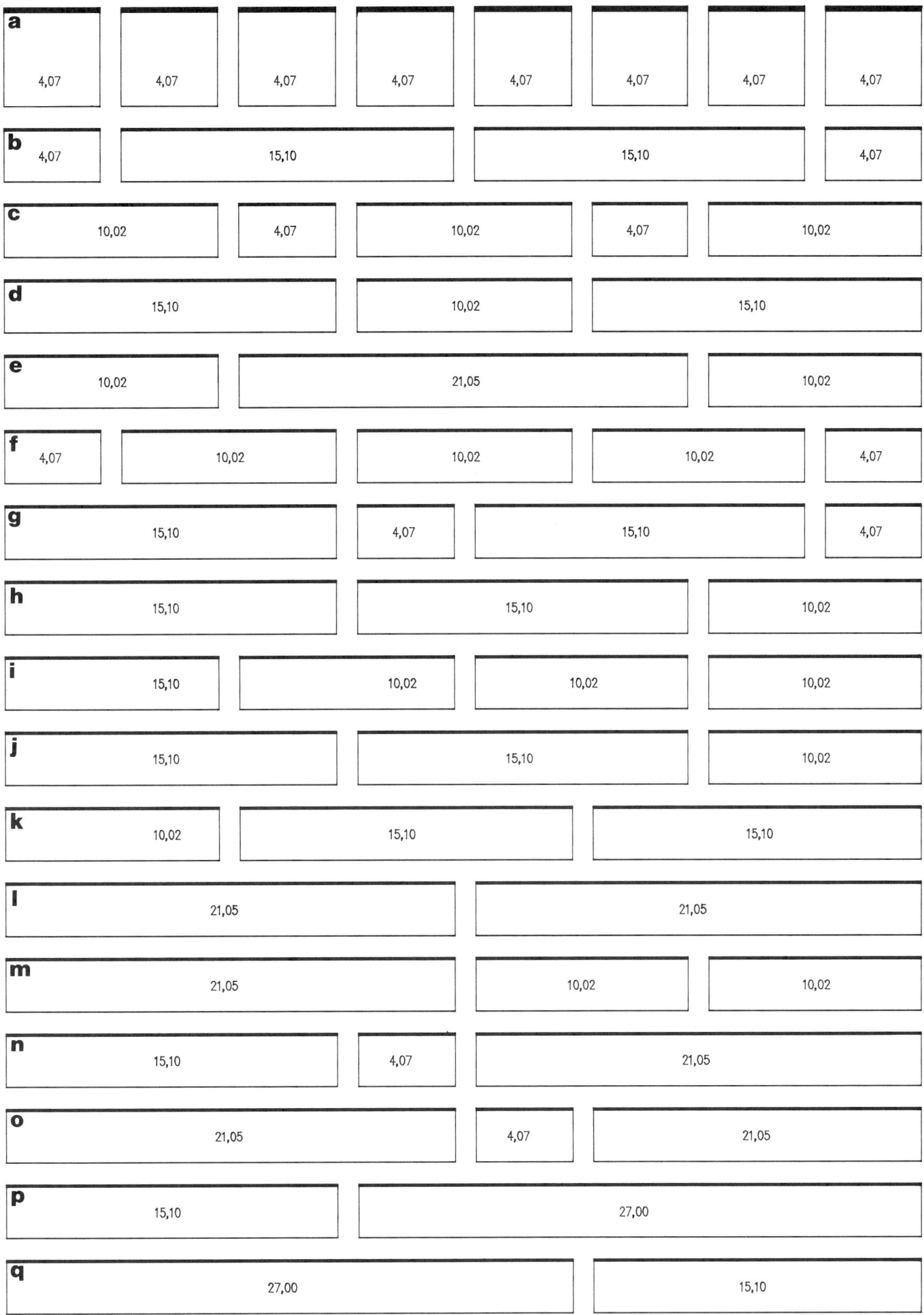

Generic Grids

## 9 Column Grid

Nine columns are useful in that these combine very easily into a standard three-equal-column format. This is one of the most popular for magazines and newsletters. Therefore, it is really a subdivision of this format. At the same time it provides a number of unequal three column formats. The thumbnail sketches show many of the variations that are possible using these nine columns. Although the majority of them are asymmetrical, they have many applications for creative layouts.

Enlarge to 235 %

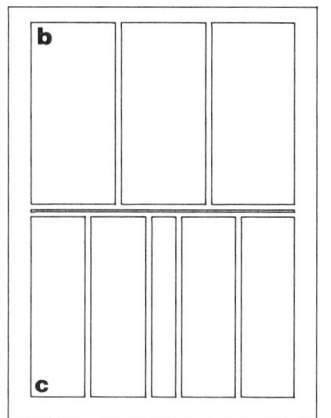 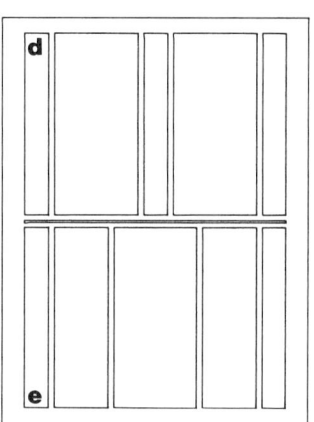 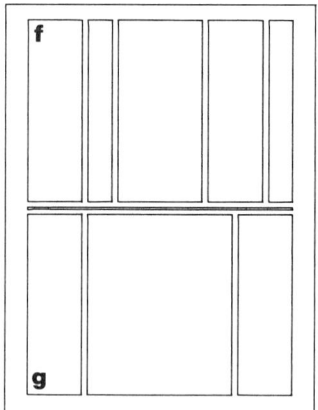 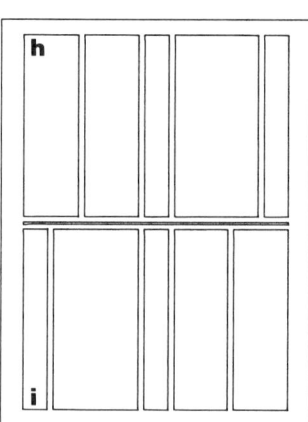

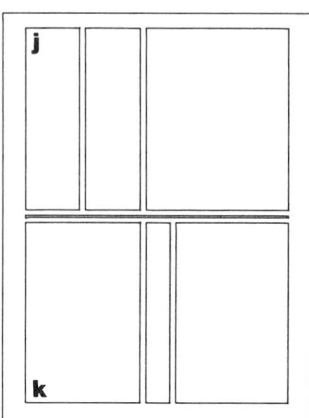 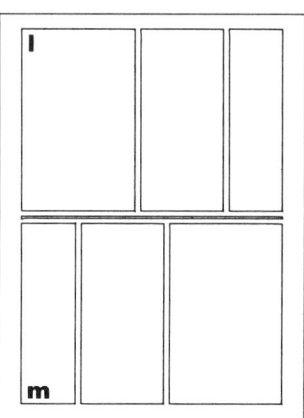 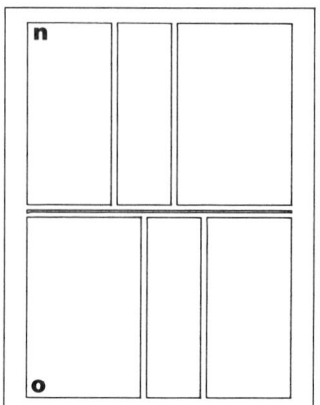 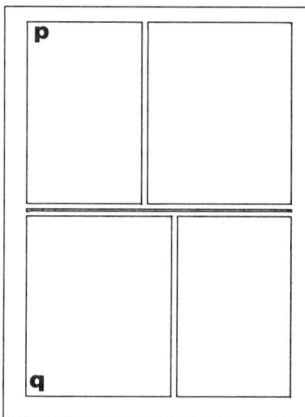

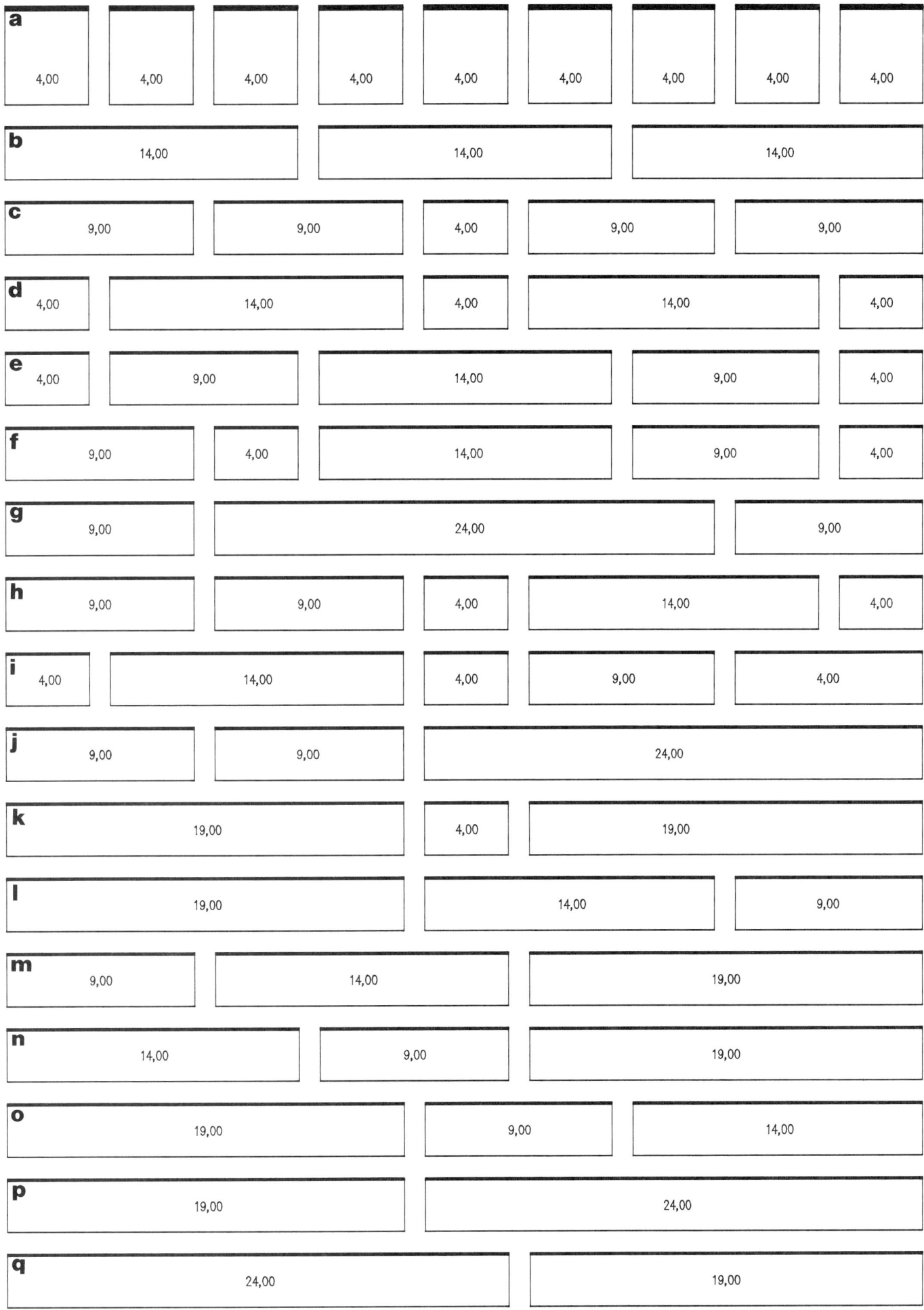

Generic Grids 105

## 10 Column Grid

A ten column grid module combines into many different formats. For example, two columns and five columns represent equal formats. But there are many unequal formats, such as a variety of three, four and six column arrangements. These are all asymmetrical formats, to be sure, but they provide an excellent opportunity to combine wide columns with narrow ones. The thumbnails only show a fraction of the various combinations that are possible using the ten column grid.

Enlarge to 235 %

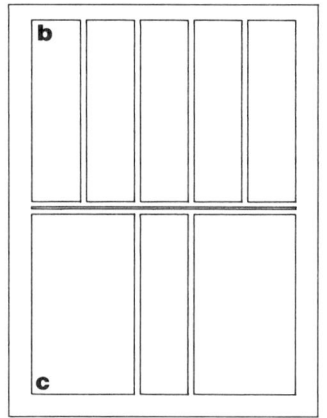 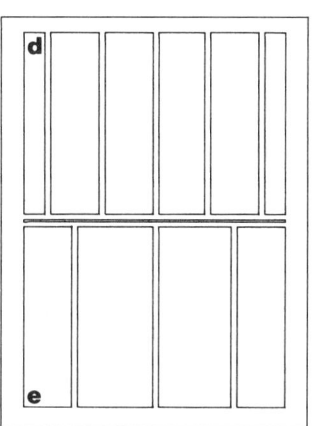 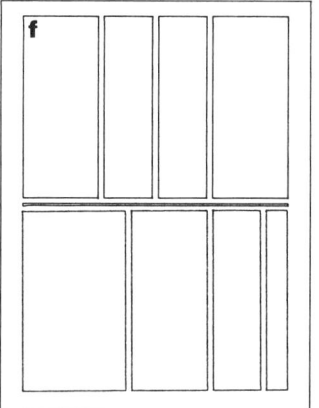 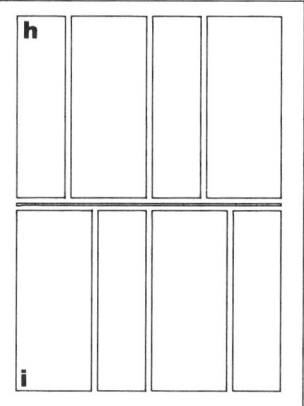

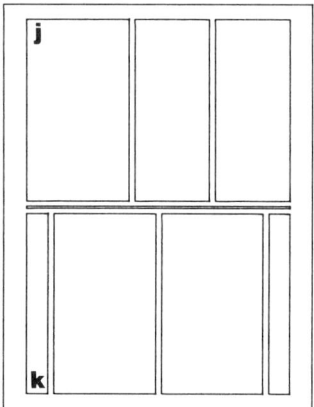 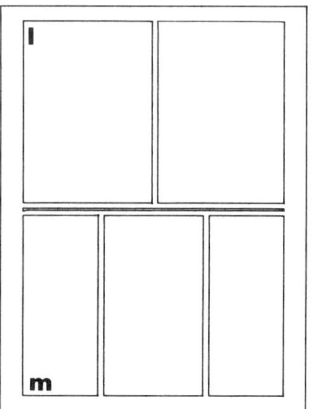  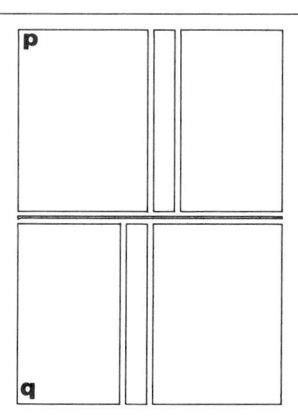

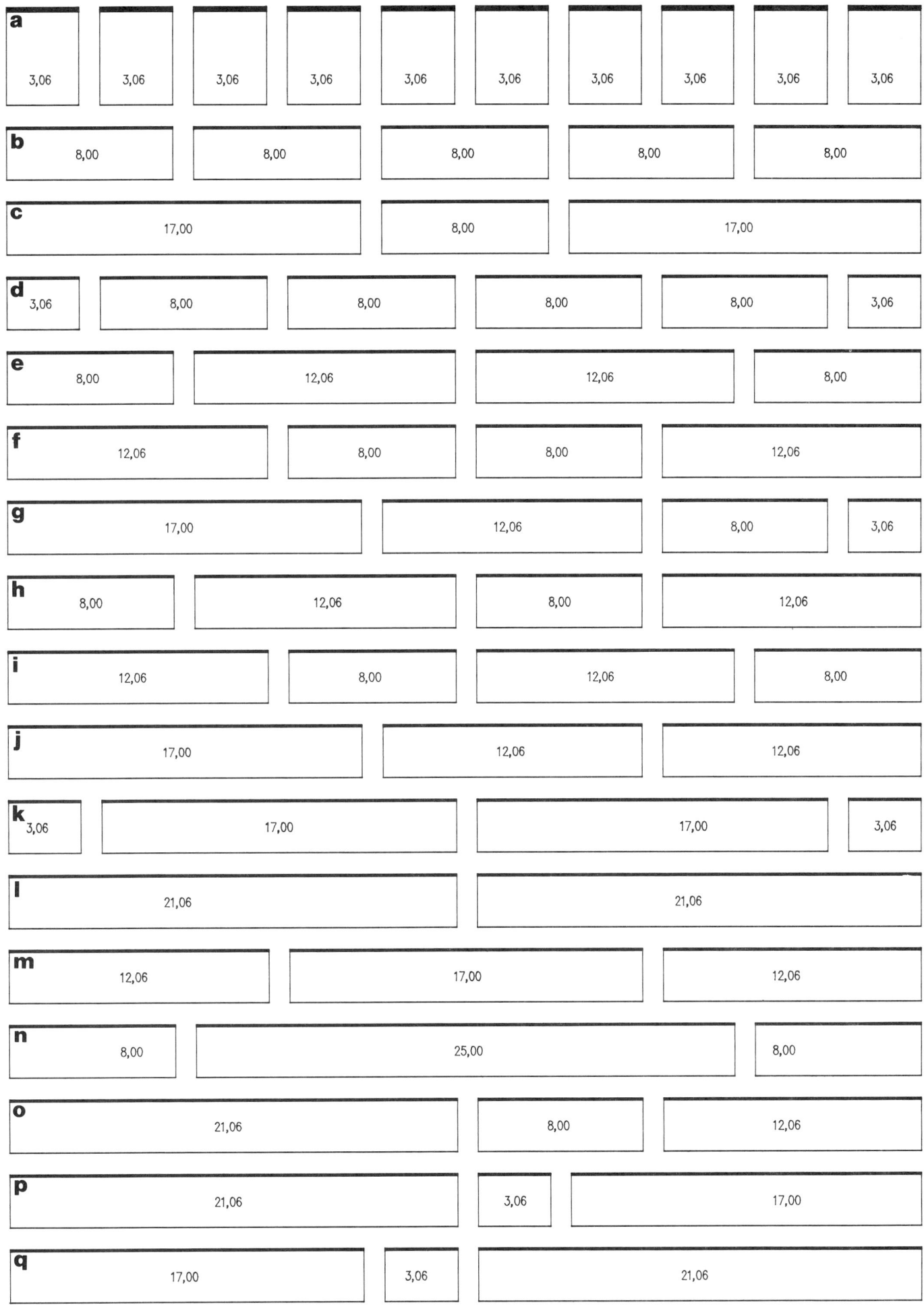

Generic Grids 107

## 11 Column Grid

The eleven column grid module is the origin of many unequal and asymmetrical column arrangements. For example, there are many variations of the three unequal column formats, as the number of combinations also increases. The thumbnails show a few of these combinations. Since the column width is so small, the most useful combinations are those with wider columns. However, the columns can be used to introduce wide margins of white space between the columns.

Enlarge to 235 %

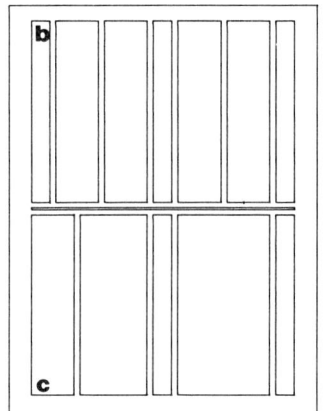 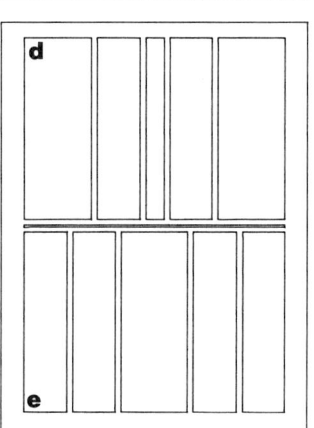 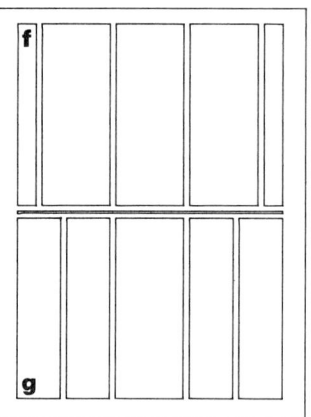 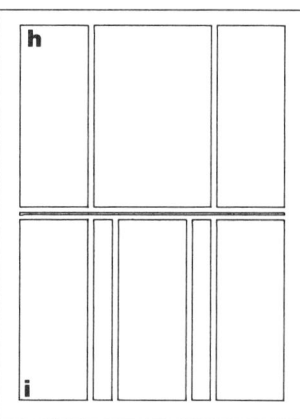

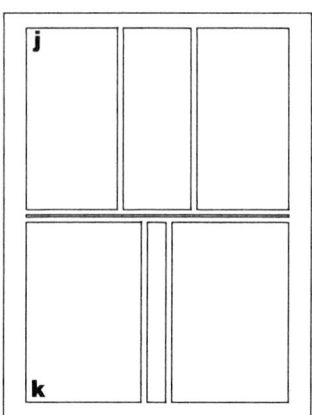   

| row | values |
|---|---|
| a | 3,01 — 3,01 — 3,01 — 3,01 — 3,01 — 3,01 — 3,01 — 3,01 — 3,01 — 3,01 — 3,01 |
| b | 3,01 — 7,02 — 7,02 — 3,01 — 7,02 — 7,02 — 3,01 |
| c | 7,02 — 11,04 — 3,01 — 11,04 — 7,02 |
| d | 11,04 — 7,02 — 3,01 — 7,02 — 11,04 |
| e | 7,02 — 7,02 — 11,04 — 7,02 — 7,02 |
| f | 3,01 — 11,04 — 11,04 — 11,04 — 3,01 |
| g | 7,02 — 7,02 — 11,04 — 7,02 — 7,02 |
| h | 11,04 — 19,06 — 11,04 |
| i | 11,04 — 3,01 — 11,04 — 3,01 — 11,04 |
| j | 15,04 — 11,04 — 15,04 |
| k | 19,06 — 3,01 — 19,06 |
| l | 19,06 — 11,04 — 11,04 |
| m | 11,04 — 11,04 — 19,06 |
| n | 19,06 — 7,02 — 19,06 |
| o | 15,04 — 7,02 — 19,06 |
| p | 11,04 — 11,04 — 19,06 |
| q | 19,06 — 11,04 — 11,04 |

Generic Grids 109

## 12 Column Grid

The twelve column grid provides the basis for many formats that can be coordinated within a publication. There is a wide array of three and four column formats since they are both multiples of twelve. Surprisingly, there are several five and seven column possibilities also. The range of asymmetrical column combinations is almost limitless, and gives a whole new dimension to what we normally think of as a page grid. Using the twelve column grid can be a rewarding design experience.

Enlarge to 235 %

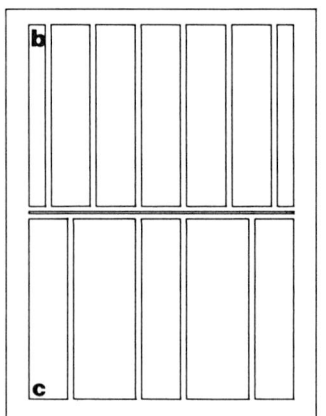
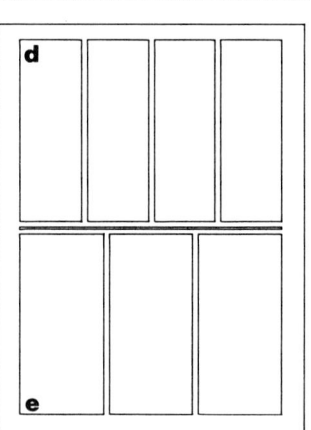
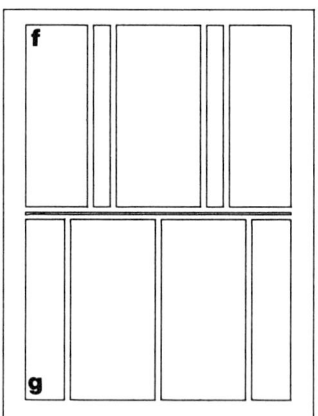
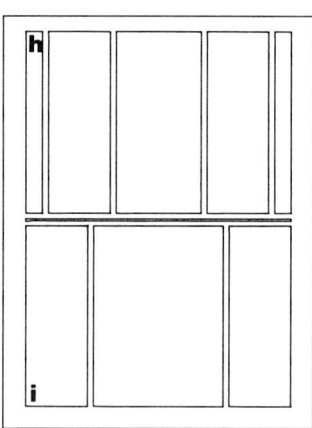
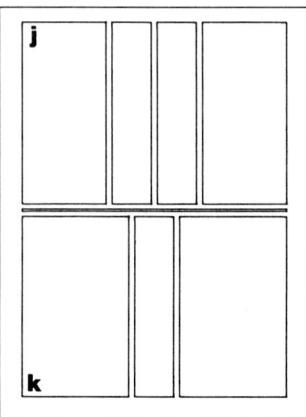
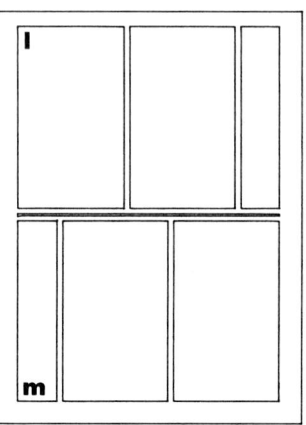
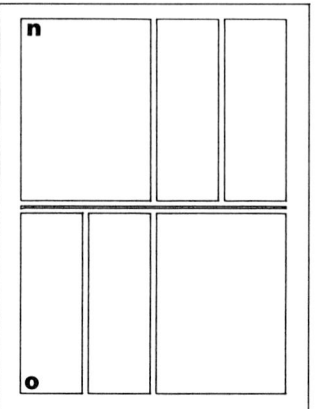
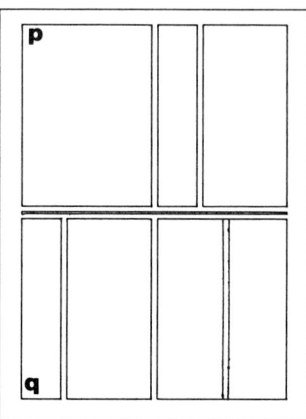

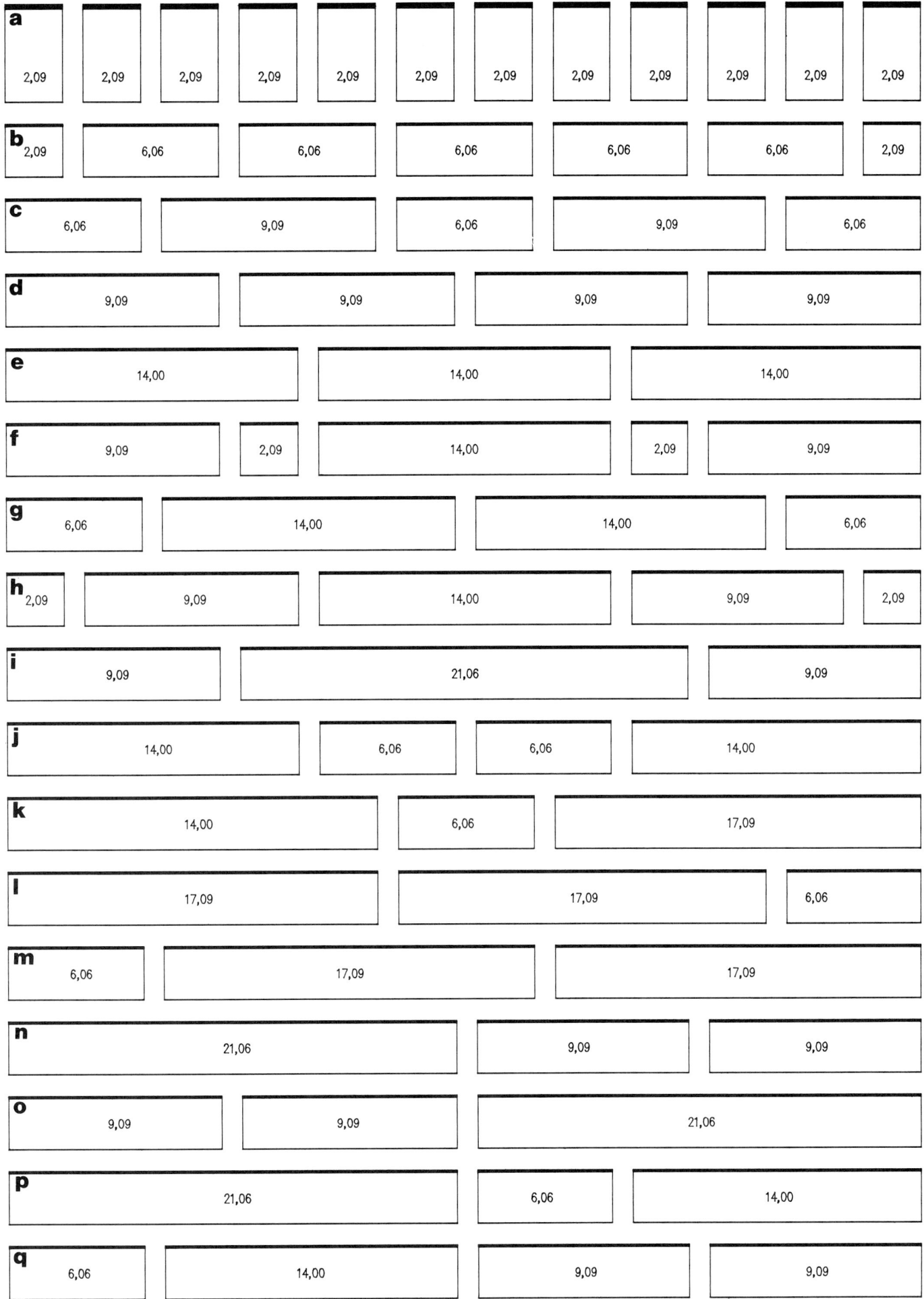

## Multi - Column Grids

Actually the page can be subdivided into as many columns as desired, and the applications become limitless. The number of divisions should be related to the particular publication and to the need for refinement within these divisions. The divisions shown here are for thirteen columns at 2 picas each, eighteen columns at 1 1/2 pica each and twenty three columns at 1 pica each. The grid on the facing page is a field grid of 1 pica square units, 44 picas wide by 60 picas high.

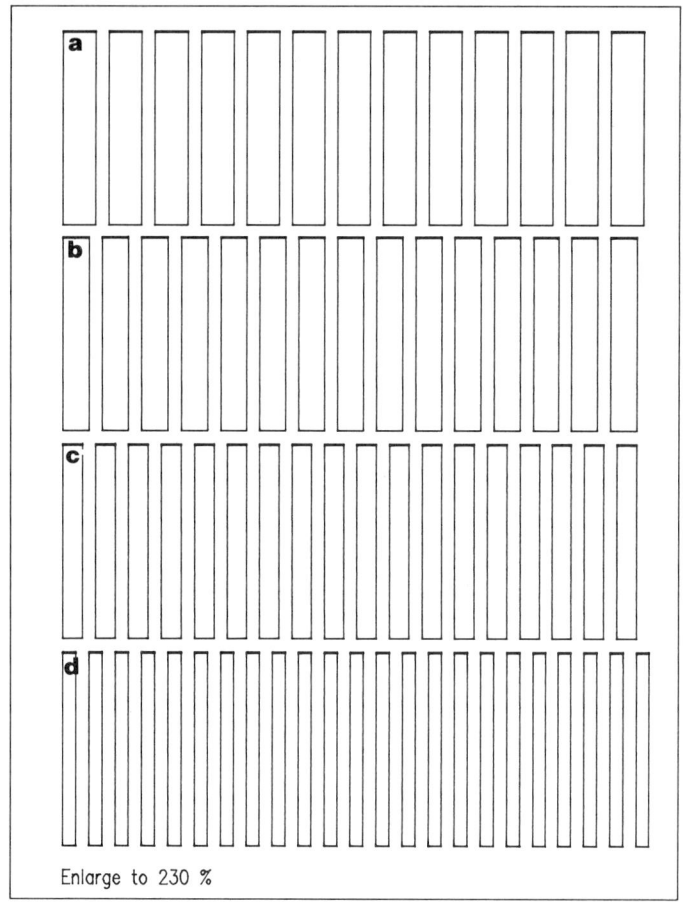

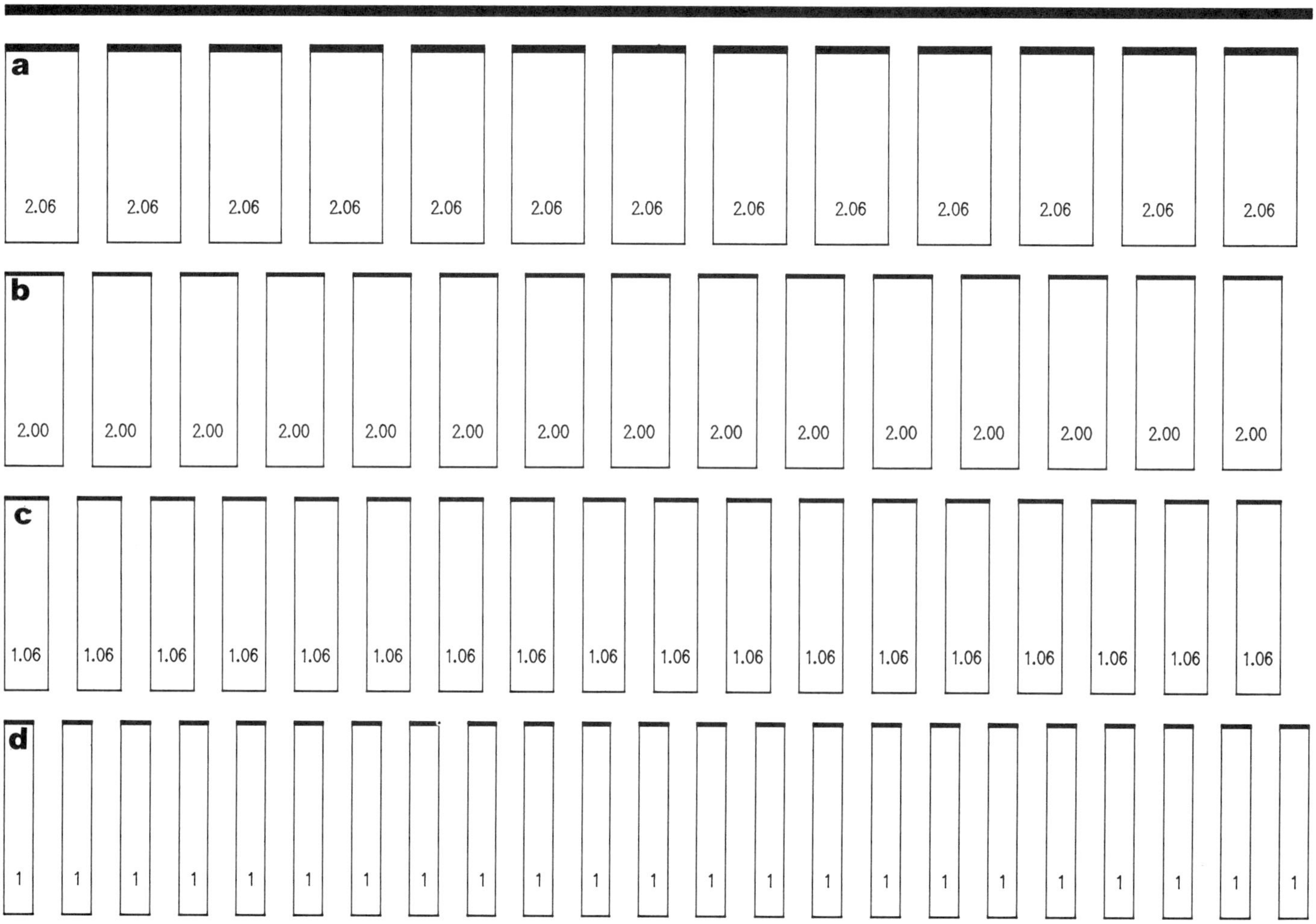

Enlarge to 230 %

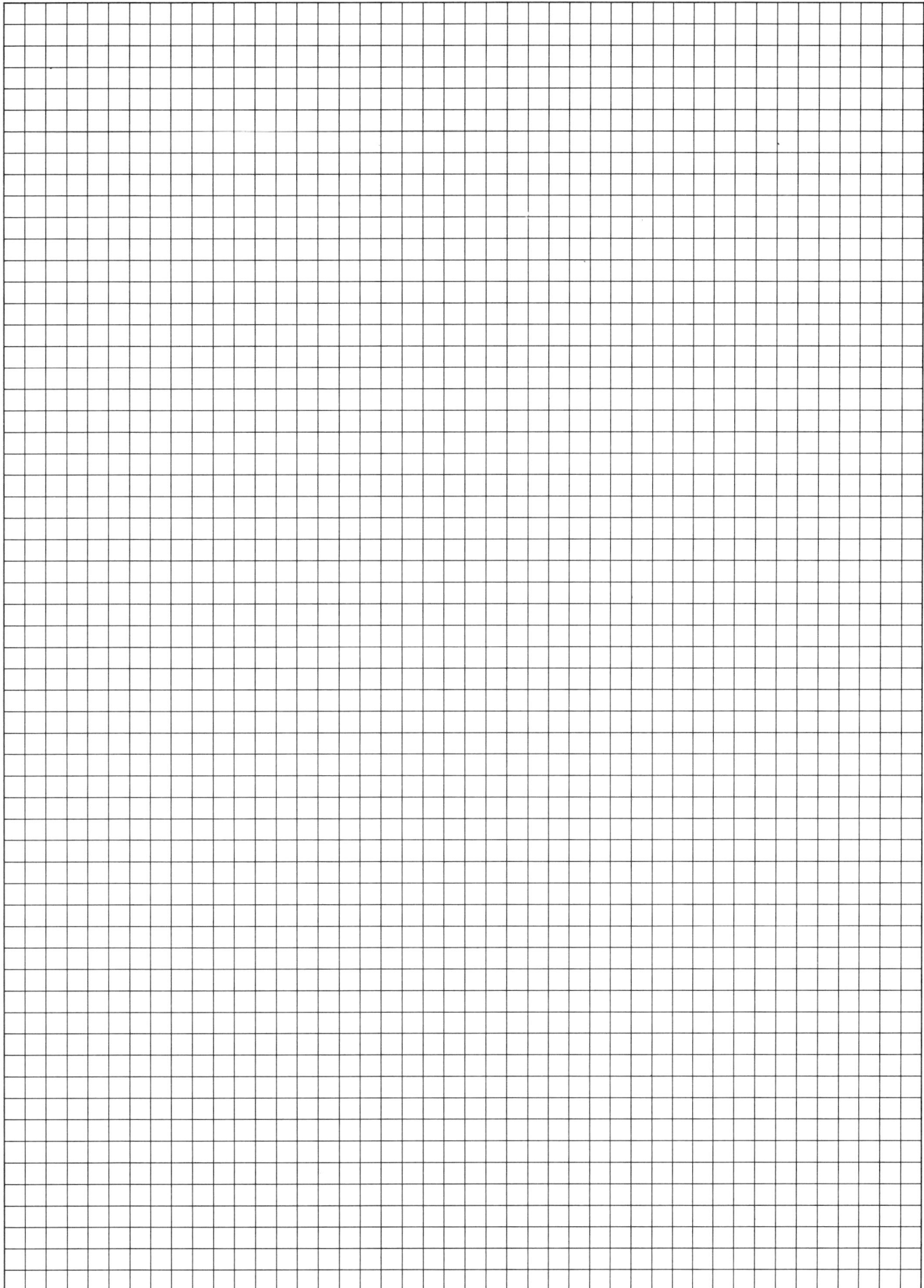

# 4 Case Study Formats

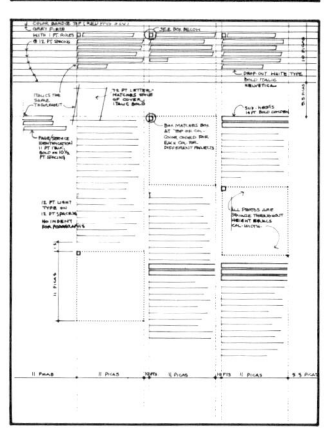

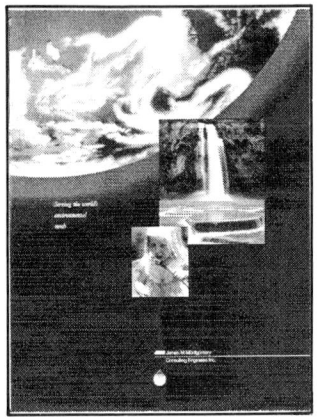

## Case Study 1

Aspects of environmental concerns are expressed in the cover photo showing the earth from space. Fresh air and clean water are both emphasized by the two photo inserts. Inside, this pristine quality is maintained through the sparse use of photographs and minimal text. Around each photograph is a rule, made up of dotted lines to appear lighter. On each spread there is one close-up head shot of someone in the firm, with a quotation under the photo. Text is arranged in a column tilted at 82 degrees

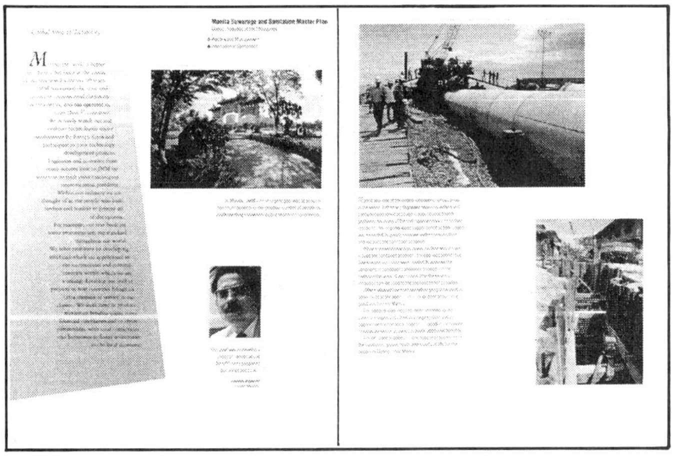

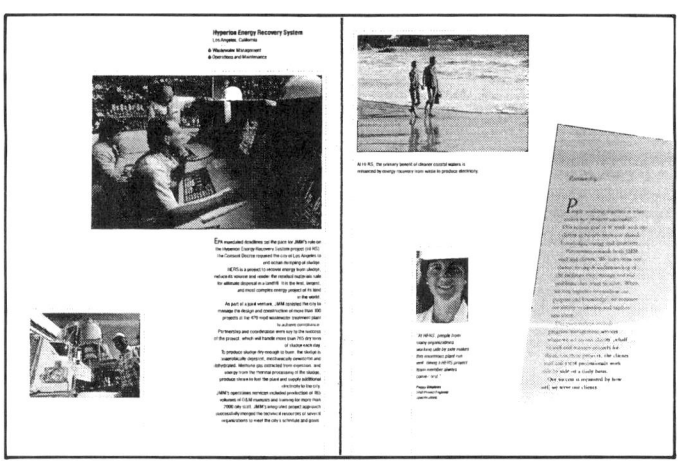

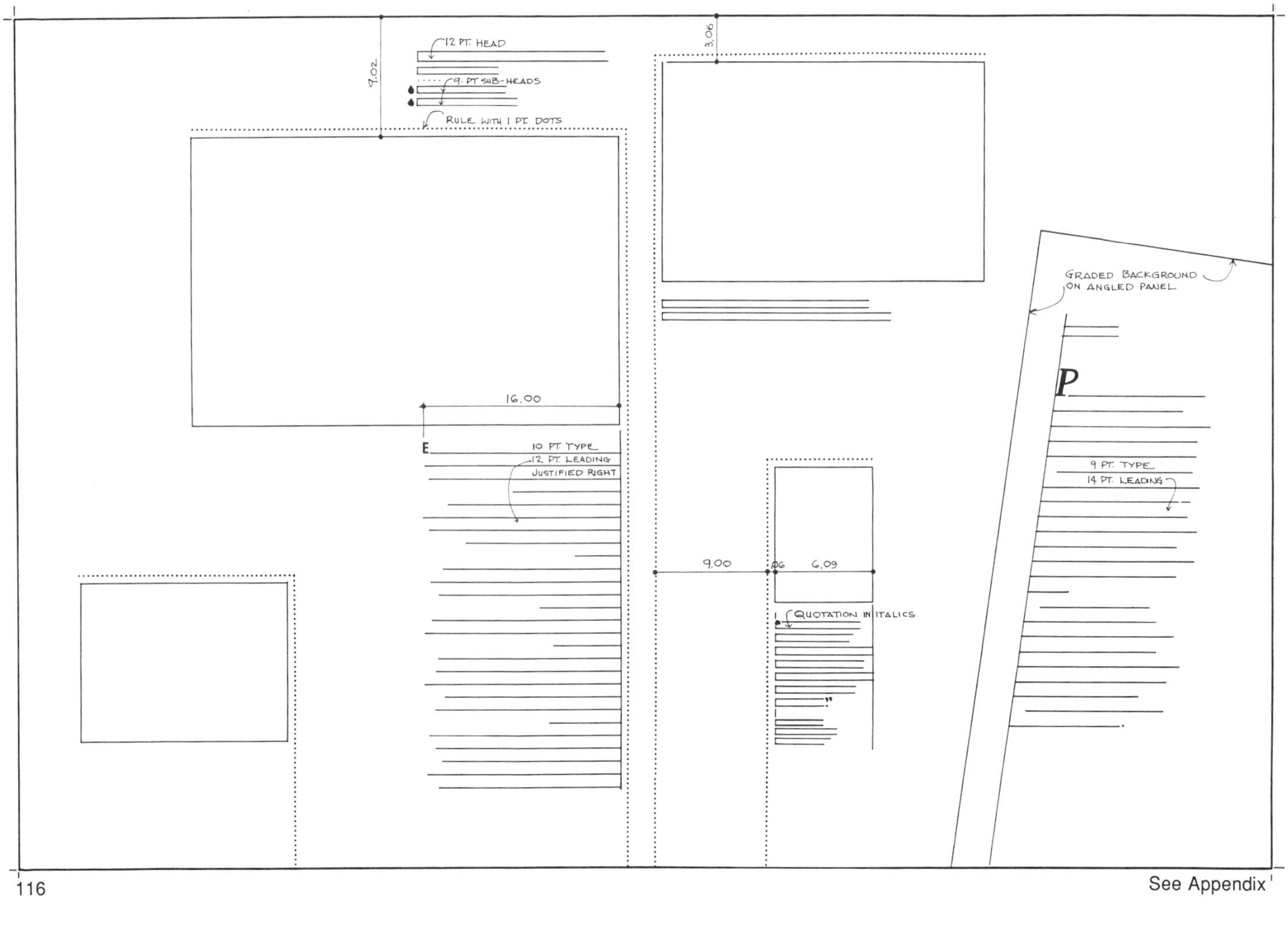

See Appendix

## Case Study 2

Each spread in this transportation brochure is laid out using the same "field grid." It consists of a left-hand page with a single column of bold type, widely spaced. A single head and shoulders photograph appears on this page with a quotation under it. It is one of the "users" of the transportation system. The opposite page is controlled by a strong color bar with drop-out type, under which appears a single column of descriptive text. Alongside is a single large photograph which contains smaller

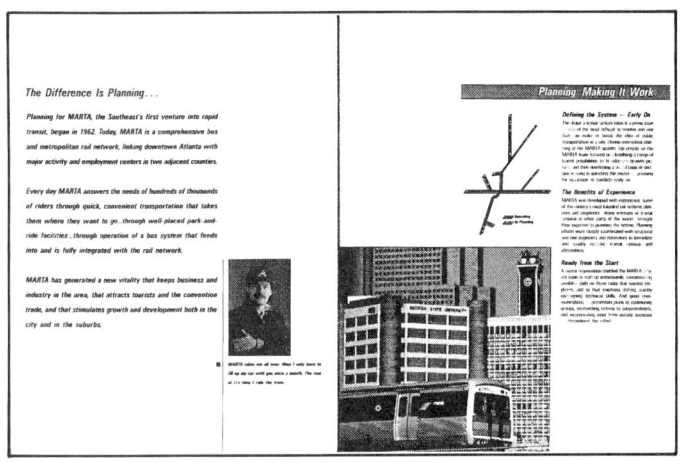

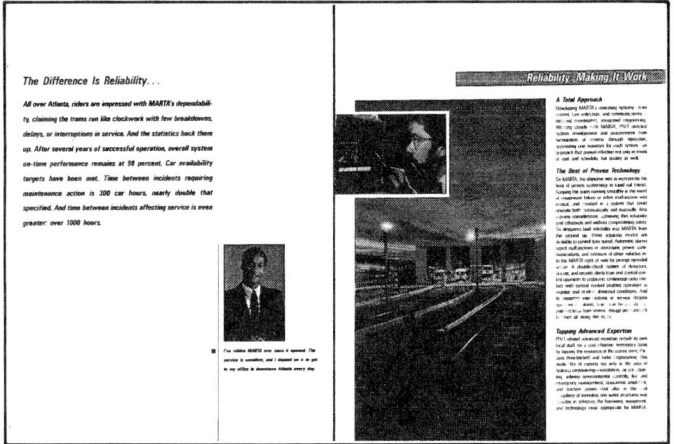

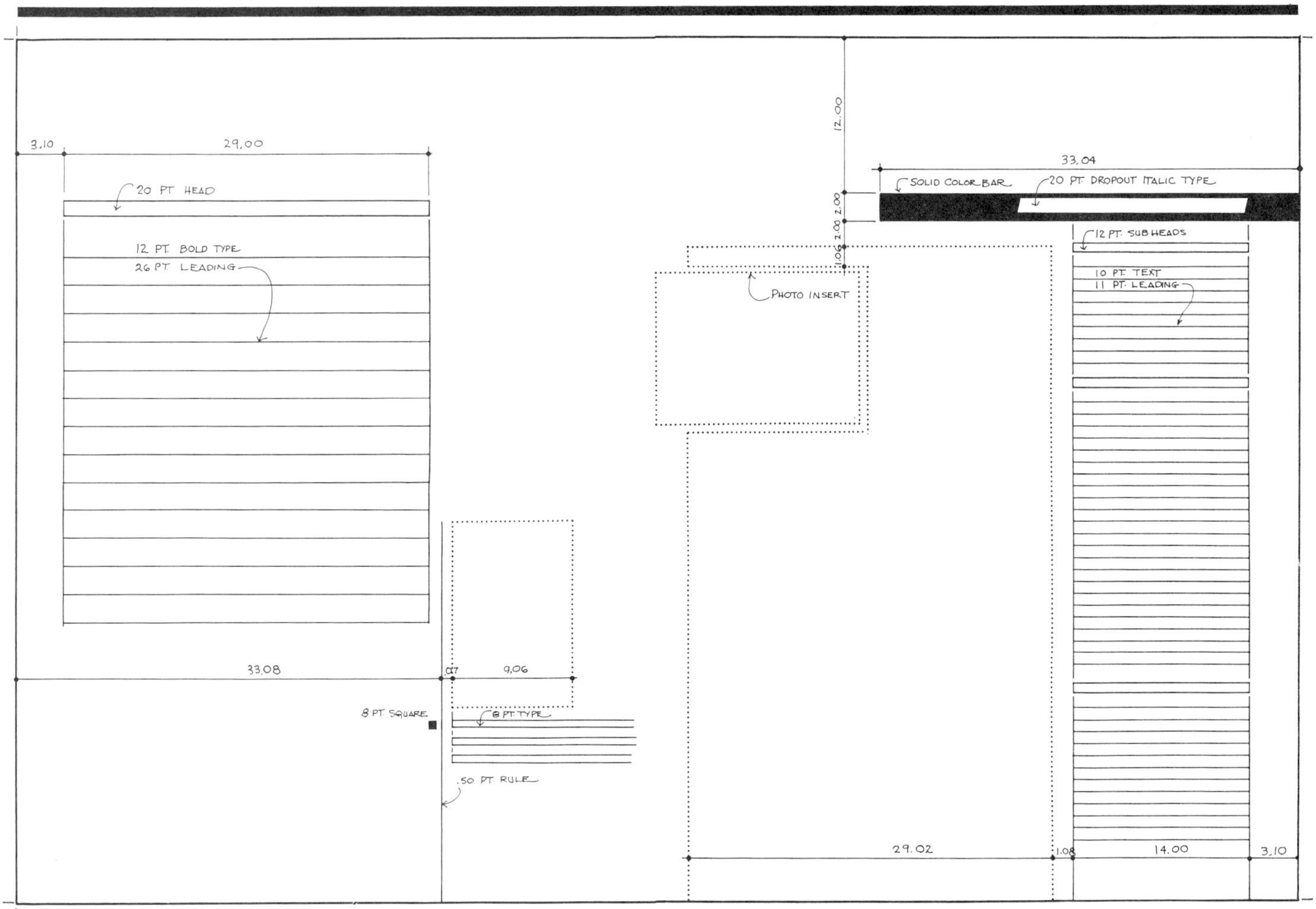

See Appendix

Case Study Formats 117

## Case Study 3

There is little indication on the cover of this brochure of the imaginative layouts within. While there is a consistency of design concept in the layout, there is no repeated grid. The elements that are repeated are the curved dotted lines circumscribing arcs across each spread. The insert photos are very carefully chosen and the total effect is one of high-end graphic style without being trendy. This innovative style can be easily adapted to many applications.

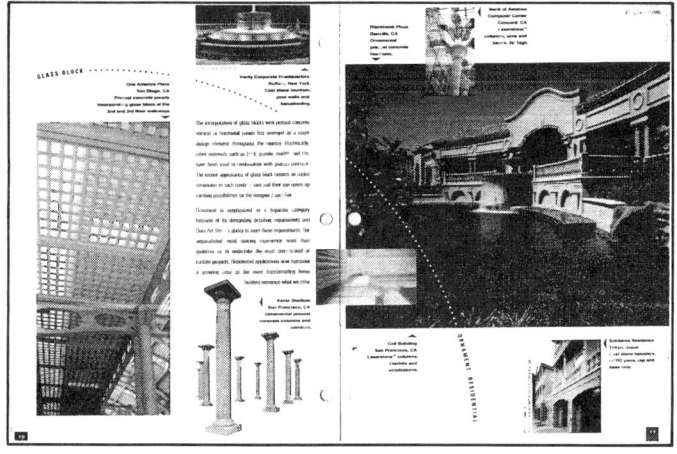

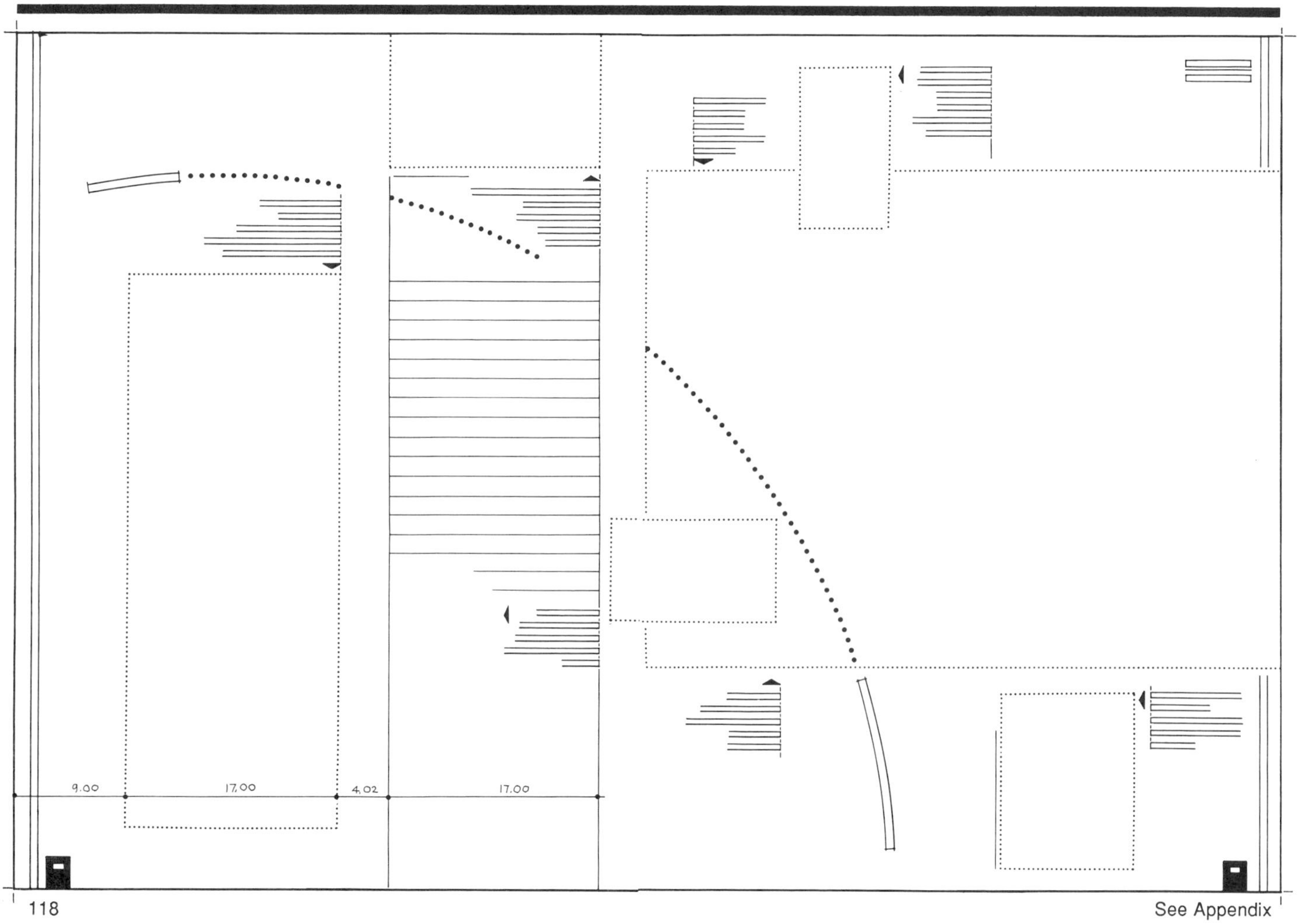

See Appendix

## Case Study 4

The underlying simplicity of design is what gives this brochure its unique appearance. There is a simplicity in the concept as well. The opening page illustrates an empty room with a simple arched opening. On the following pages the room becomes decorated with different "props" to highlight certain aspects of the firm's services. The last photo shows the room again, with the architects, the photographic equipment, and helpers carrying out the props.

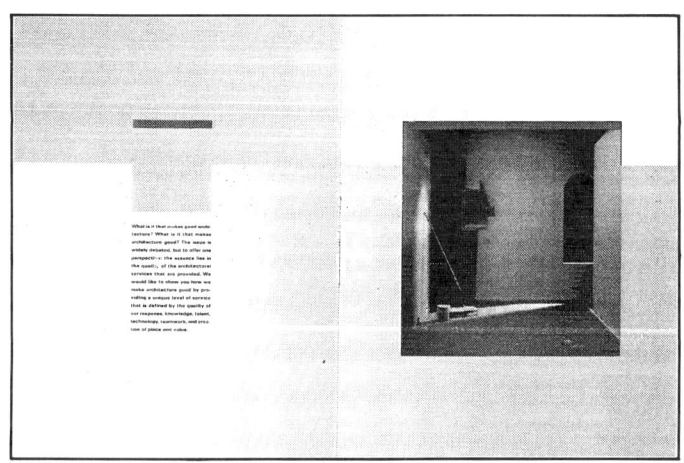

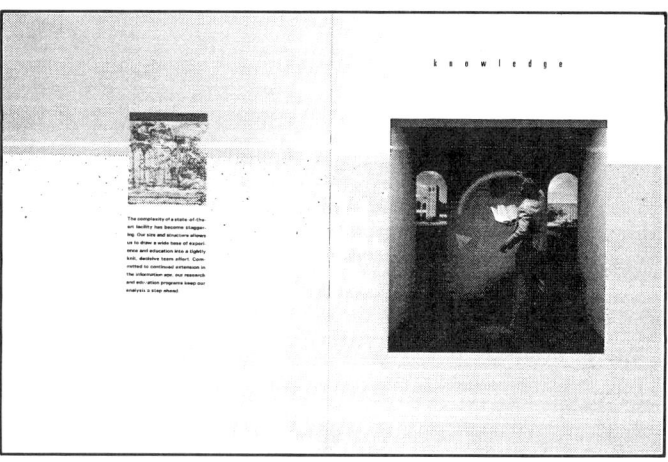

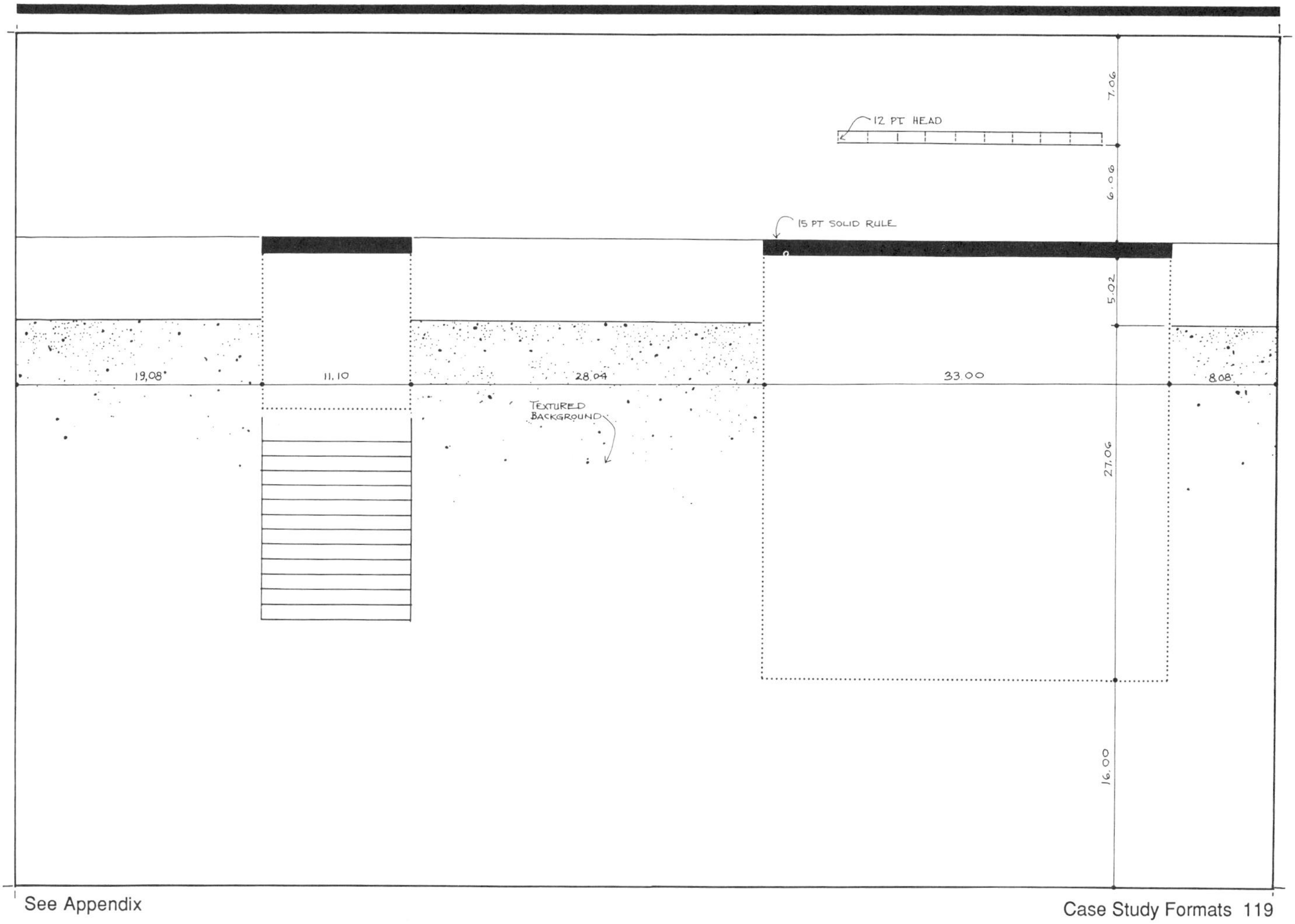

See Appendix

## Case Study 5

The distinctive feature of this construction management firm's brochure is the use of a torn paper edge. This is achieved through photography, wherein the rough textured paper used for the cover was replicated in four colors. The type is printed over this color. The remainder of the spread is a full bleed color photo of a typical installation.

See Appendix for full spread grid

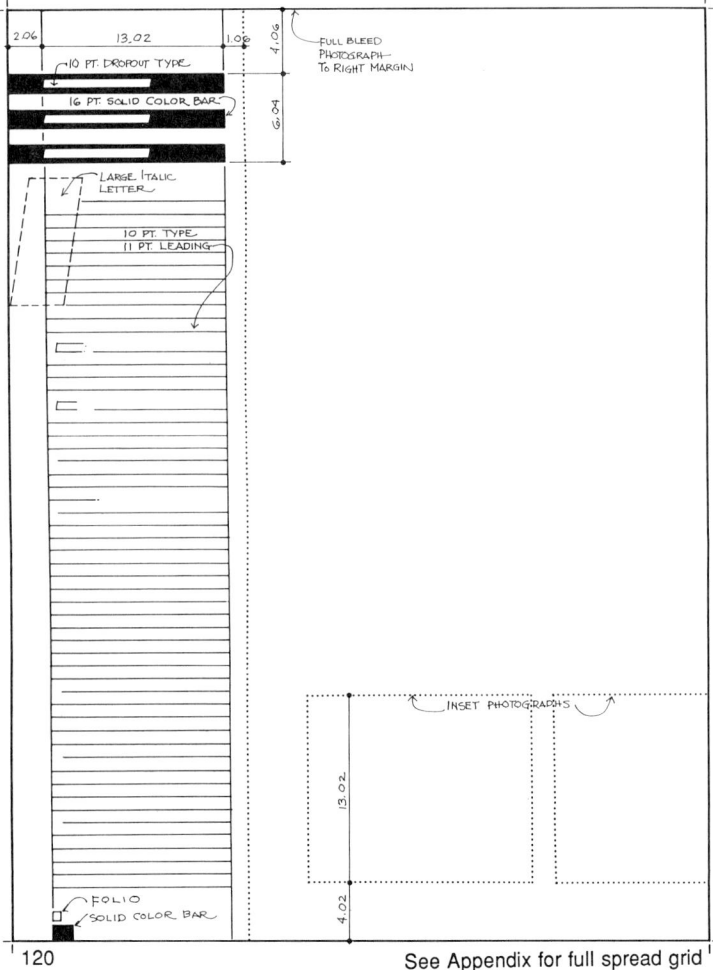

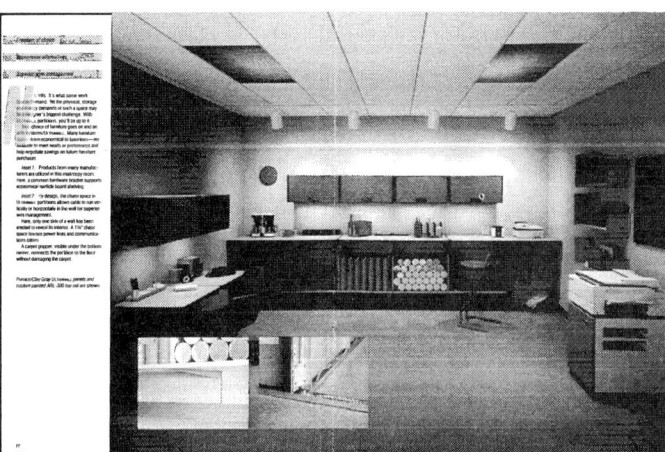

## Case Study 6

This brochure was designed with a similar concept, except that the edge of the large photograph is a clean straight line. Other smaller photos are inserted into this large one. The single column of type is started with a large italic letter above, which is a series of color bars with surprinted type.

See Appendix for full spread grid

## Case Study 7

Everything you need to know about the specialties of this firm can be found on one page. At the top is the identification of the discipline; underneath is a photograph of the key individual overseeing that division. The identification of the person and position is under the picture. The column of type has a discription of that particular service segment. At the bottom is a list of representative projects and clients. Opposite is a large single photo depicting some aspect of the service. A very simple and efficient layout.

Enlarge to 154% for actual size    Trim size 8 1/2 x 11" (51.3 x 66.4 picas)    See Appendix for full spread grid

Case Study Formats 121

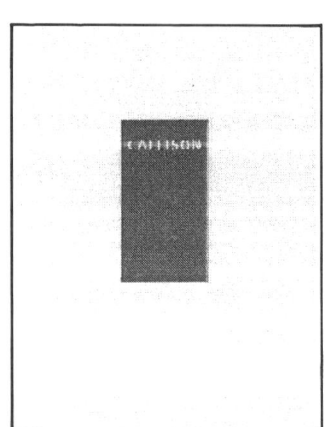

### Case Study 8

Simplicity in layouts can be achieved by several methods, and this brochure demonstrates a few of them. The entire verbal message is carried by a single narrow column of type. It is headed by a three to four line statement of service. Beneath this is a small vignette illustration, with descriptive type underneath it. At the bottom of the column is a quotation from a client. The main photos are large and have smaller photos inserted into them.

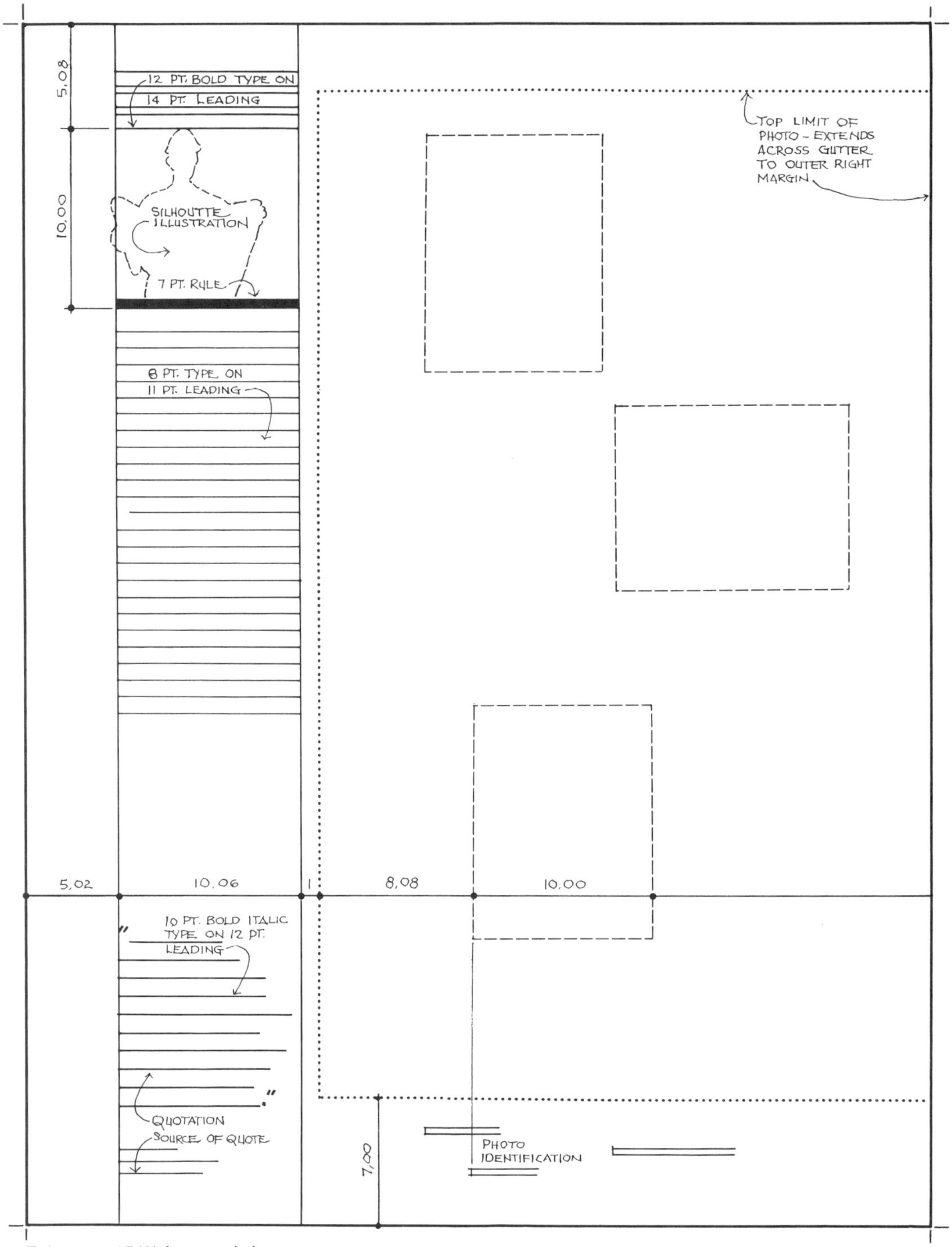

Enlarge to 154% for actual size    Trim size 8 1/2 x 11" (51.3 x 66.4 picas)    See Appendix for full spread grid

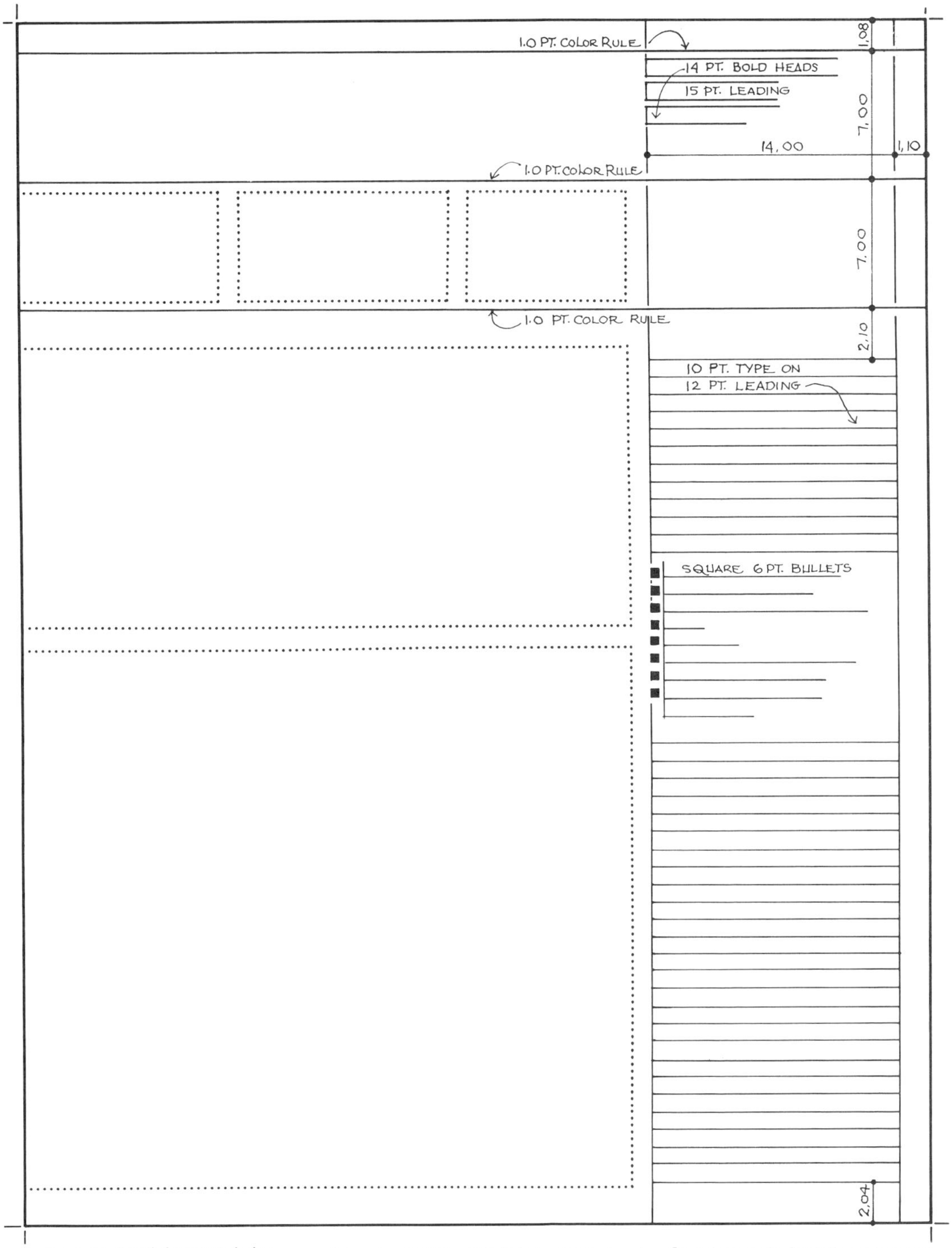

## Case Study 9

This tightly designed, sophisticated layout is controlled by a single column of type on the far corner of each spread. It is visually controlled by two tiers of photographs. The top tier appears as a narrow slot, and the small photos in this slot are wide-angle views of projects or computer models. This slot appears on the cover of the brochure, and carries the firm's name. The second tier contains large dramatic photographs of typical projects. The text is highlighted by a list of square bullets.

Enlarge to 154% for actual size    Trim size 8 1/2 x 11" (51.3 x 66.4 picas)    See Appendix for full spread grid

Case Study Formats  123

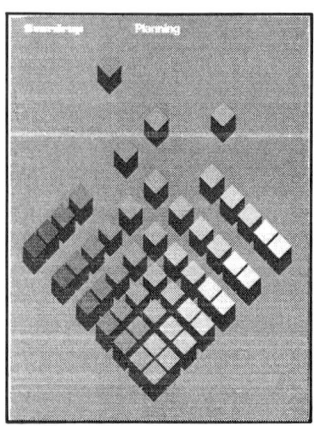

## Case Study 10

An interesting graphic design is displayed on the front cover, and the inside front cover. One element of this design is pulled out as a solo element, and appears on every inside page. It is a cube turned diagonally with a deep shadow cast vertically. This element hovers over a wide single column of type. The photo underneath it lines up with the type. The opposite page contains a very dramatic full bleed color photo with smaller photos set into it, using very thin margins around the photos.

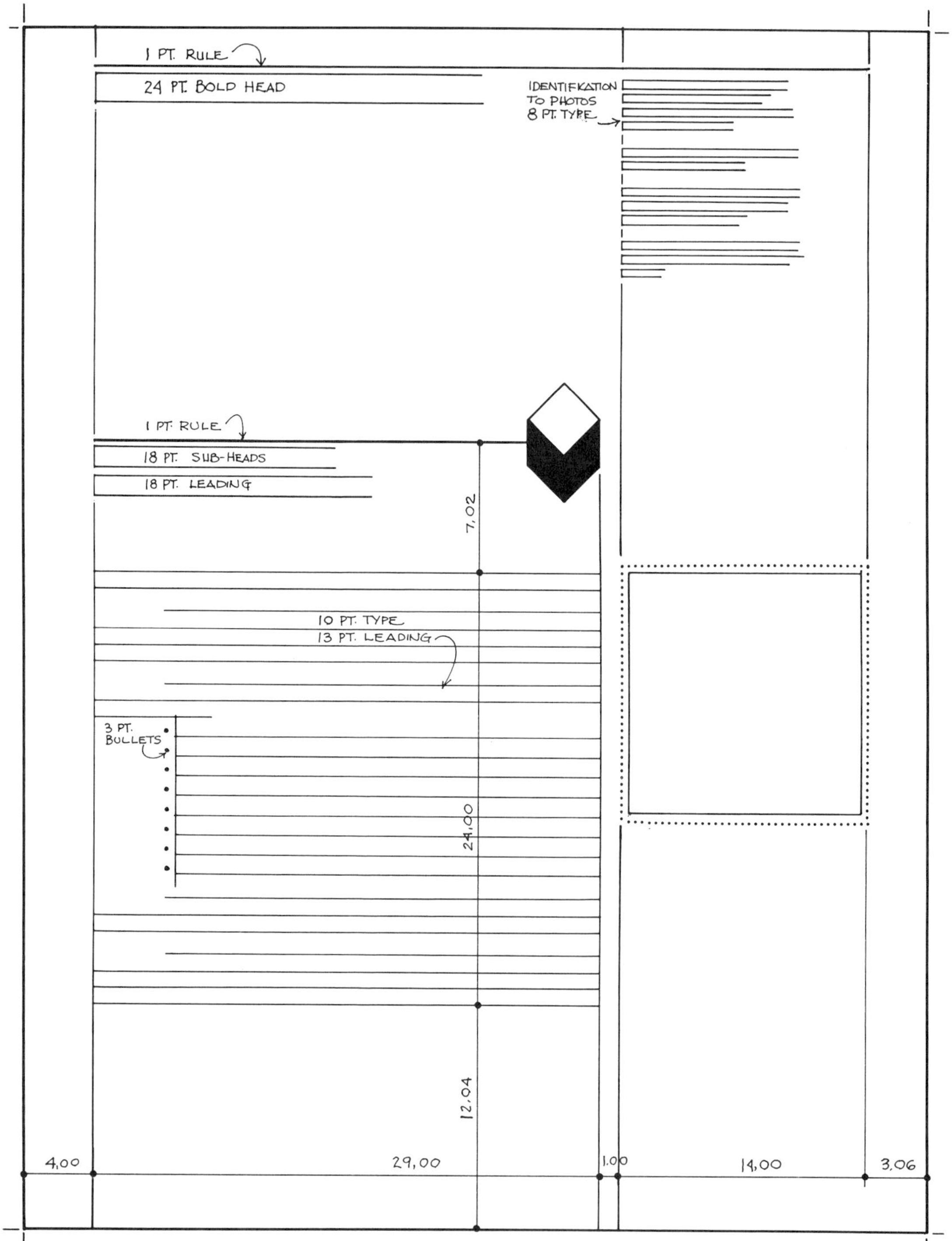

1 PT. RULE
24 PT. BOLD HEAD
IDENTIFICATION TO PHOTOS 8 PT. TYPE
1 PT. RULE
18 PT. SUB-HEADS
18 PT. LEADING
10 PT. TYPE
13 PT. LEADING
3 PT. BULLETS
7.02
24.00
12.04
4.00  29.00  1.00  14.00  3.06

Enlarge to 154% for actual size      Trim size 8 1/2 x 11" (51.3 x 66.4 picas)      See Appendix for full spread grid

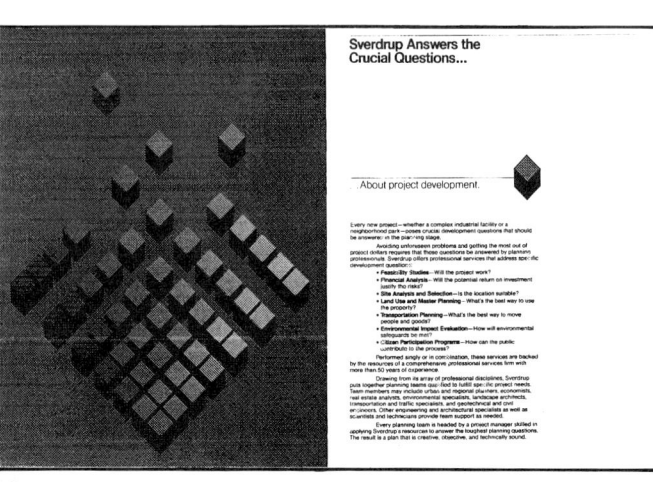

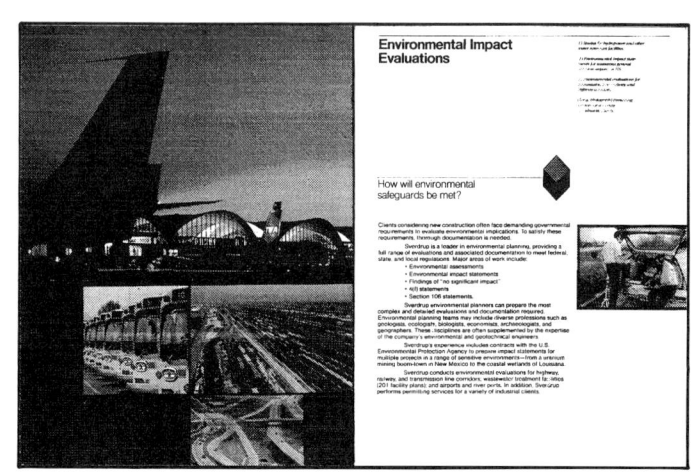

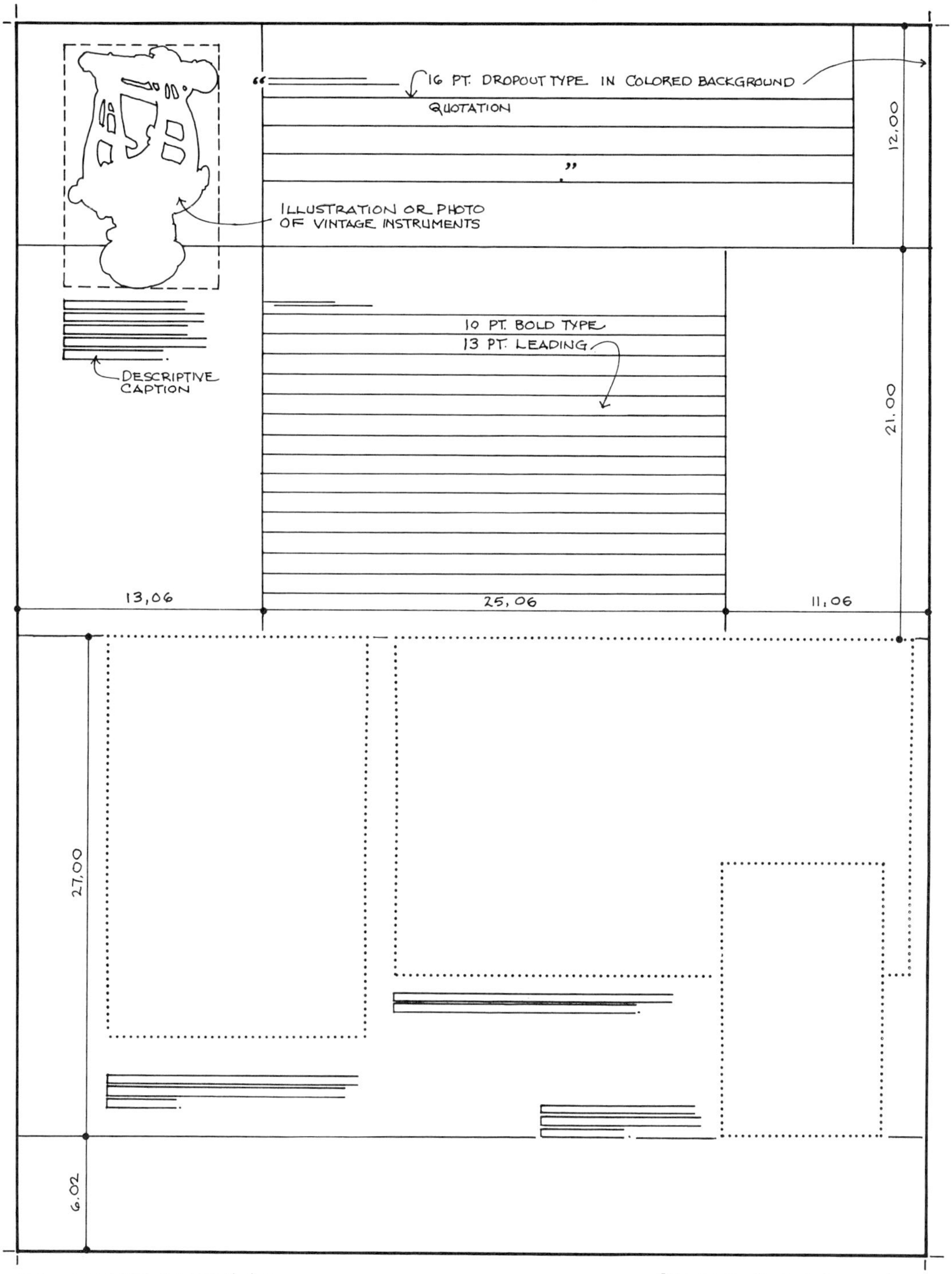

### Case Study 11

The theme of "concept to completion" dominates the cover of this firm's brochure. The background photo shows a roll of drawings on a table. It has an insert photo showing a newly completed project. On the inside comparisons are made between the firm's early methods and the modern techniques that have replaced them. The top margin includes a color band with a quotation set in drop-out type. In the top corner there is either a vintage photograph or an illustration. Below this is a wide single column of type, and below the type a panel of photographs. The facing page contains a full bleed photo with inserts.

Enlarge to 154% for actual size    Trim size 8 1/2 x 11" (51.3 x 66.4 picas)    See Appendix for full spread grid

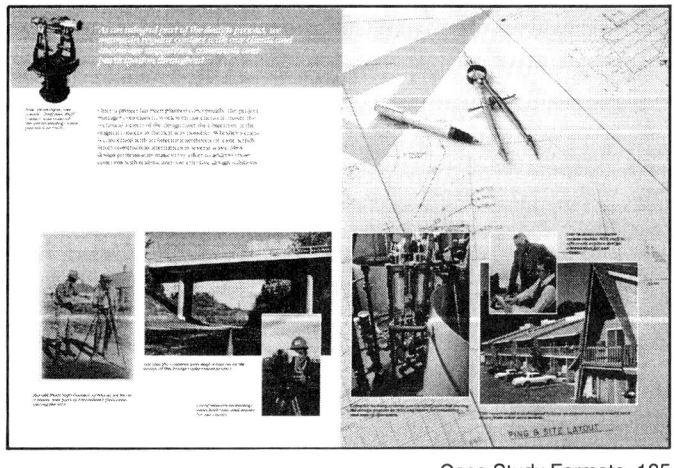

Case Study Formats  125

HUMPHREYS & HARDING, INC.

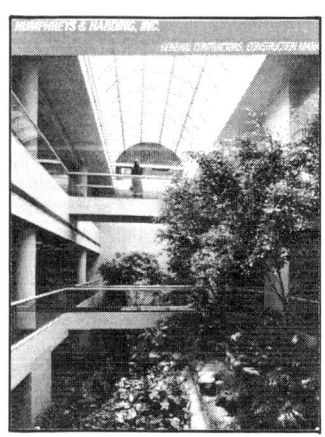

## Case Study 12

A single column of type is all that is used to offset the photo layout in this brochure. A strong solid color bar with drop-out type stands over the column of small photographs adjacent to the text. From here, a large photograph covers the remainder of the spread. Although some of the photos show workers and installation procedures, the main photograph is a dramatic view of a completed project.

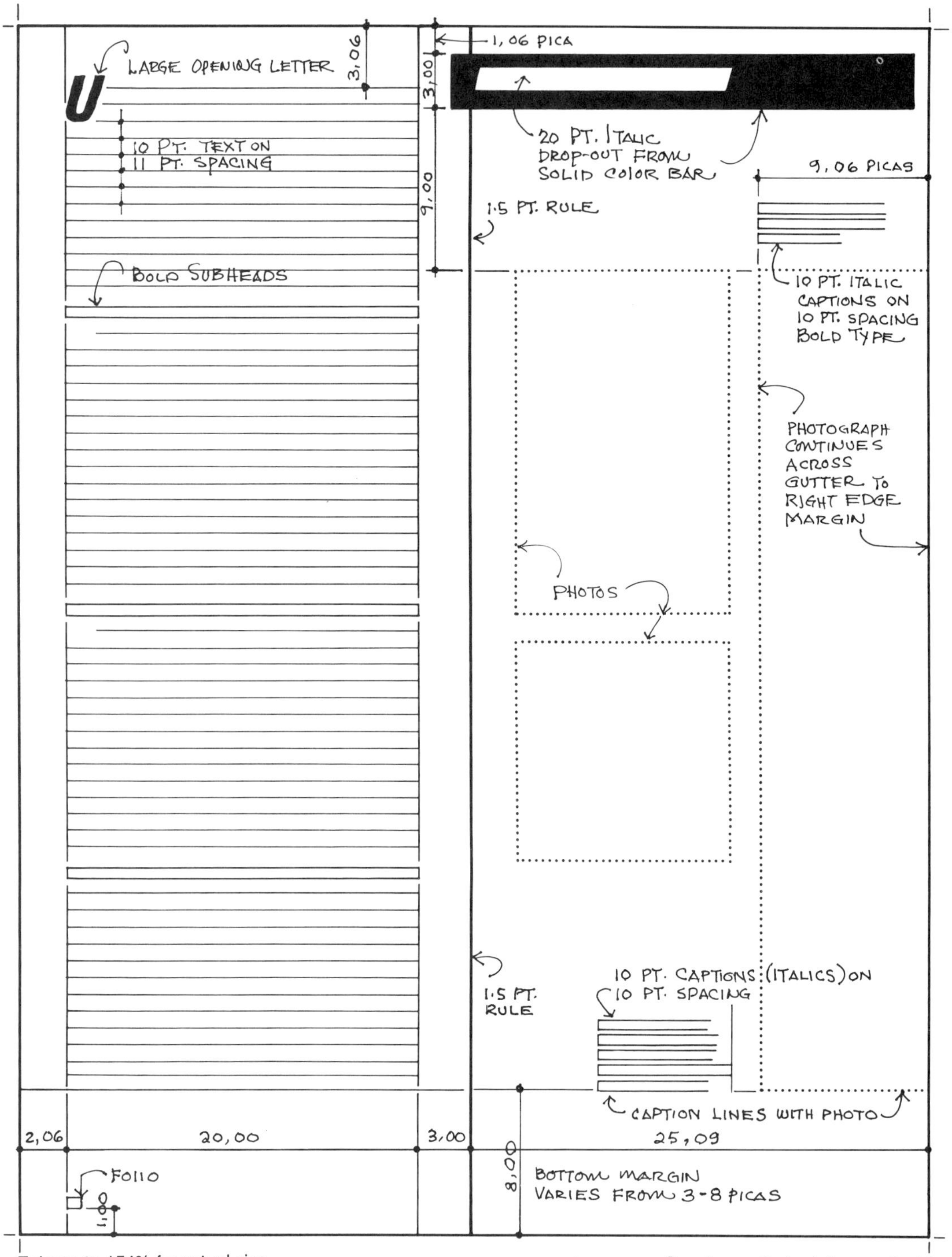

U
LARGE OPENING LETTER
3,06
10 PT. TEXT ON
11 PT. SPACING
BOLD SUBHEADS
1,06 PICA
3,00
9,00
20 PT. ITALIC
DROP-OUT FROM
SOLID COLOR BAR
9,06 PICAS
1.5 PT. RULE
10 PT. ITALIC
CAPTIONS ON
10 PT. SPACING
BOLD TYPE
PHOTOGRAPH
CONTINUES
ACROSS
GUTTER TO
RIGHT EDGE
MARGIN
PHOTOS
1.5 PT.
RULE
10 PT. CAPTIONS (ITALICS) ON
10 PT. SPACING
CAPTION LINES WITH PHOTO
2,06
20,00
3,00
25,09
FOLIO
1
8,00
BOTTOM MARGIN
VARIES FROM 3-8 PICAS

Enlarge to 154% for actual size

Trim size 8 1/2 x 11" (51.3 x 66.4 picas)

See Appendix for full spread grid

HUMPHREY'S & HARDING
A DIVERSIFIED CONSTRUCTION ORGANIZATION
A HISTORY OF EXCELLENCE AND INNOVATION
TABLE OF CONTENTS

EXCELLENCE IN ALTERATIONS

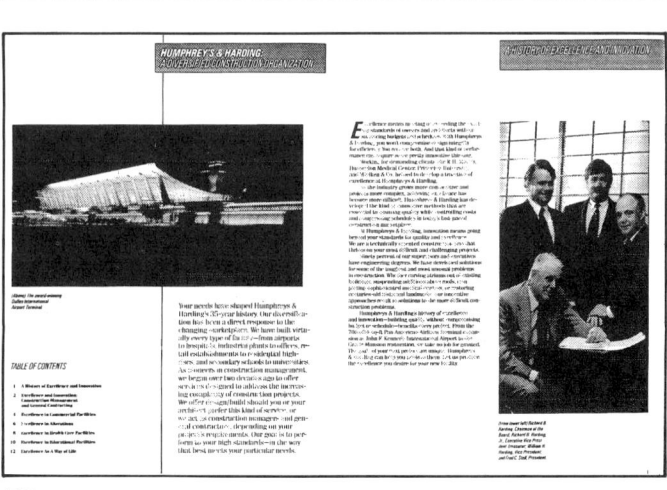

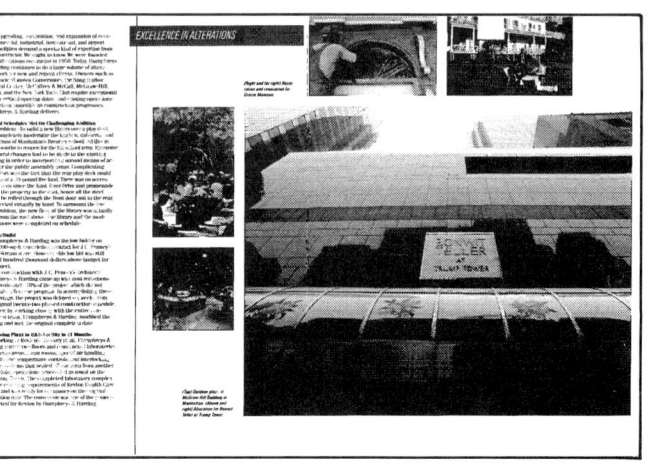

## Case Study 13

The stair-stepped design that frames the cover photograph of this manufacturer's catalog continues onto the back cover. Inside, the step design is repeated in miniature at the top of each column of type. The type is set in reverse and the photographs are inserted into the black background. Rather than show the raw product, they have chosen to showcase recent installations. On each side of the spread there is a wide band of color, replicating the texture of the product.

Enlarge to 154% for actual size   Trim size 8 1/2 x 11" (51.3 x 66.4 picas)   See Appendix for full spread grid

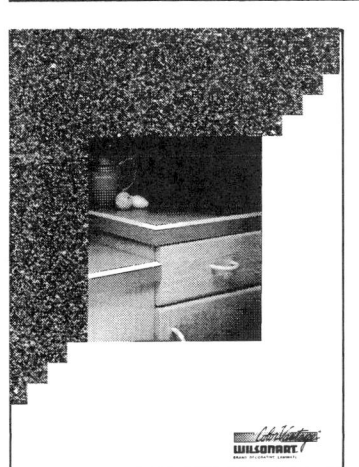  

## Case Study 14

An elaborate head margin provides the primary focus for this specialty brochure. This is expressed first in the ruled lines going across the top of the cover. Inside, this head margin has its own grid, which is independent of the material below it. The single column of type is flanked by a single column of large photographs. In many cases there are explanatory graphic illustrations below the photographs. The firm's logo appears on each page in a light tone.

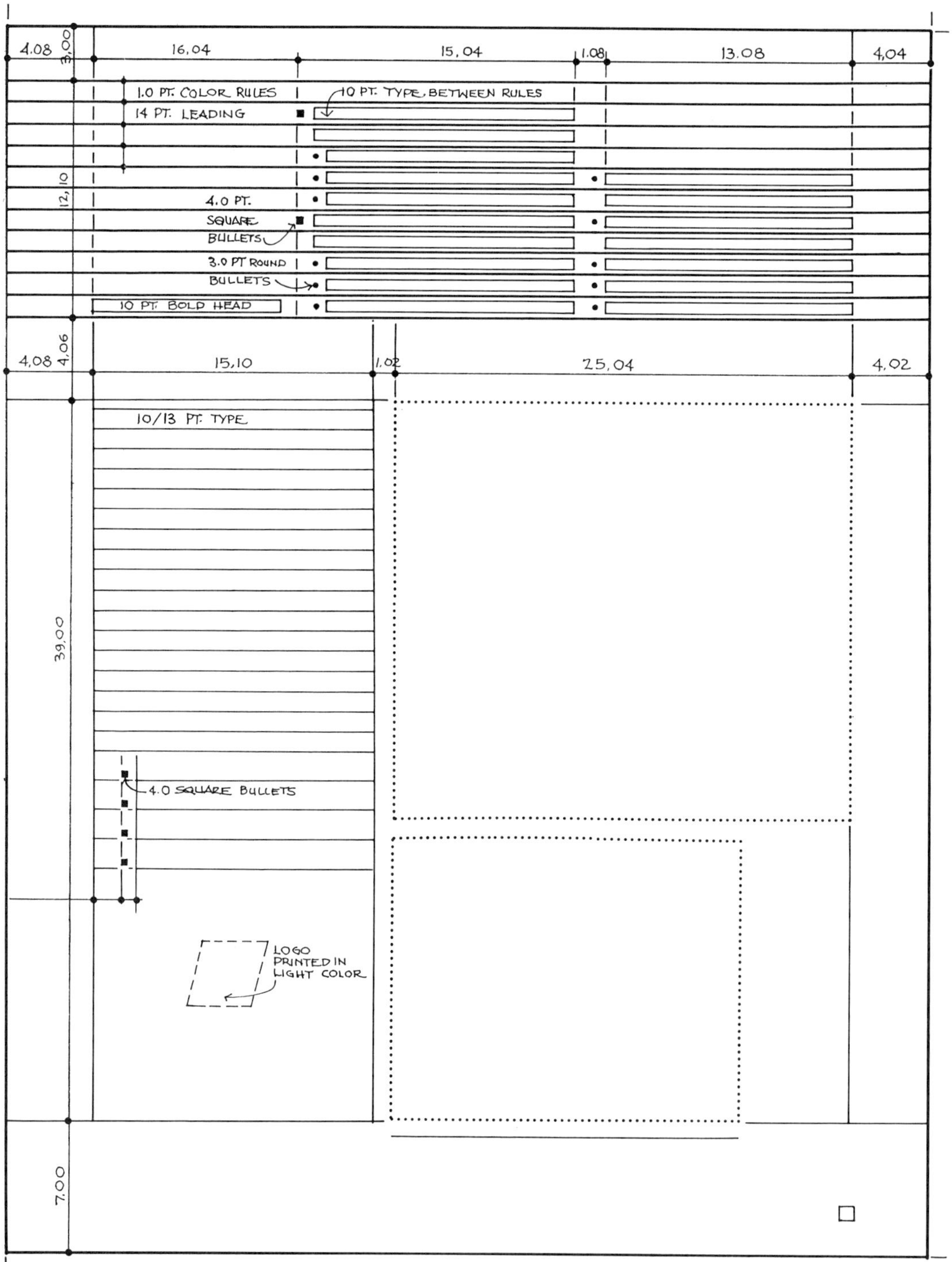

Enlarge to 154% for actual size    Trim size 8 1/2 x 11" (51.3 x 66.4 picas)    See Appendix for full spread grid

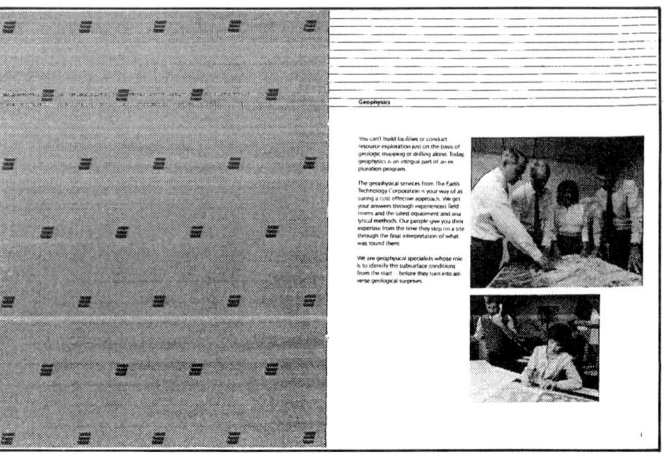

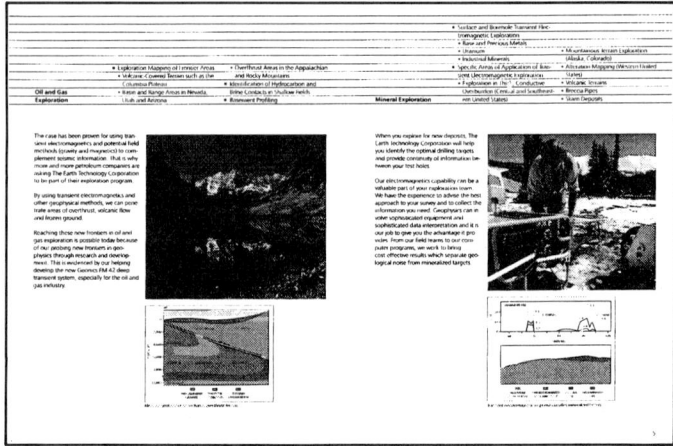

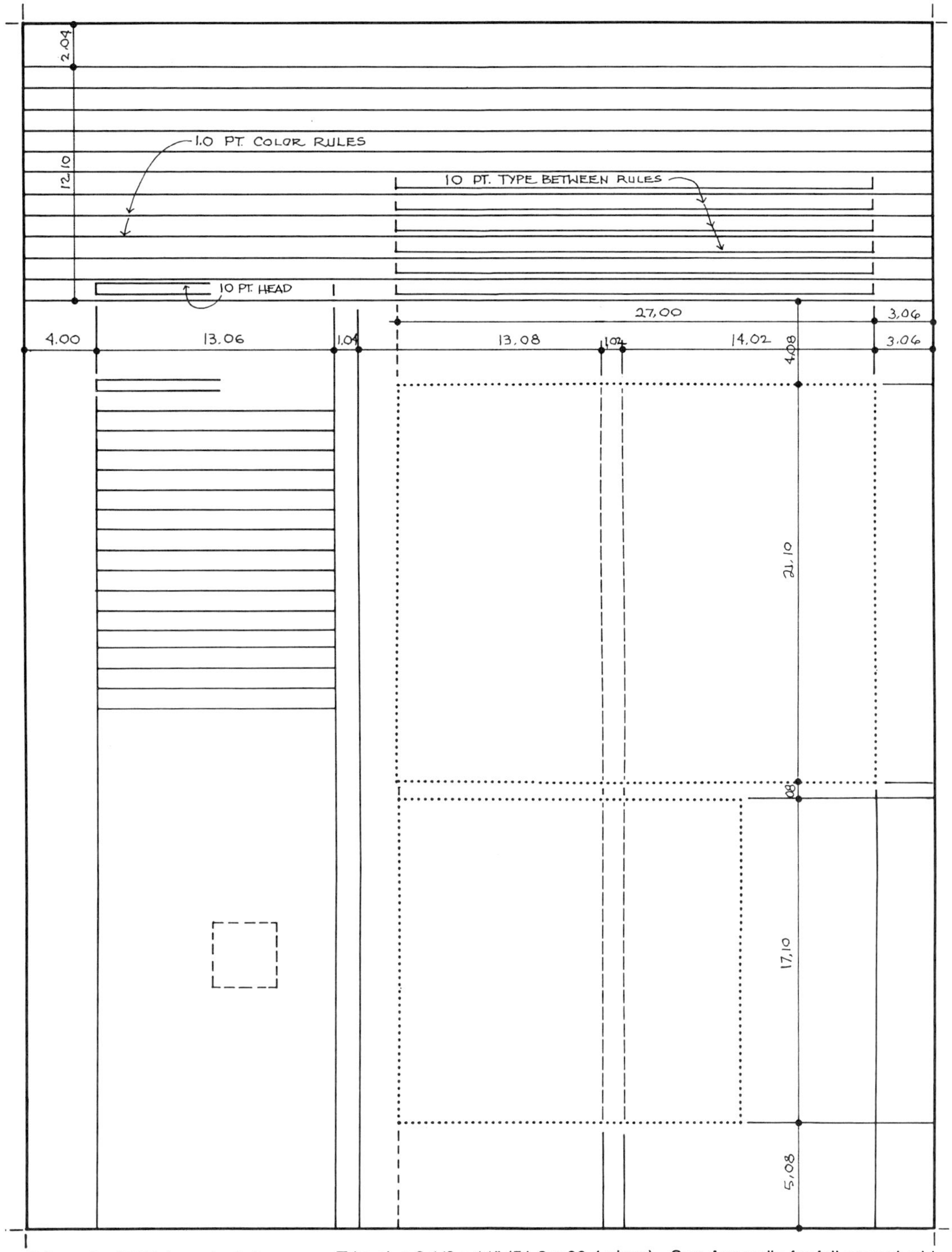

## Case Study 15

The rules in the wide head margin of this specialty brochure are similar to the previous one, in that they show first on the cover. However, the type between these rules lines up with the photo column below. In essence it is a three column grid wherein the photographs occupy two of the columns in most cases. The firm's logo is printed in the lower corner of each column of type in a light tone.

Enlarge to 154% for actual size    Trim size 8 1/2 x 11" (51.3 x 66.4 picas)    See Appendix for full spread grid

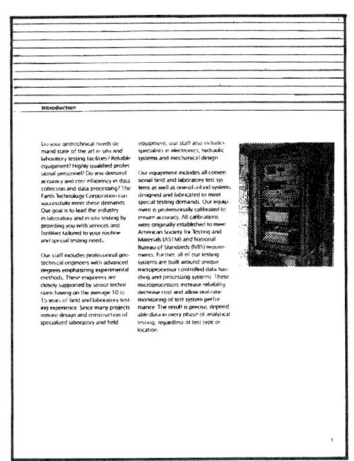
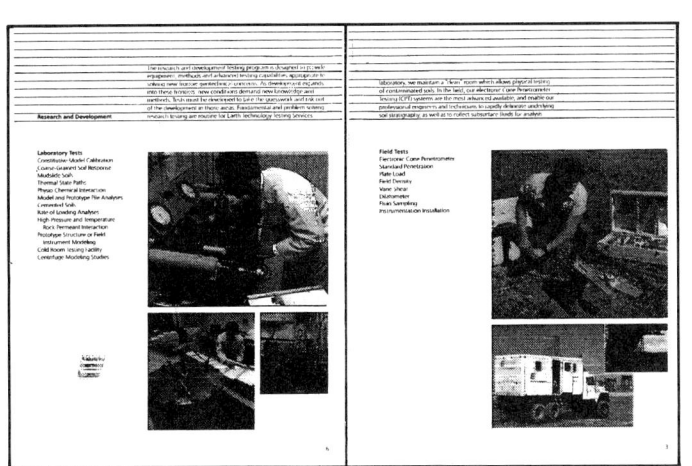
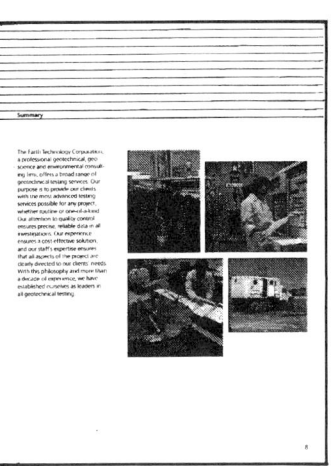

Case Study Formats 129

## Case Study 16

The brochures for this world renowned industrial design firm were designed as spreads, each featuring a different aspect of the practice. Within each aspect there was a combination of photographs of products and a historical perspective. This was achieved by showing a vintage design in the lower left corner of each page. Each spread is dominated by a large photograph, and supplemented by several smaller ones. They are consistent within a very simple grid.

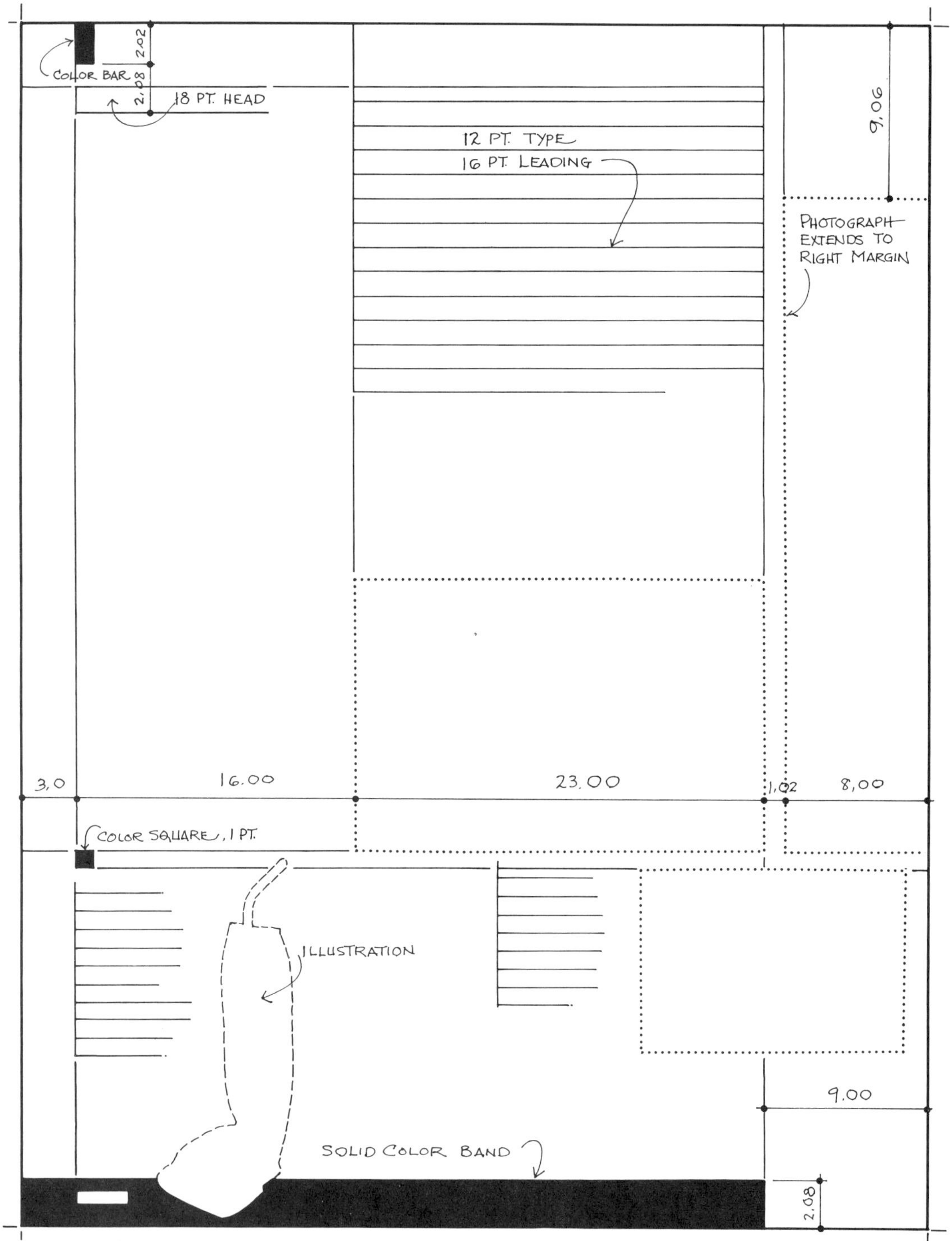

Enlarge to 154% for actual size   Trim size 8 1/2 x 11" (51.3 x 66.4 picas)   See Appendix for full spread grid

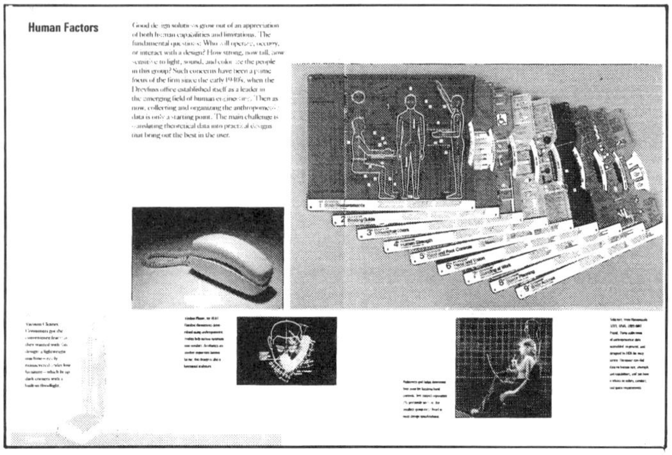

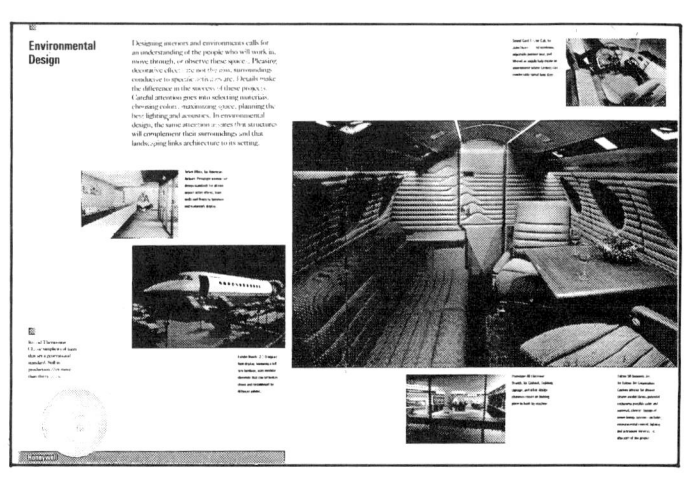

## Case Study 17

The format for this annual report for a major magazine publishing company has a traditional two-column grid combined with a presentation of miniature cover designs running across the top of the spread. Each spread in the report features a different division of the company program. The left page features a large photograph, and the rules underlining the subheads continue across the gutter and into the photograph as a drop-out rule.

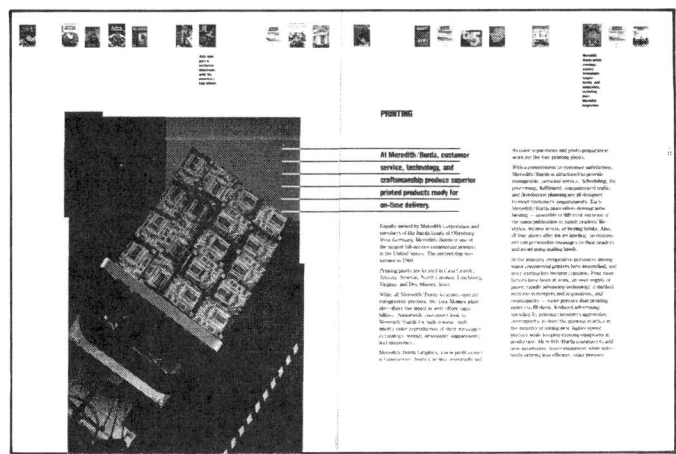

See Appendix for full spread grid   Trim size 8 1/2 x 11" (51.3 x 66.4 picas)

## Case Study 18

Each spread is dominated by a large photo which straddles the gutter. It leaves a wide single column on one side and a narrow column on the other. At the top of the narrow column is a featured product. Across the bottom is a chronology illustrating the development of high-tech instruments. The years are separated by thin rules.

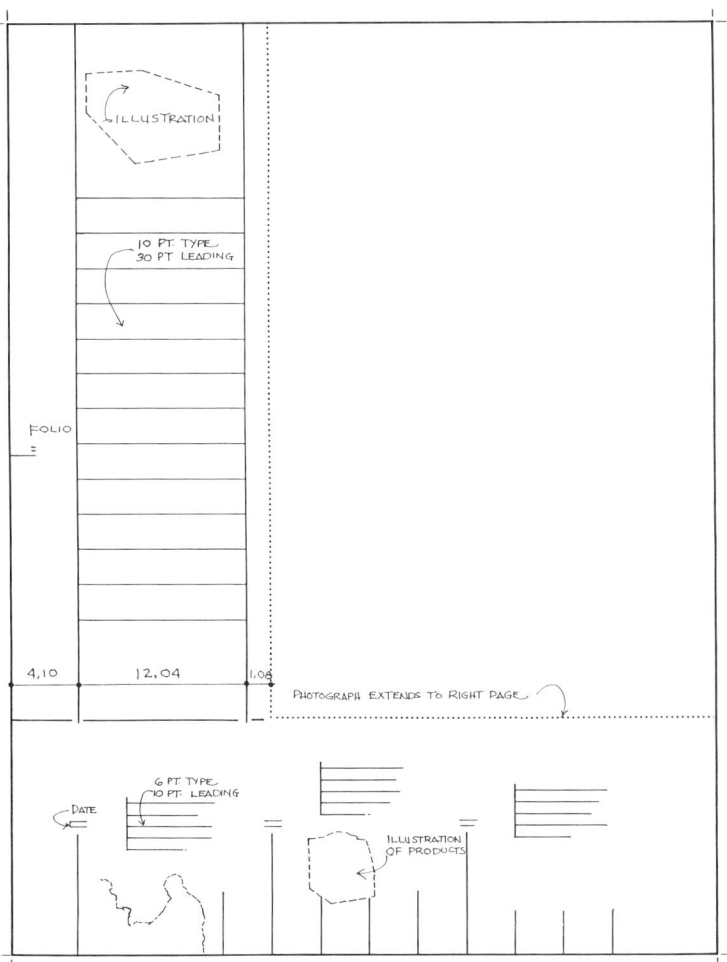

See Appendix for full spread grid   Trim size 8 1/2 x 11" (51.3 x 66.4 picas)

**Case Study 19**

For most firms this would be a trendy brochure design, but it comes alive in both the graphic design and cleverly written text. It is based more on storytelling than plain descriptions of projects. The lively spot illustrations support this theme. In addition the innermost column of text is interrupted by vignette sketches of the floor plans and sections of the featured structures. Each spread is dominated by one large photograph which crosses the gutter.

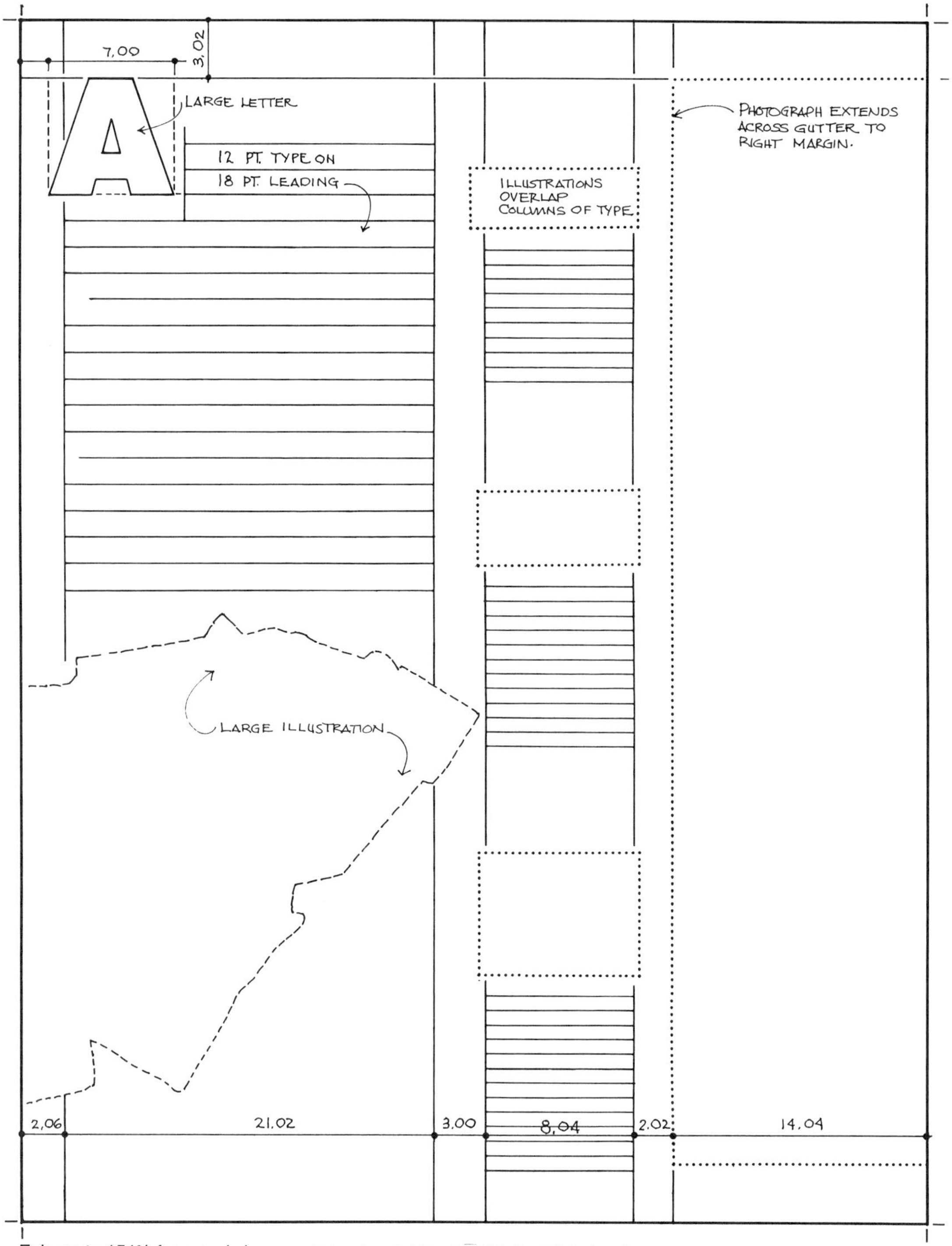

— LARGE LETTER
12 PT. TYPE ON
18 PT. LEADING
ILLUSTRATIONS OVERLAP COLUMNS OF TYPE.
PHOTOGRAPH EXTENDS ACROSS GUTTER TO RIGHT MARGIN.
LARGE ILLUSTRATION

7,00  3,02
2,06  21,02  3,00  8,04  2,02  14,04

Enlarge to 154% for actual size     Trim size 8 1/2 x 11" (51.3 x 66.4 picas)     See Appendix for full spread grid

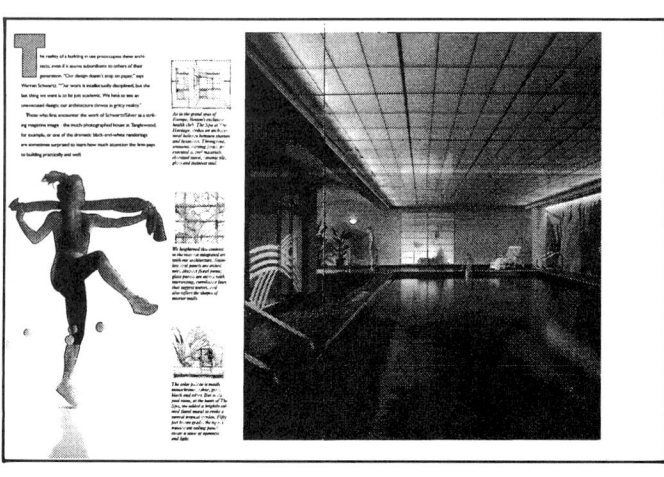

132

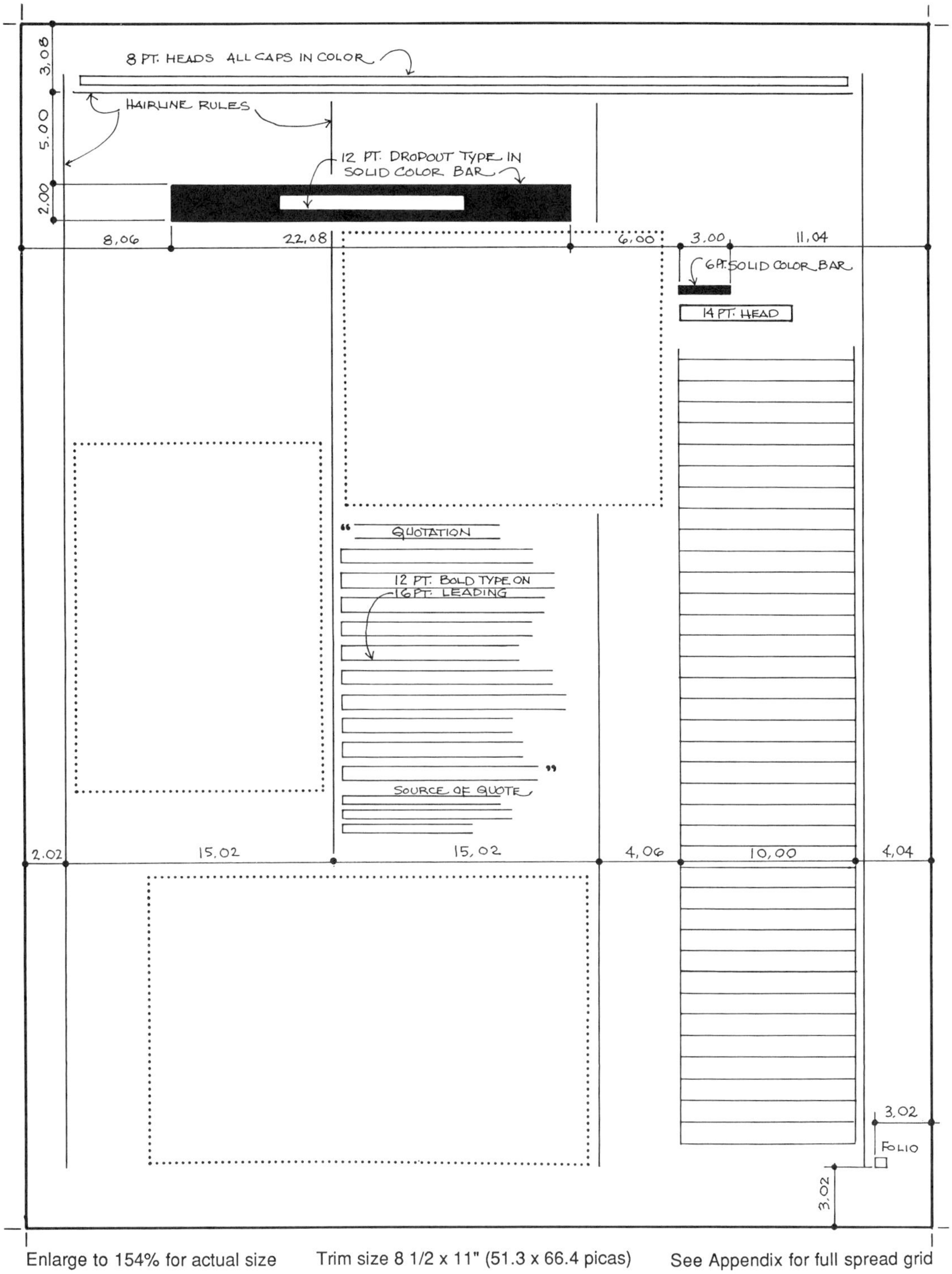

### Case Study 20

The grid for this project brochure is a simple three column grid. However the application and use of the photographs and text make it appear more complex. This is due to the overlapping of photographs outside of the columns. Also, the main column of text is justified and narrower than the grid, allowing for a more flexible arrangement. The descriptive text is set in the full column width, either ragged right or ragged left depending on the position of the photograph.

Enlarge to 154% for actual size   Trim size 8 1/2 x 11" (51.3 x 66.4 picas)   See Appendix for full spread grid

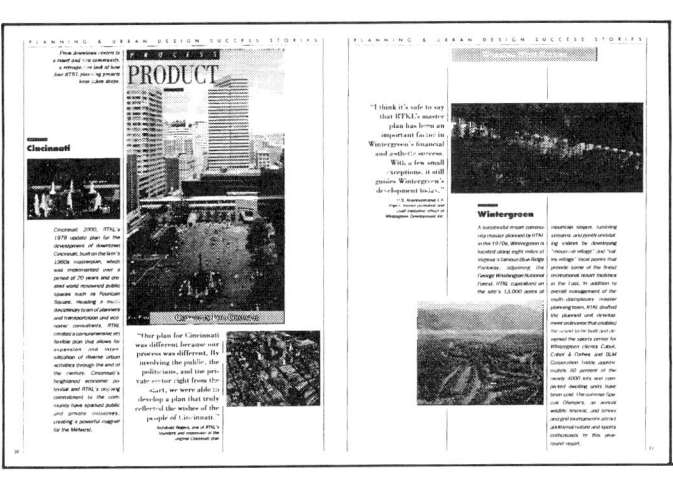
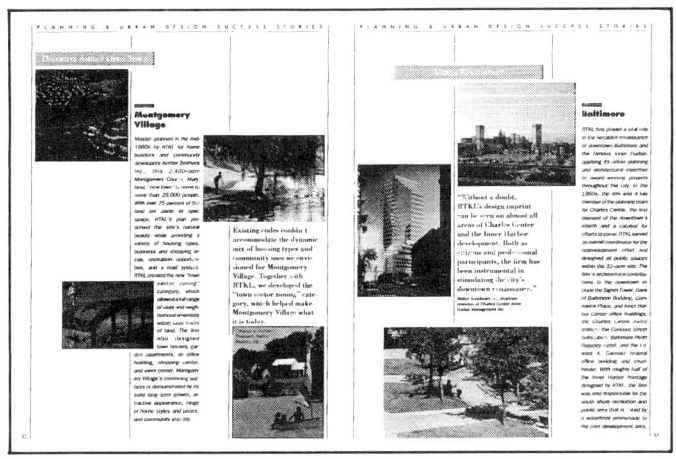

Case Study Formats  133

### Case Study 21

This magazine's masthead becomes a design element that is carried inside to all the left-hand pages in the form of a color bar with drop-out type. Below and in line with this bar is a panel of photographs, and below that are two equal columns of type. Each double column begins with a large decorative letter, set inside a box. Since the products are best depicted in photographs, they are shown as large as possible, sometimes nearly covering the entire spread. Each project shown includes a list of design and construction credits.

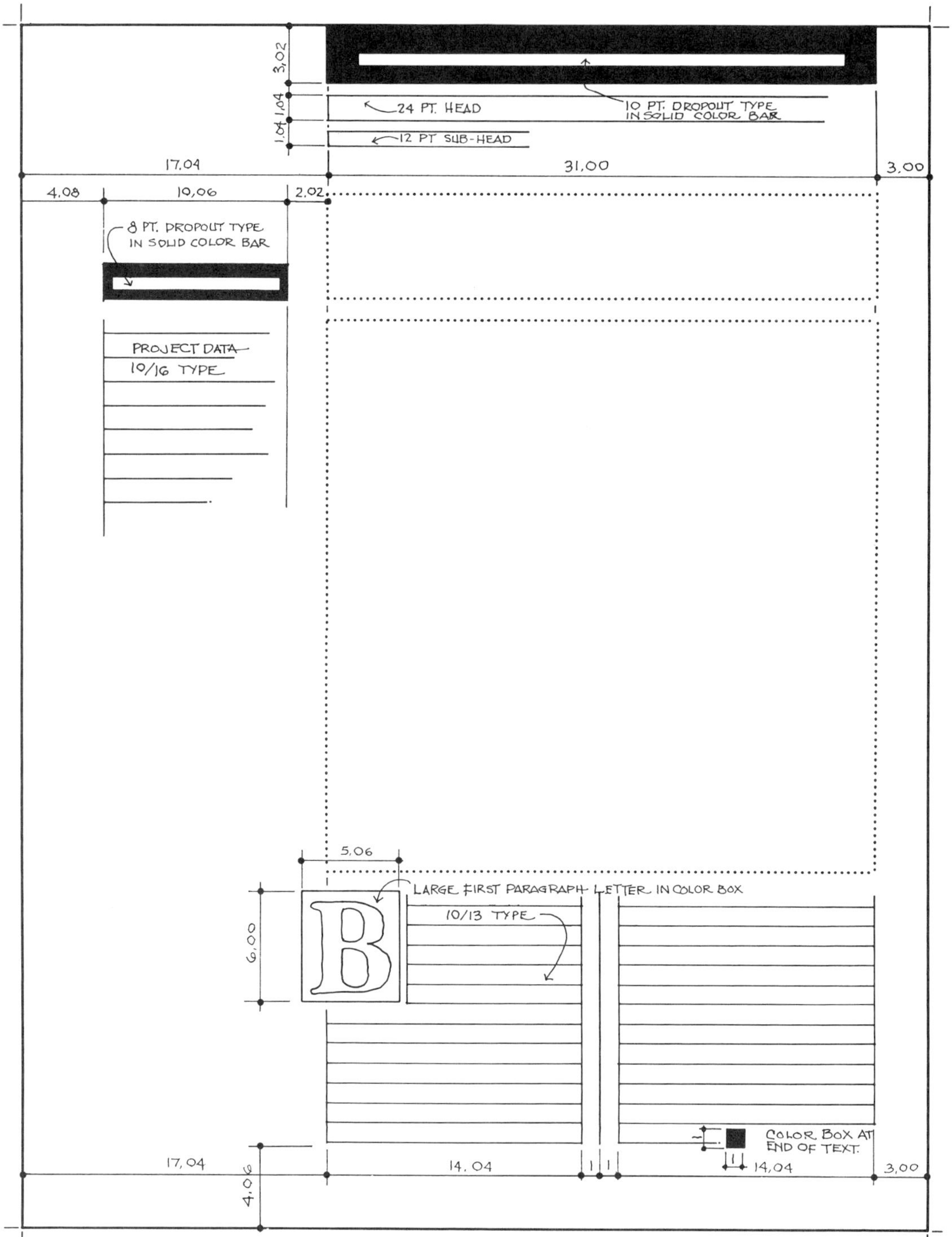

Enlarge to 154% for actual size   Trim size 8 1/2 x 11" (51.3 x 66.4 picas)   See Appendix for full spread grid

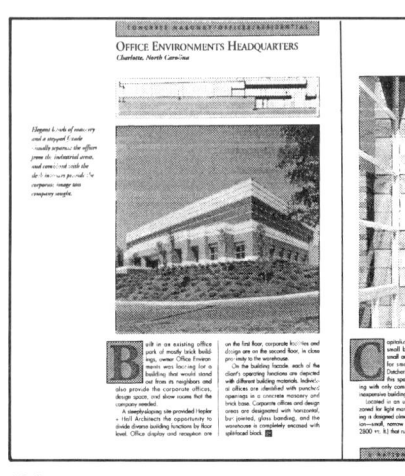

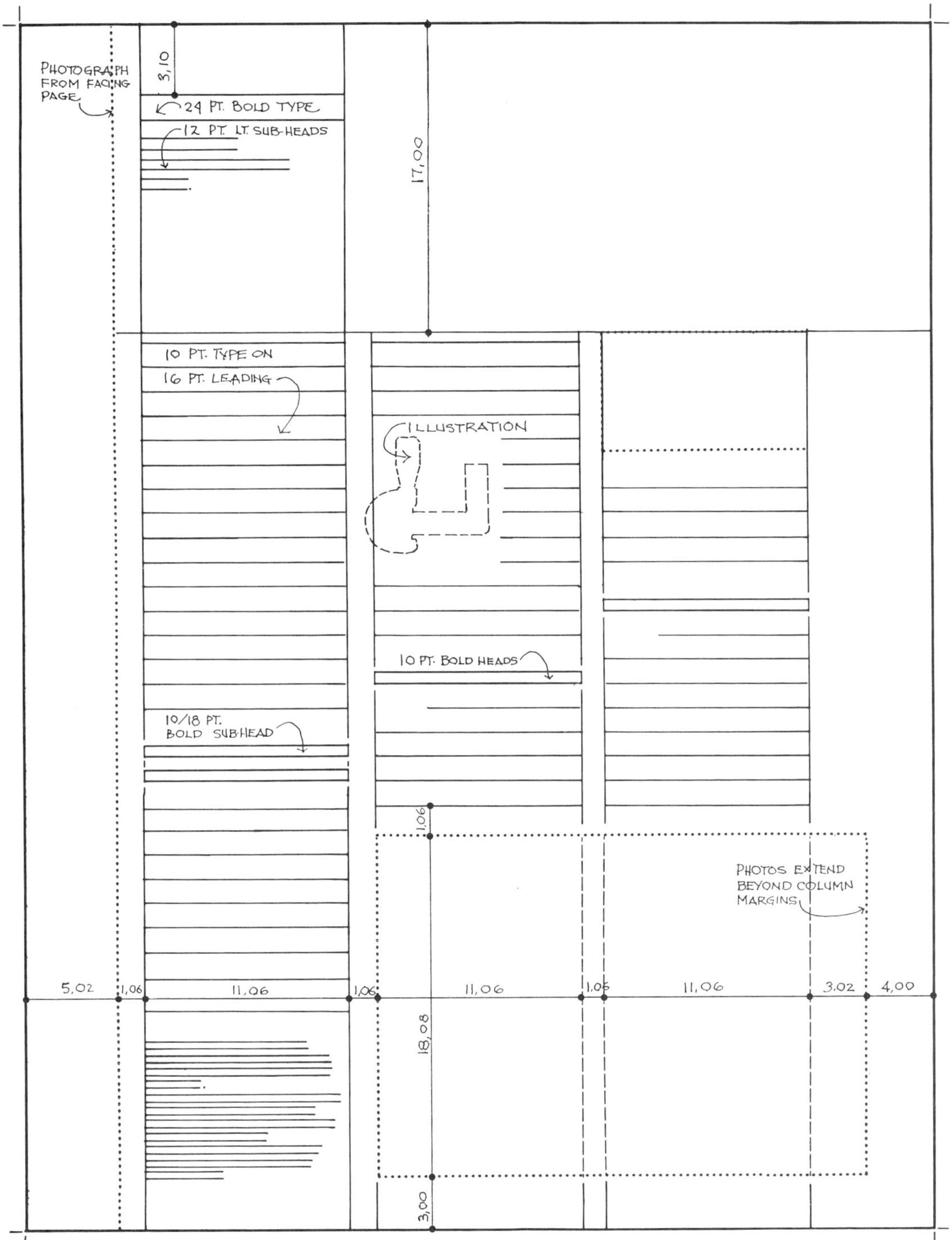

### Case Study 22

Large full bleed photos and a simple three column format gives this brochure a clean appearance. Some of the full bleed pages have small insert photos. The head margin is very deep, and the text spacing is open, which provides an uncluttered look. Spot illustrations are used throughout the text to break the reading into smaller pieces. All this makes a simple but pleasing and easy to read layout.

Enlarge to 154% for actual size   Trim size 8 1/2 x 11" (51.3 x 66.4 picas)   See Appendix for full spread grid

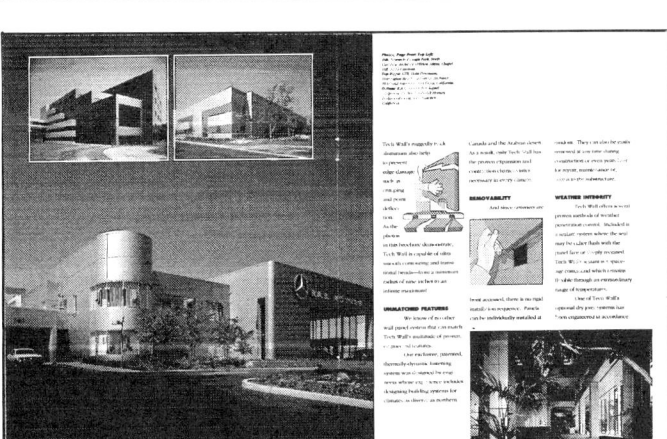
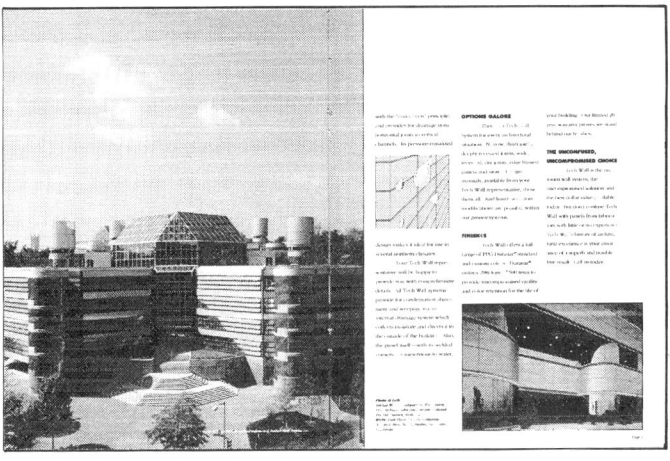

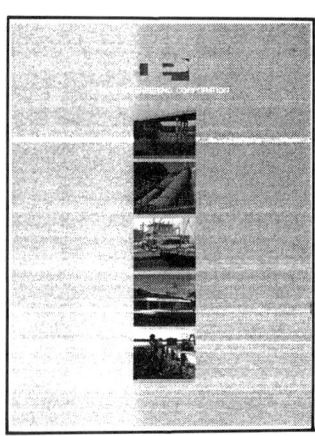

## Case Study 23

The five color photographs that are stacked in a column on the cover are repeated again as theme photos on each of the five spreads. This is an "active" grid in that the elements stay the same from page to page, yet slide to different locations on the page. The main elements are the theme photos, which slide across the top margin in relation to the solid bar which contains drop-out type. The large solid vertical bar, also with drop-out type, moves from side to side. The text appears in one of the three remaining columns.

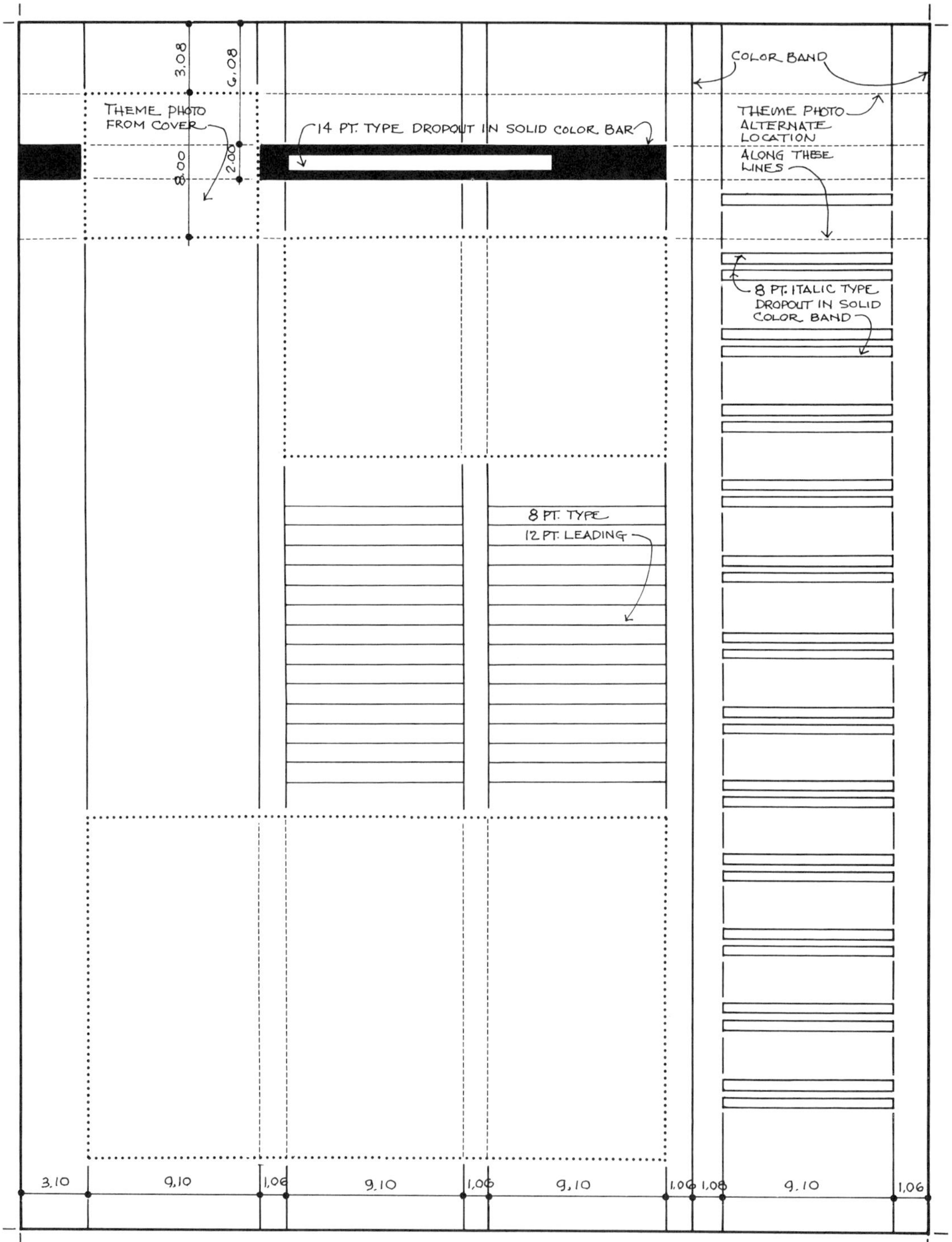

Enlarge to 154% for actual size    Trim size 8 1/2 x 11" (51.3 x 66.4 picas)    See Appendix for full spread grid

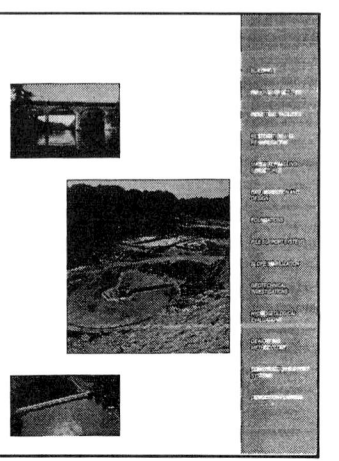

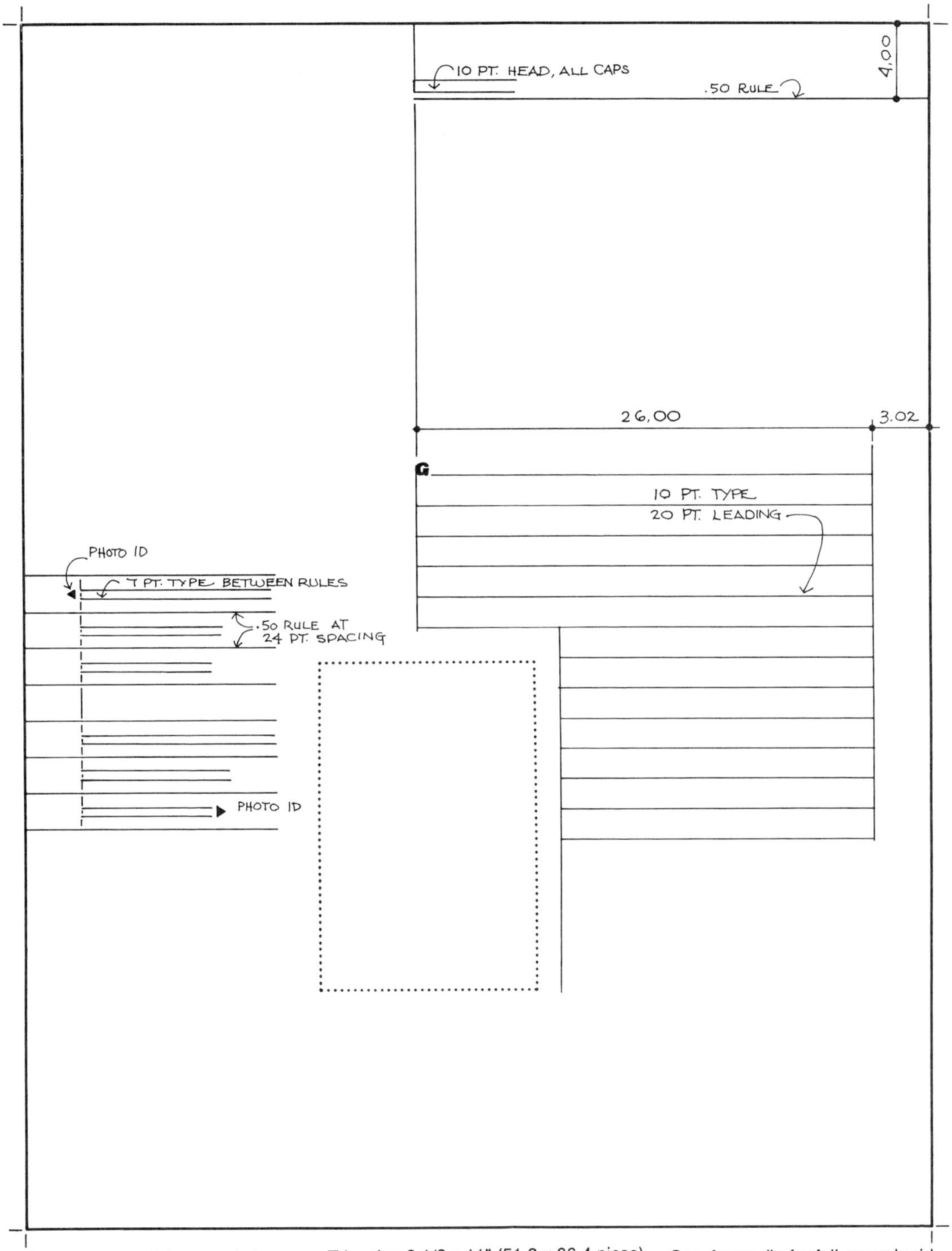

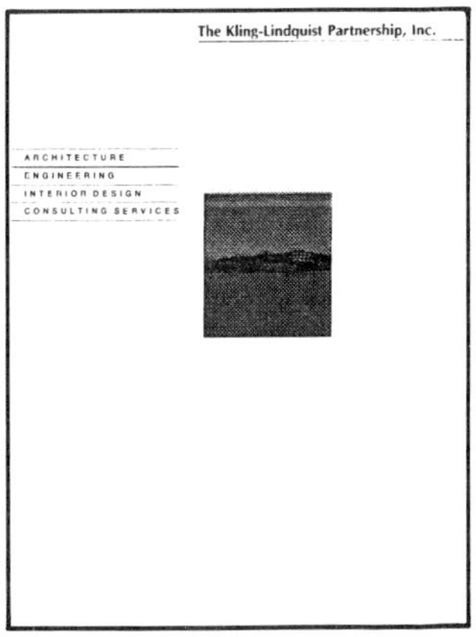

## Case Study 24

The sparseness of the cover layout is indicative of the inside layouts in this brochure. Thin rules separate minimal headlines at the side and middle of each page. The left pages hold only one large photograph, and the right pages hold a smaller photo, around which the text is wrapped. The pristine quality of this brochure is heightened by the use of high quality photography, featuring dramatic views of large scale projects. The text is bold and set wide for maximum openness.

Enlarge to 154% for actual size  Trim size 8 1/2 x 11" (51.3 x 66.4 picas)  See Appendix for full spread grid

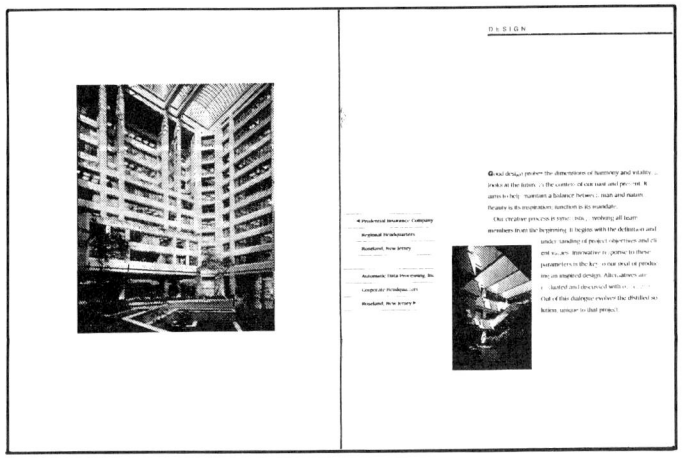
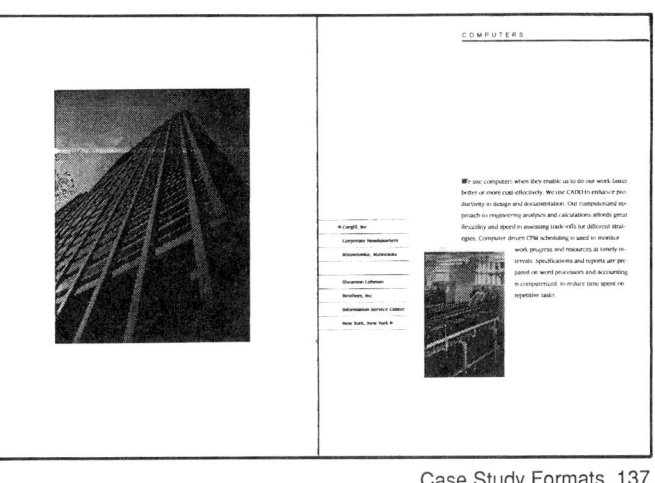

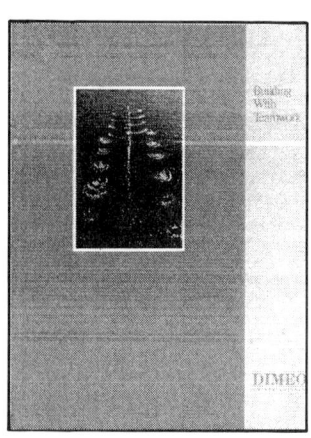

### Case Study 25

Team spirit and competitiveness are the themes set forth on the cover of this brochure. It shows an aerial view of a racing crew. This scene also identifies the locale of this contracting firm's work. Each page is introduced by a head shot of key people with a quotation directly underneath. There is one large project photograph and two columns of type above it. Opposite, there is a single mood photograph of the racing team in action. The connection between the images and the services of the firm is explicit in the text.

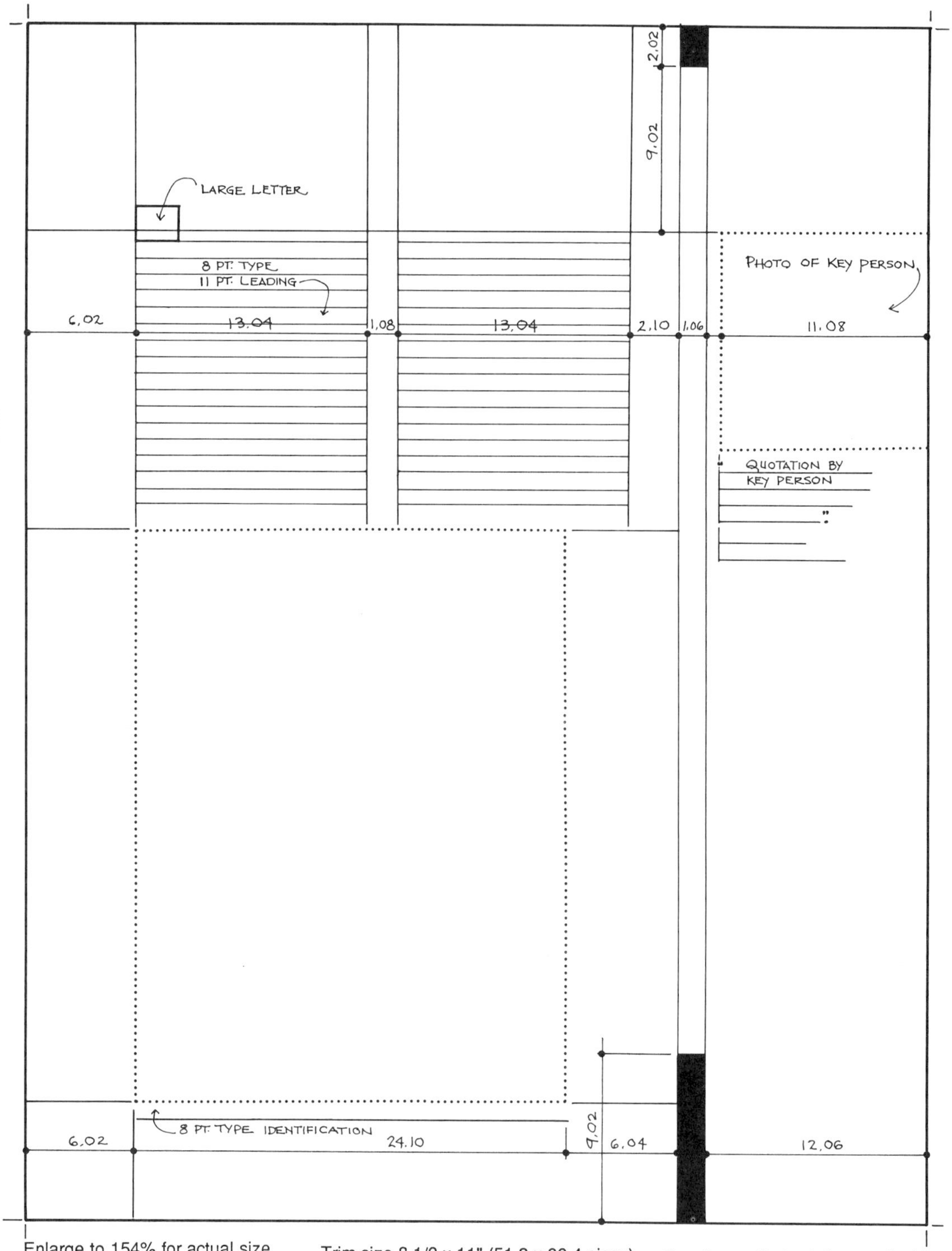

Enlarge to 154% for actual size    Trim size 8 1/2 x 11" (51.3 x 66.4 picas)    See Appendix for full spread grid

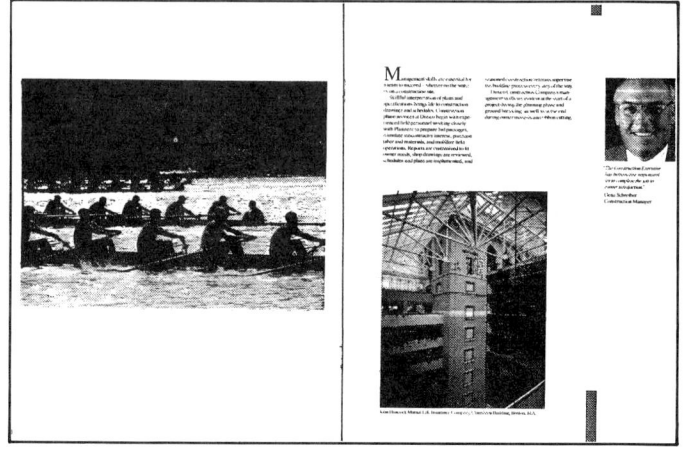

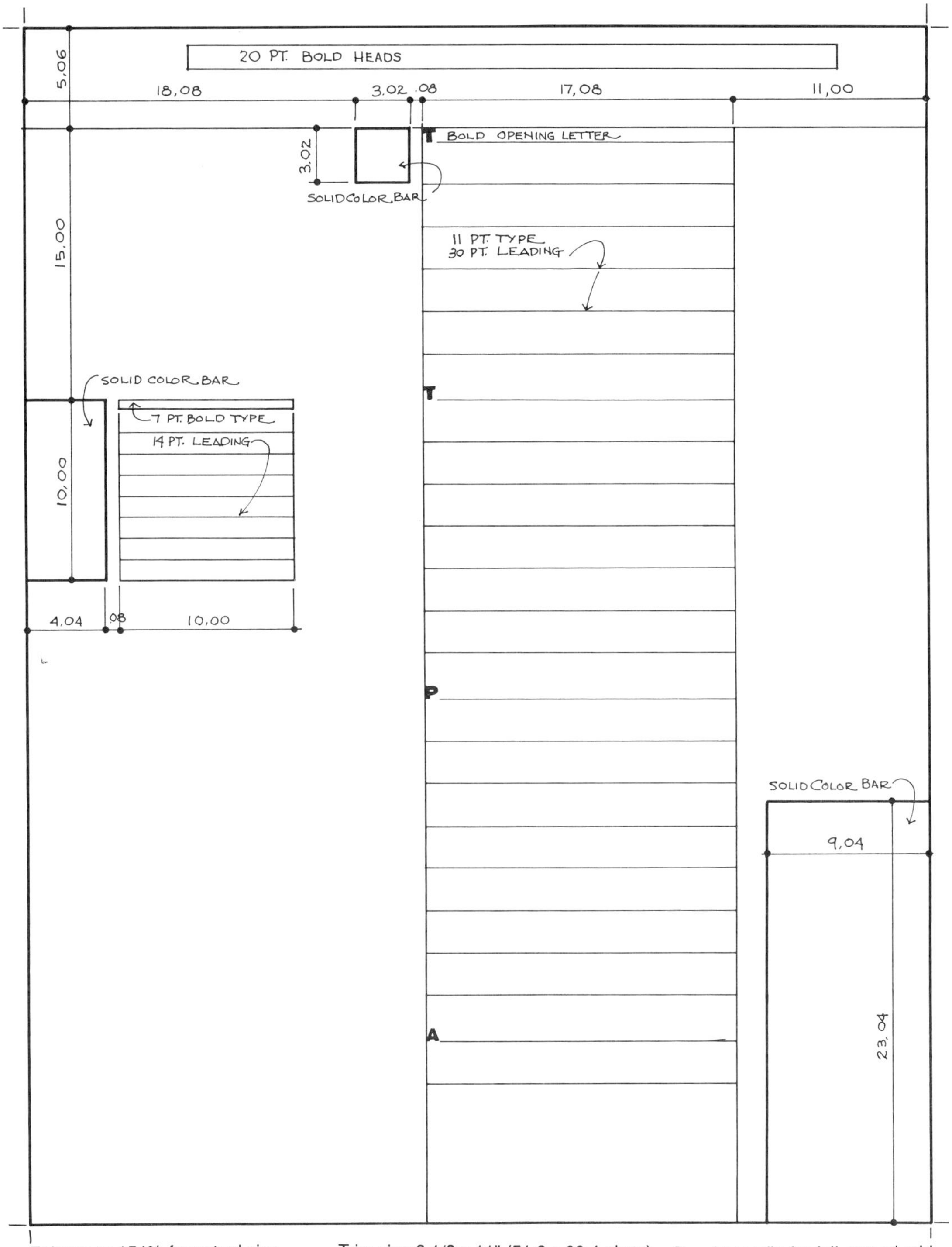

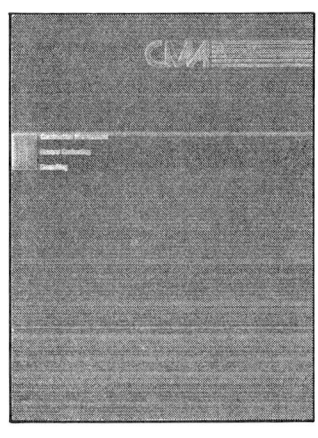

## Case Study 26

A hint at the inside theme can be seen on the embossed heavy paper cover, where a silver bar highlights the firm's name. Inside, the bar becomes expressed on every page. It appears on the right side of the gutter margin, where an equal size bar is removed from the photo. The bar continues to get wider in the lower right corner as well. The reason is that each page is trimmed slightly smaller to expose the bar on the page beyond. In the back the bar and photo theme is carried to a conclusion.

Enlarge to 154% for actual size    Trim size 8 1/2 x 11" (51.3 x 66.4 picas)    See Appendix for full spread grid

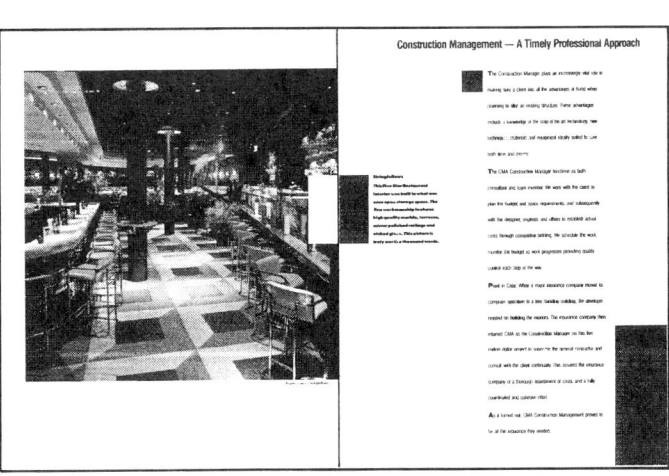
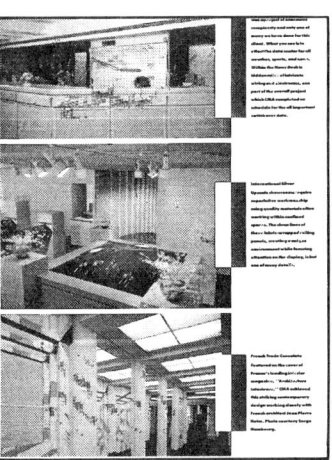

Case Study Formats 139

### Case Study 27

A symmetrical one column grid can sometimes lead to an uninteresting layout. This grid overcomes this problem by varying the elements within the grid, and by careful coordination of these elements. The column is quite wide, but the line spacing is also wide, which counteracts it. The headline is bold and centered. The main photo is centered, flanked by two columns of bold type, one justified left and one justified right. All this helps to create interest and order within the single column format.

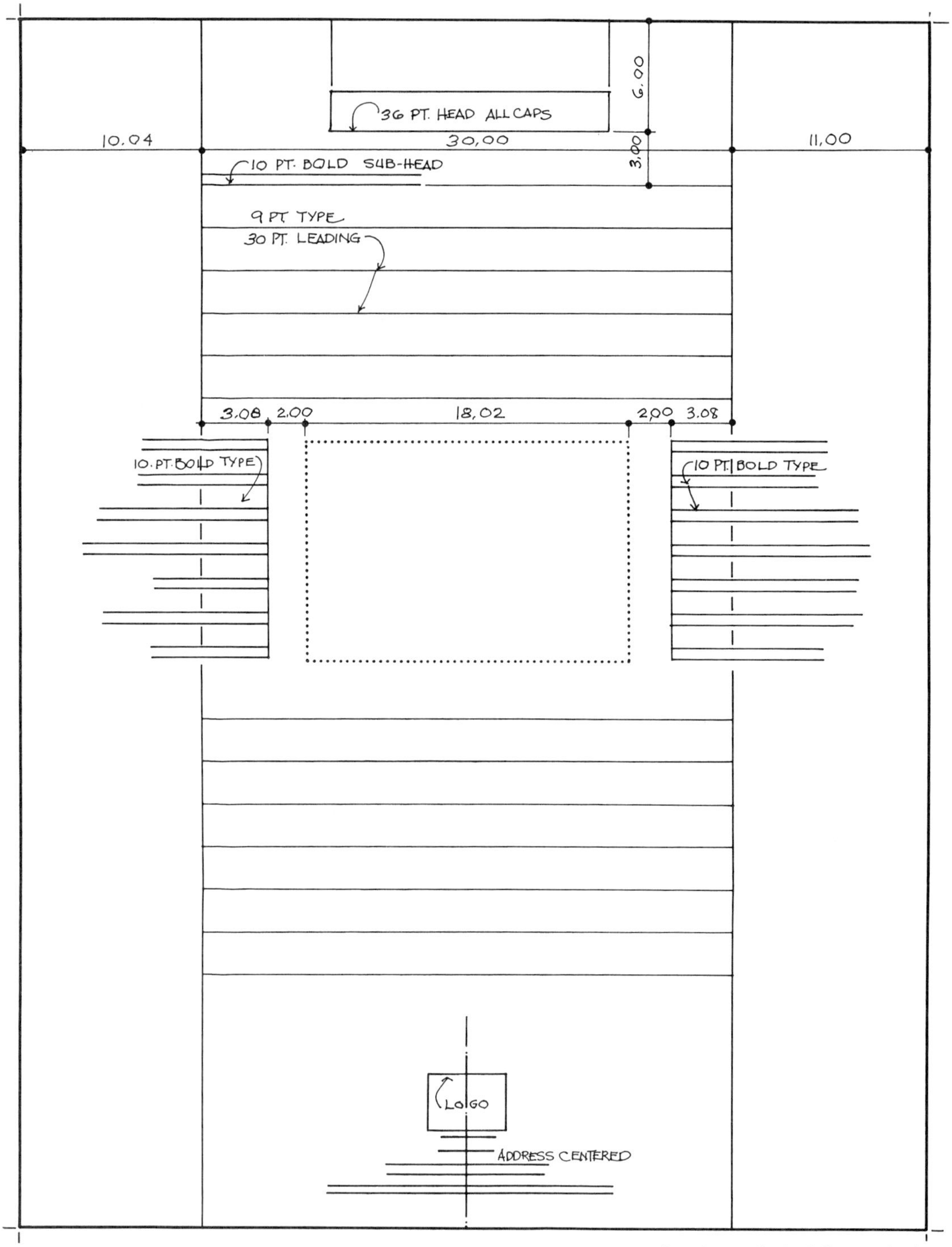

Enlarge to 154% for actual size   Trim size 8 1/2 x 11" (51.3 x 66.4 picas)   See Appendix for full spread grid

140

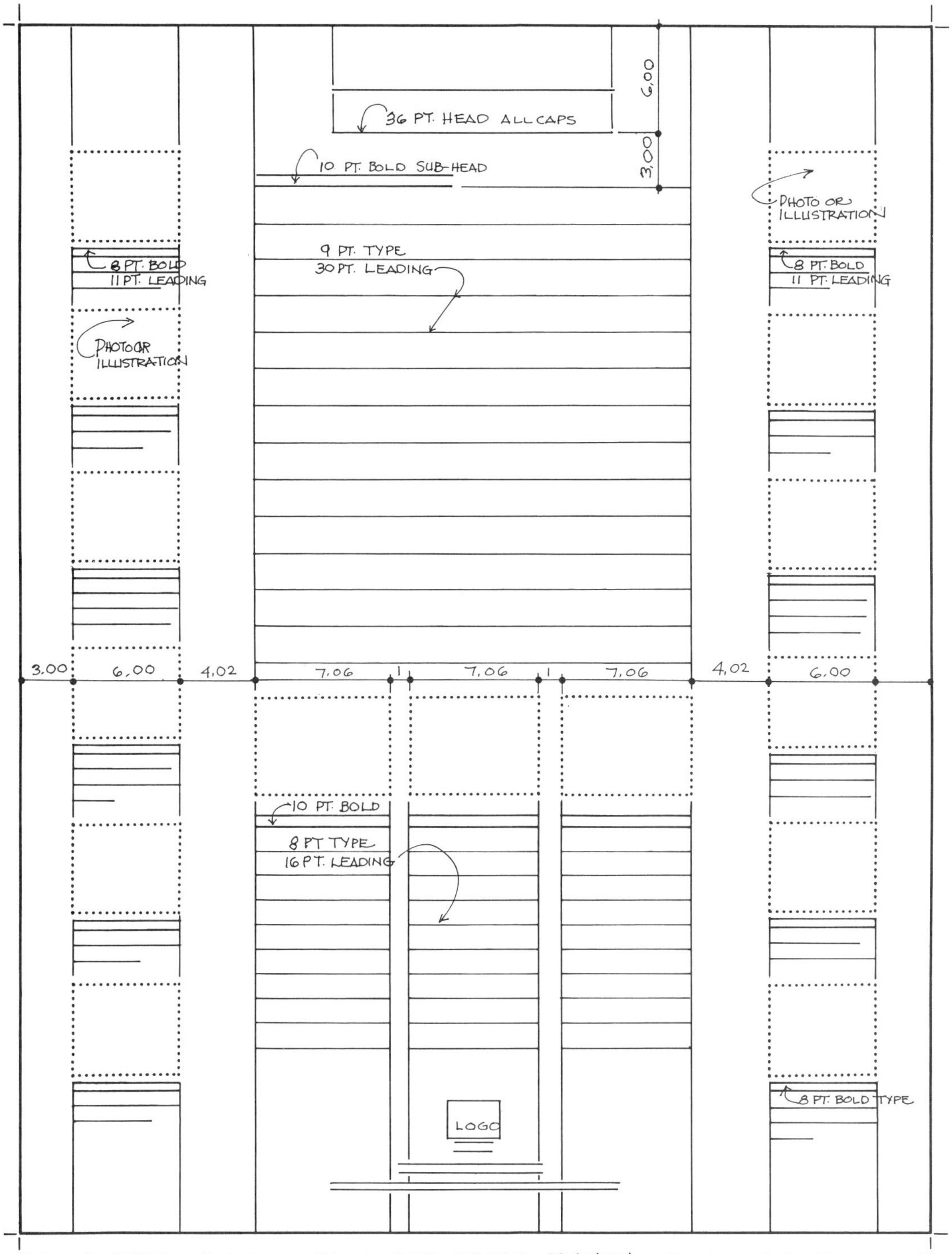

## Case Study 28

Other symmetrical grids follow in the same publication. This one shows a wide center column flanked by two narrow ones. The side columns feature small images, such as those captured off a computer screen or video monitor. This format is typical of those found in computer software publications and advertisements, but can be adapted to many other applications where several small format photos are appropriate.

Enlarge to 154% for actual size    Trim size 8 1/2 x 11" (51.3 x 66.4 picas)    See Appendix for full spread grid

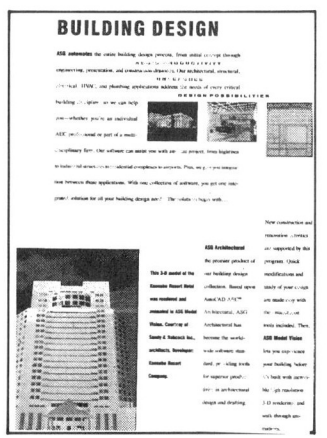

**Case Study 29**

Rarely does a symmetrical grid arrangement have as much interest as this one. It was designed for a landscape and planning firm's brochure. Centered at the top of the page is a small square dot. Under it is a question, set off in quotes. The three columns of type under it are symmetrically arranged. The widest one is in the center flanked by narrower ones on each side for descriptive captions for the photos. Some of the material in these columns are also set in quotations, sometimes as answers to the question at the top, or as comments from clients.

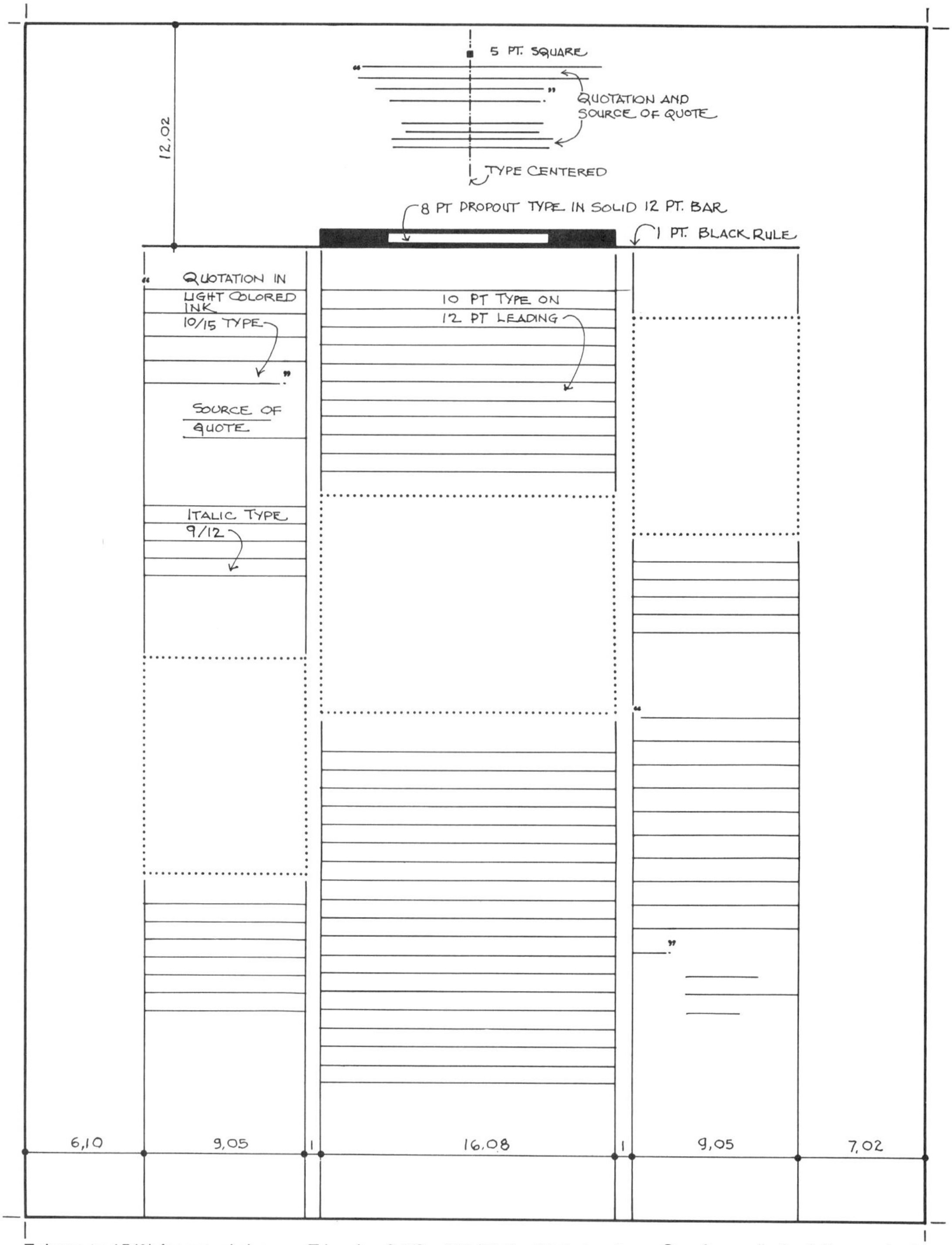

Enlarge to 154% for actual size   Trim size 8 1/2 x 11" (51.3 x 66.4 picas)   See Appendix for full spread grid

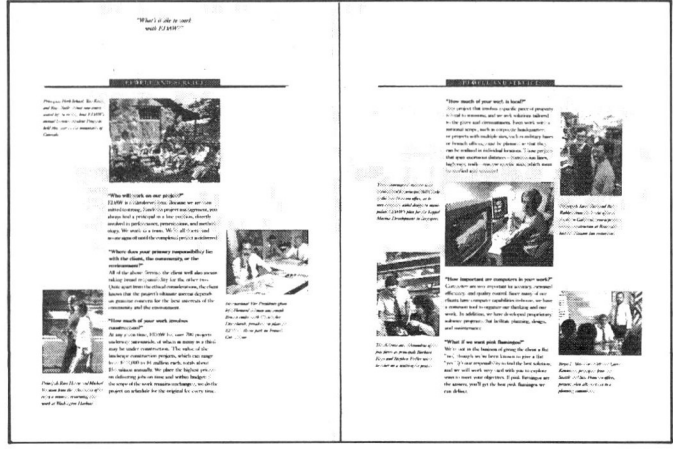
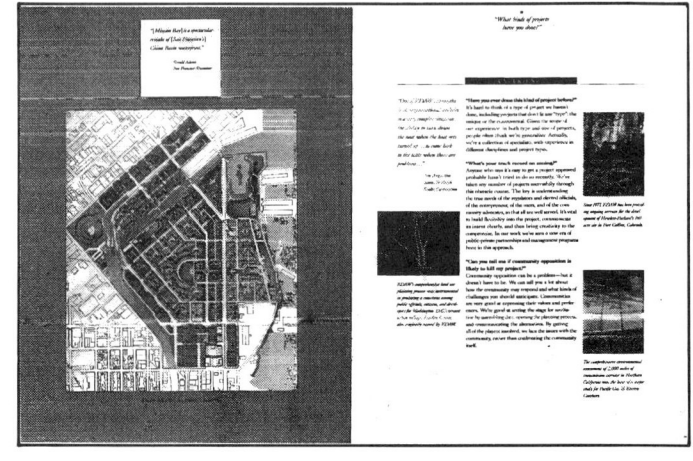

142

## Case Study 30

This annual report utilized a three column grid wherein the wide outer columns carry the text and the narrower middle column features charts, photos and statistical information. There is a decorative initial on the cover, which is repeated inside at the beginning of the text. On the facing page, there is a large photograph, framed by white borders, which are wider than the ones for the text pages. Both the head and foot margins are generous, which gives this report an open look considering the amount of text on the page.

Enlarge to 154% for actual size   Trim size 8 1/2 x 11" (51.3 x 66.4 picas)   See Appendix for full spread grid

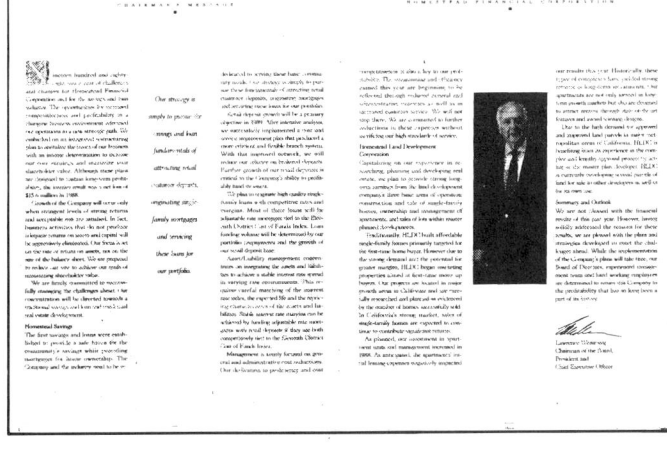
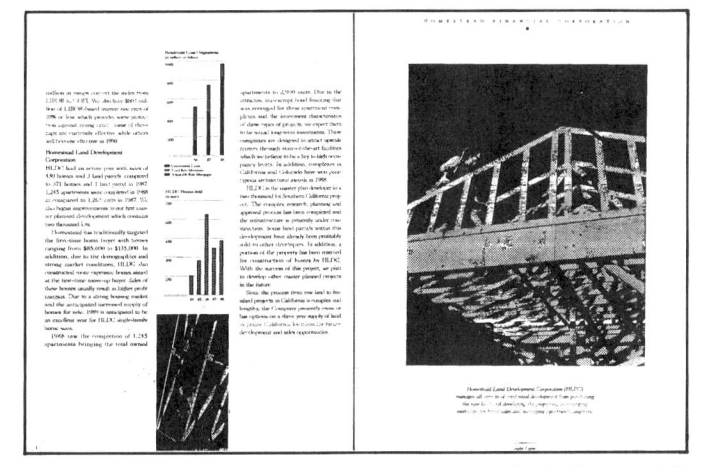

Case Study Formats 143

### Case Study 31

The grid for this brochure on planning services is expressed on the cover by using a large photograph to fill the entire grid area. Inside, one finds a 3 column format in which one of the columns is used for the text and description of the project photographs. There is also a strict horizontal grid, making the resultant grid units fit into a square pattern. Photos are then either within these square grids or oblong, using two grids across. The descriptive text is minimal.

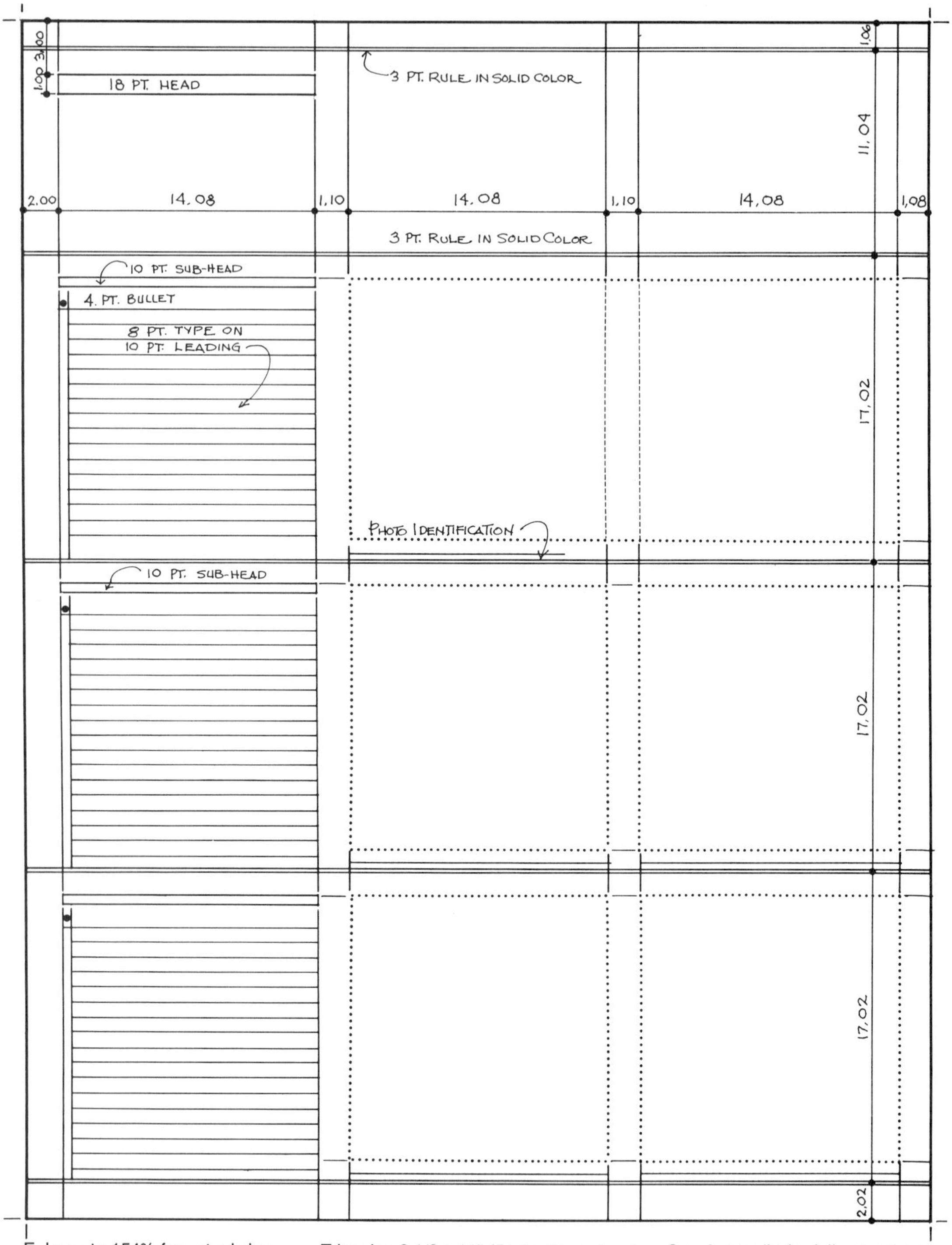

Enlarge to 154% for actual size   Trim size 8 1/2 x 11" (51.3 x 66.4 picas)   See Appendix for full spread grid

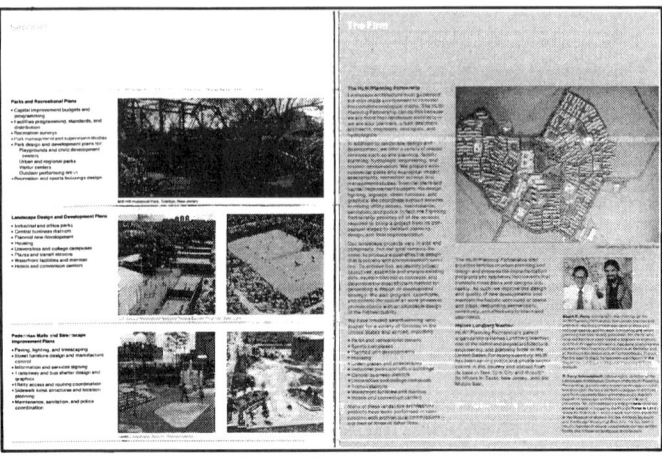
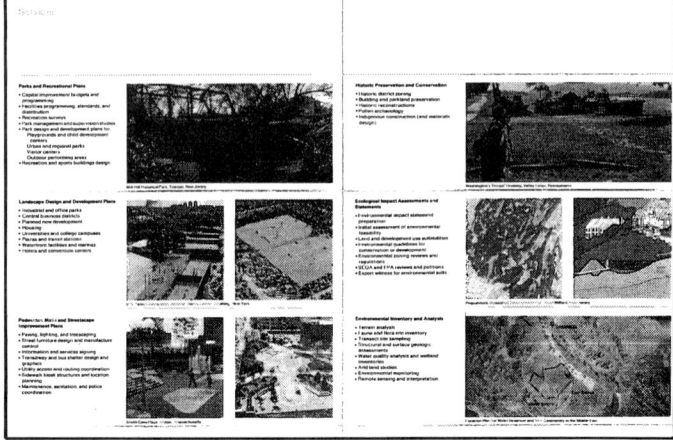

144

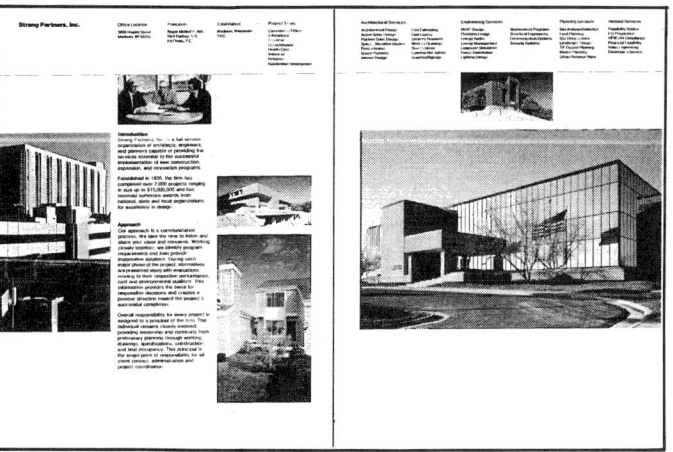
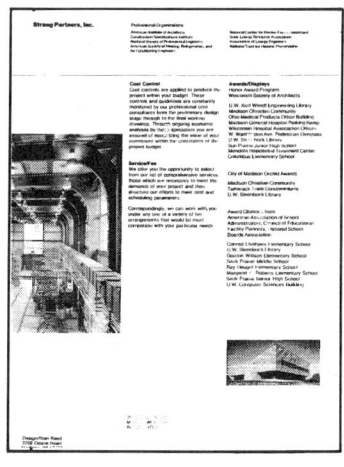

Enlarge to 154% for actual size    Trim size 8 1/2 x 11" (51.3 x 66.4 picas)

## Case Study 32

The grid used for this brochure consists of a combination of three and six columns. It also establishes a horizontal pattern at the top of the page for a column of descriptive information about the project illustrated below. The overall effect of the page is one of simplicity, due in part to the organizing design of its grid.

See Appendix for full spread grid

Case Study Formats 145

### Case Study 33

This anniversary piece was designed as a continuous grid spread across three fold-out pages. At each section there is a page of solid color with dropout type in white (below left). When opened the grid spreads across three pages. It is a 3 column grid with wide margins. It is a tiered design vertically with the top tier featuring dates of important stages of the firms growth and development. The middle tier consists of photographs illustrating major key projects. The lowest tier also features a time line depicting events that happened concurrently with the development of the firm.

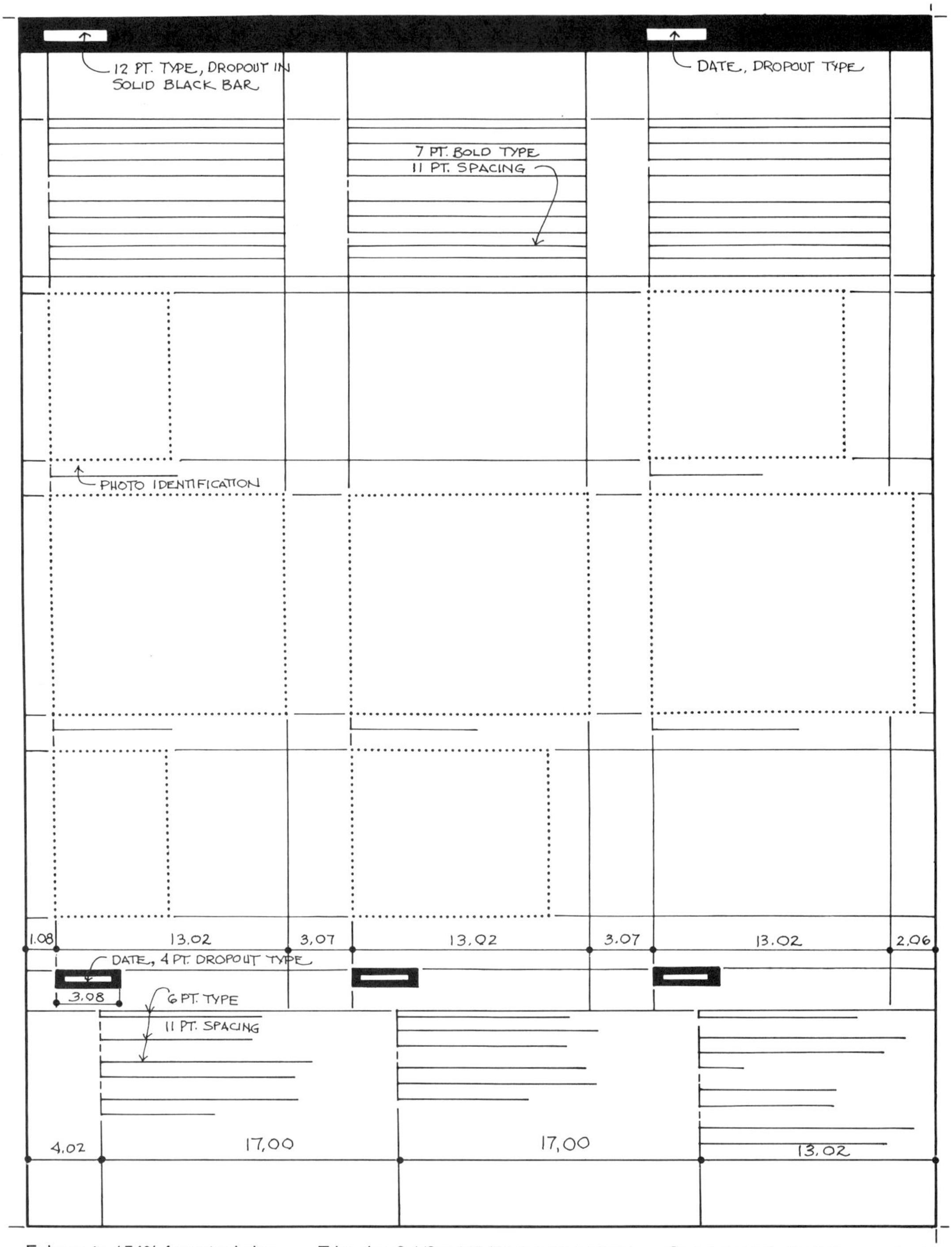

Enlarge to 154% for actual size    Trim size 8 1/2 x 11" (51.3 x 66.4 picas)    See Appendix for full spread grid

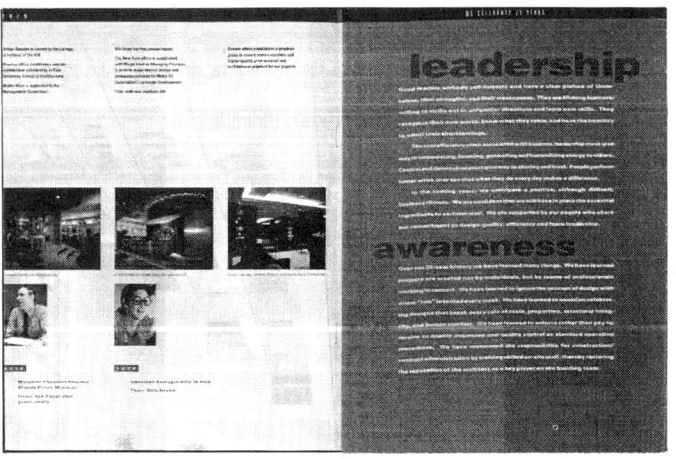

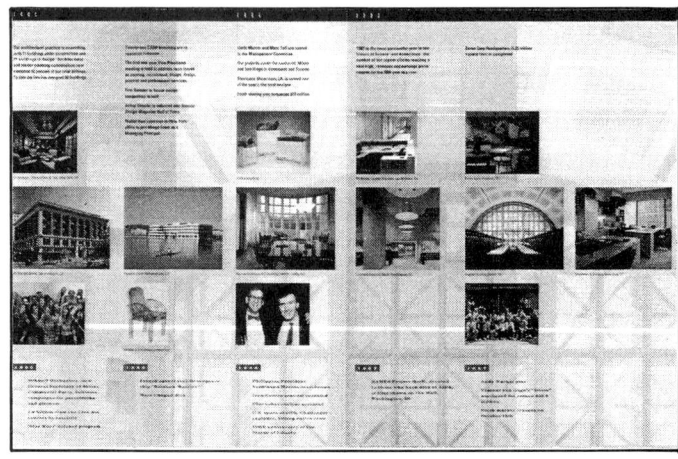

## Case Study 34

The grid for this anniversary piece is an open structure across a piece which unfolds to 30" in length. The grid does not relate to the folds at all. There is a yearly timeline at the top, and a large panel of photographs which make up the bulk of the page. Each photograph is captioned and the project is described at the bottom. At the end of the foldout is a statement signed by the president and a ruled design in the corporate colors. At the top of this ruled column is the numeral 30. At the bottom of the rules is the firm's logo.

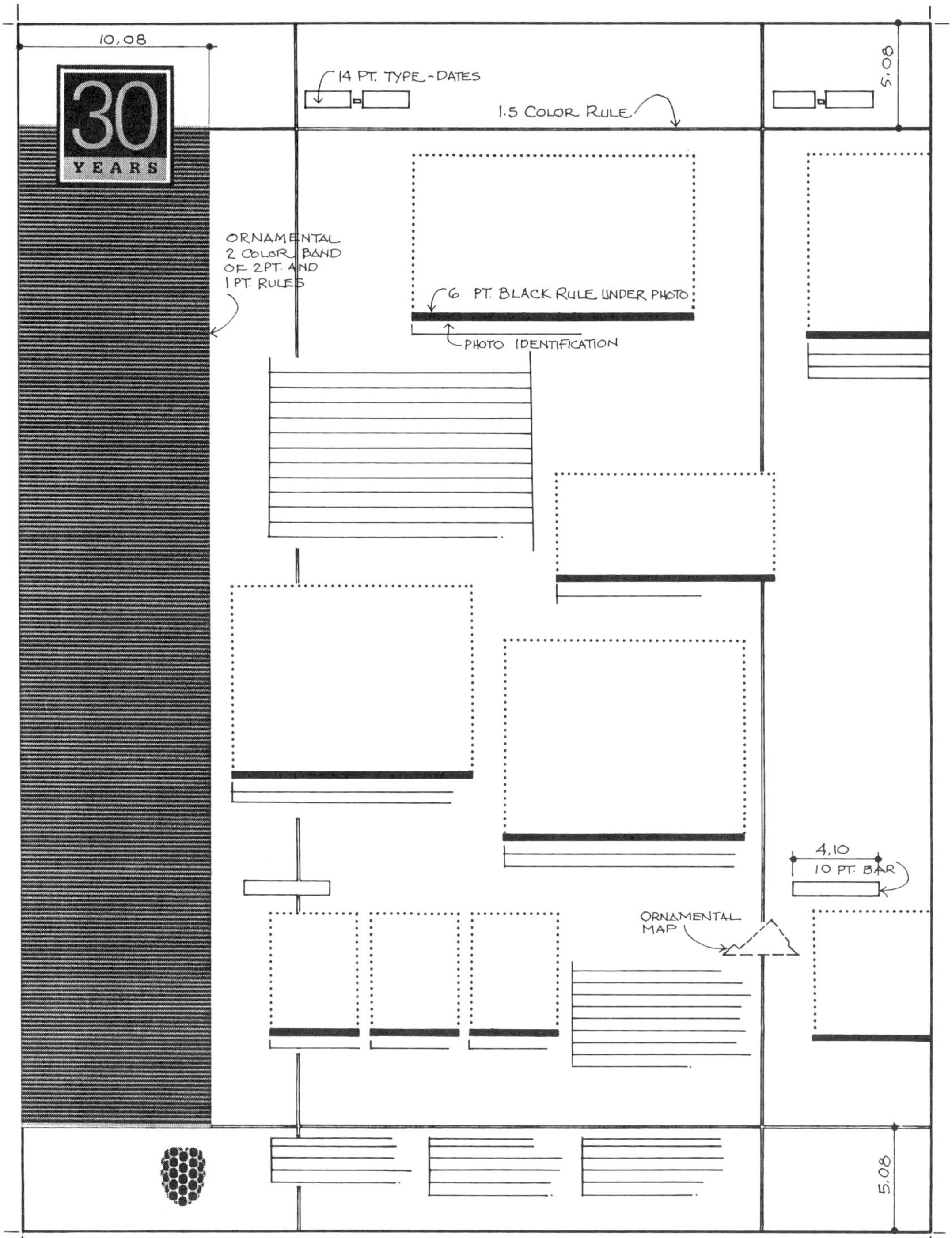

Enlarge to 154% for actual size   Trim size 8 1/2 x 11" (51.3 x 66.4 picas)   See Appendix for full spread grid

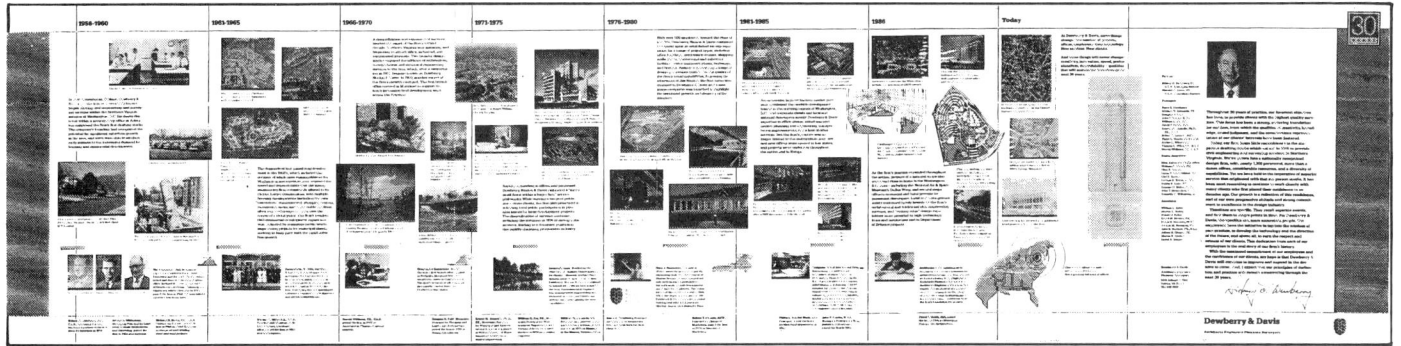

Case Study Formats 147

## Case Study 35

The design of this brochure is based on a three column grid, although all three columns are not expressed in the text. The photos follow a rigid format within the grid. The larger ones appear on the lower section of the page. Smaller photos, text, and a solid color bar appear on the top portion of the page.

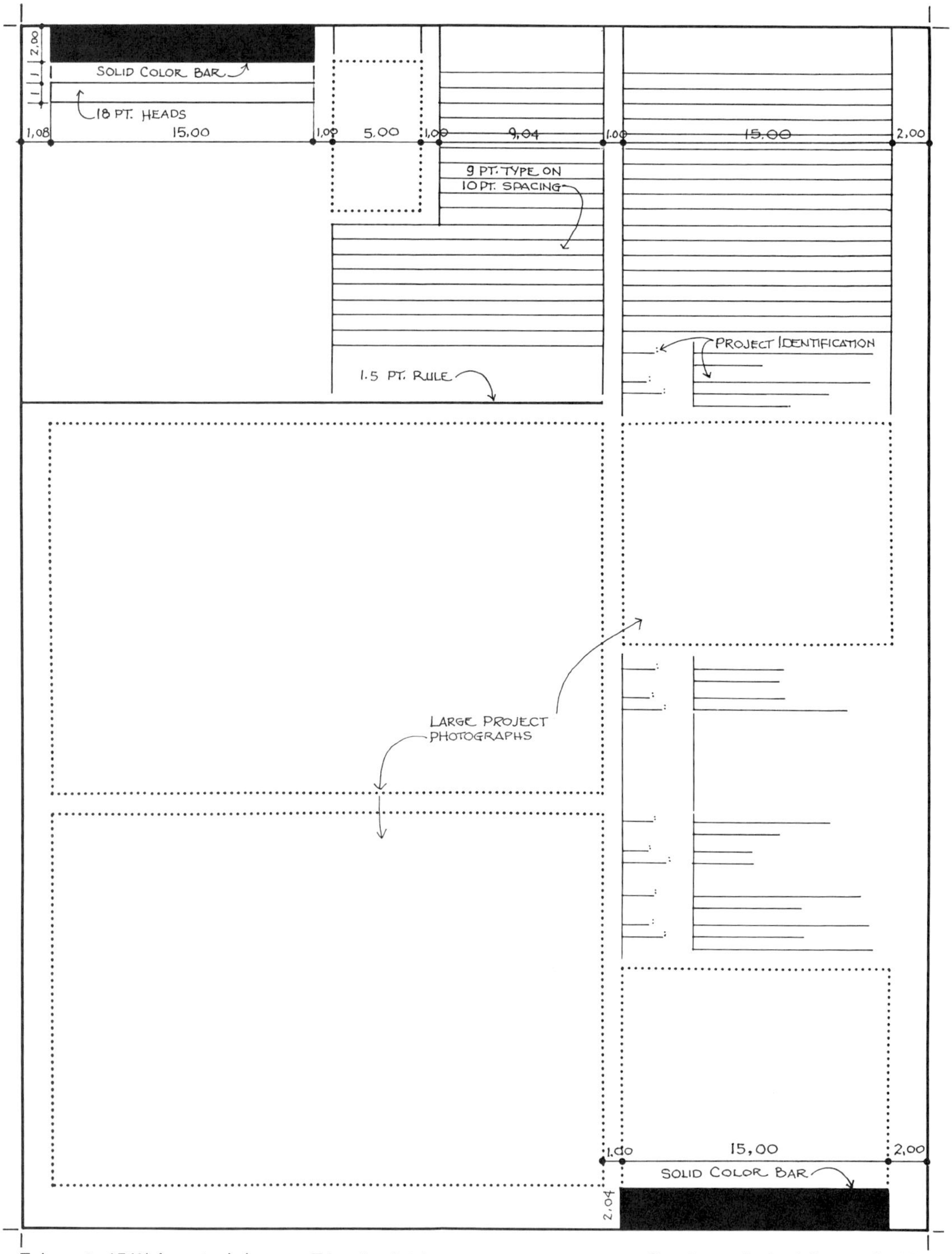

Enlarge to 154% for actual size   Trim size 8 1/2 x 11" (51.3 x 66.4 picas)   See Appendix for full spread grid

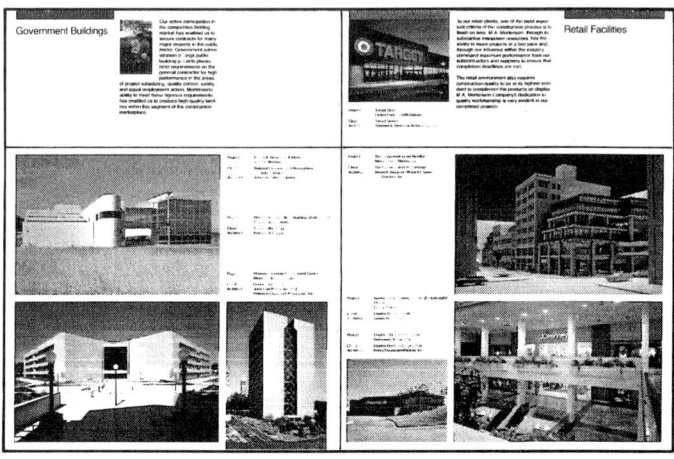

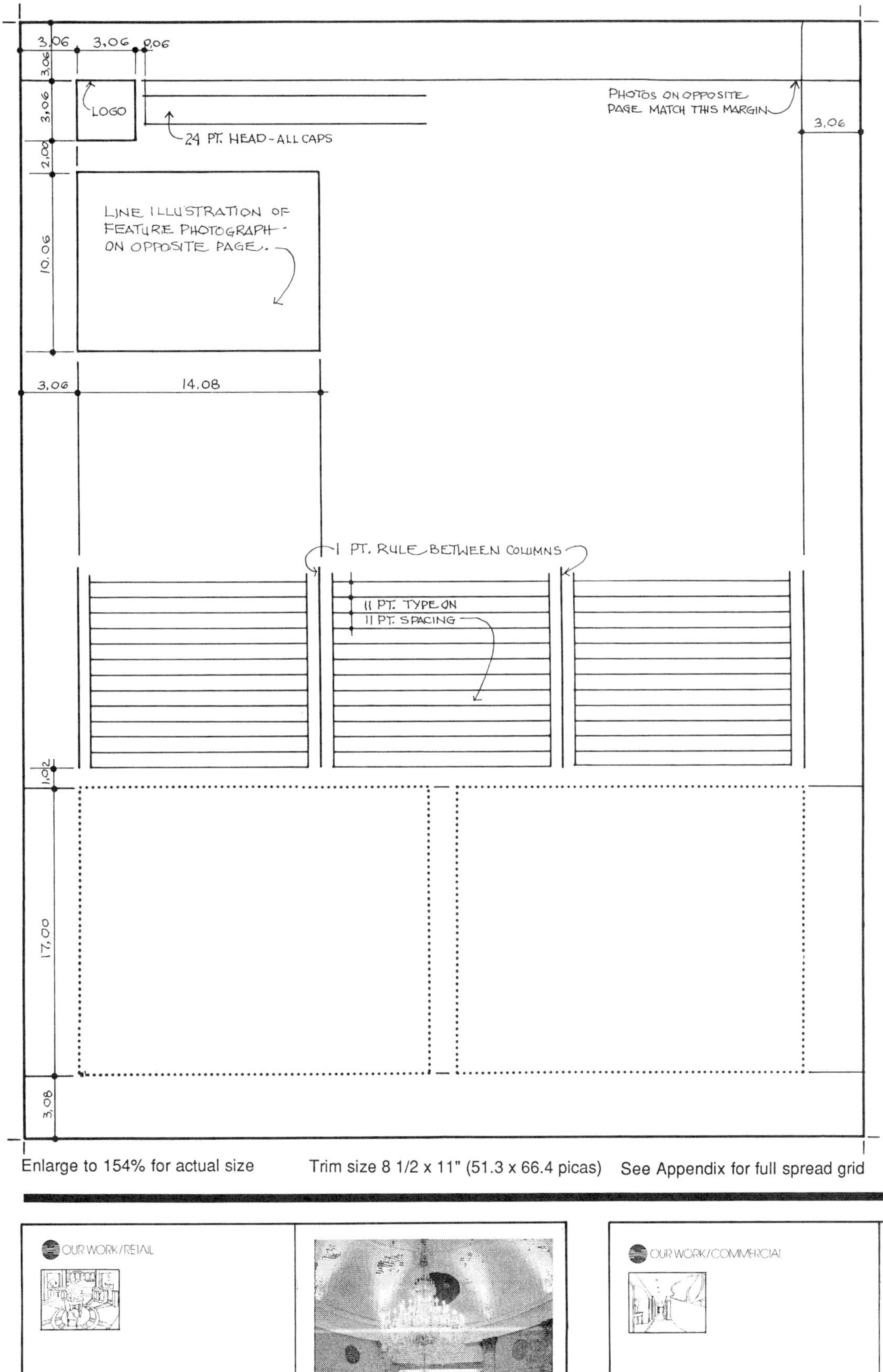

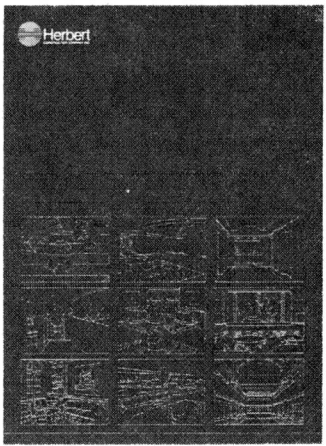

## Case Study 36

The cover of this construction firm's brochure features a series of nine drawings, grouped together, printed in a light blue color on a black background. These same images appear as line drawings on each page inside the brochure, to introduce the various aspects of this firm's work. The illustrations are line tracings over key photographs which are shown on the opposite page, at a much larger scale. The grid for the text is a simple three column arrangement separated by thin rules.

Enlarge to 154% for actual size    Trim size 8 1/2 x 11" (51.3 x 66.4 picas)    See Appendix for full spread grid

Case Study Formats 149

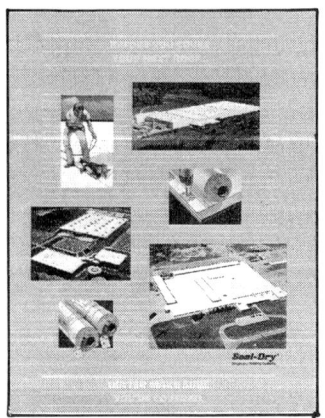

## Case Study 37

This brochure describes the benefits of a particular roofing system using a step-by-step approach. It is a gatefold design which unfolds to reveal four panels of pages. On the far left the headline copy asks a question. It is then answered and elaborated in the panel of nine boxes. Then, in a three column format per page, each numeral is explained in detail in the columns of text. The boxes and drop-out numerals are in a gray color which matches the color background at the top panel.

See Appendix

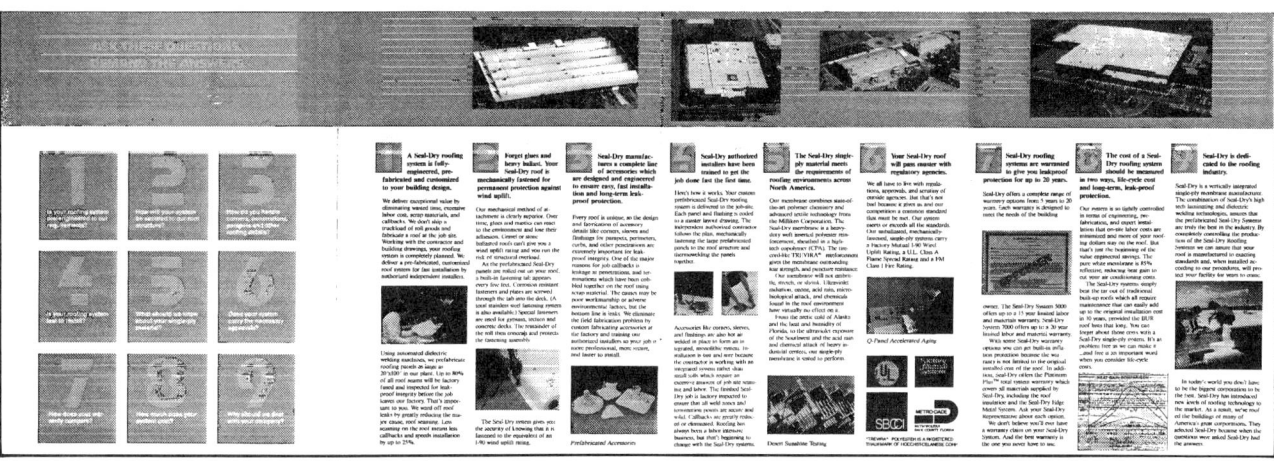

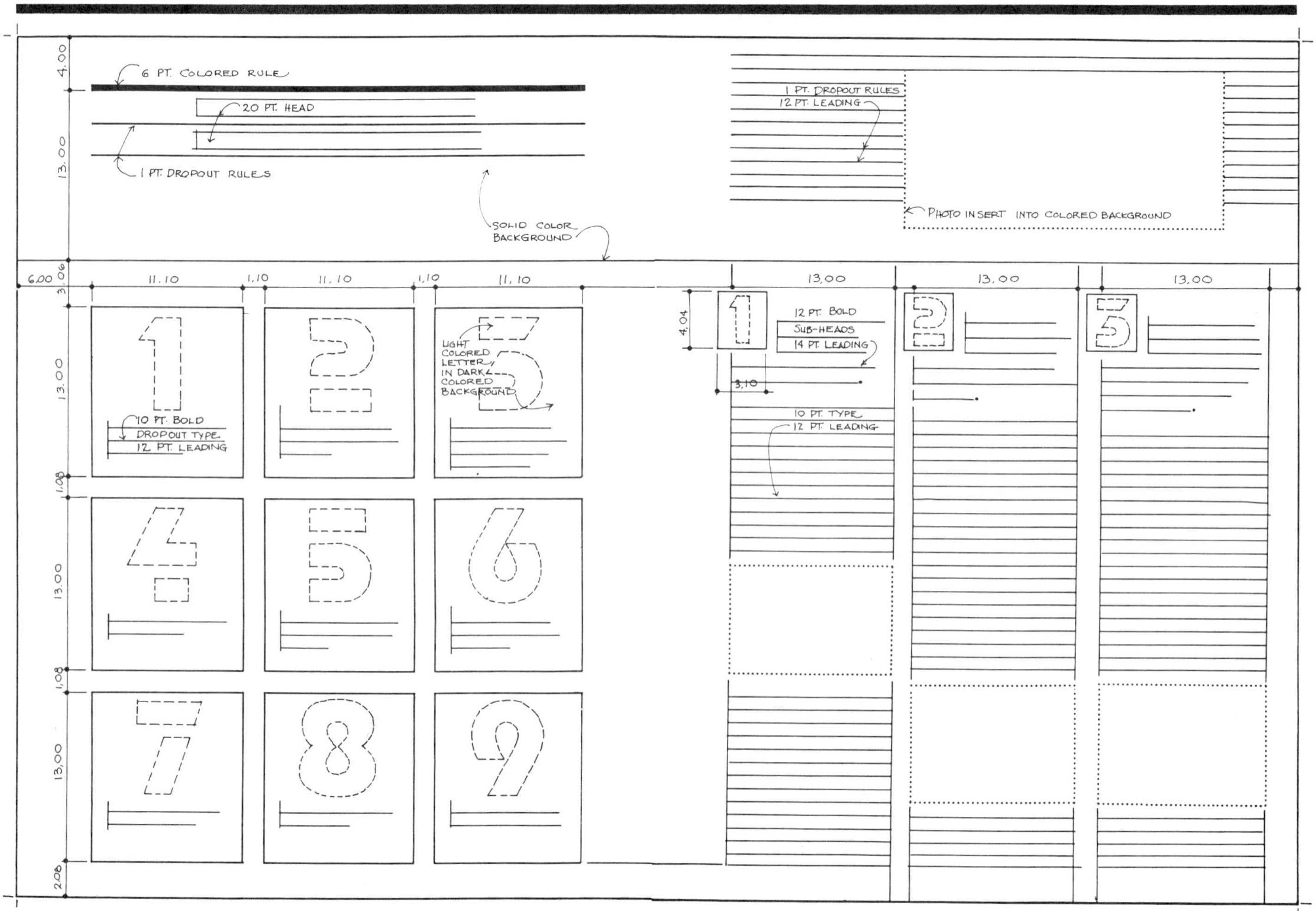

## Case Study 38

In this brochure for a specialty service, the entire design to construction process is outlined and explained. The cover is a glossy blue; inside, there is a linear numbered sequence at the top of all pages. These outline the elements in the three column grid format below. The lower panels are separated first by columns of descriptive text and below by drawings and/or photographs of the particular stage of service.

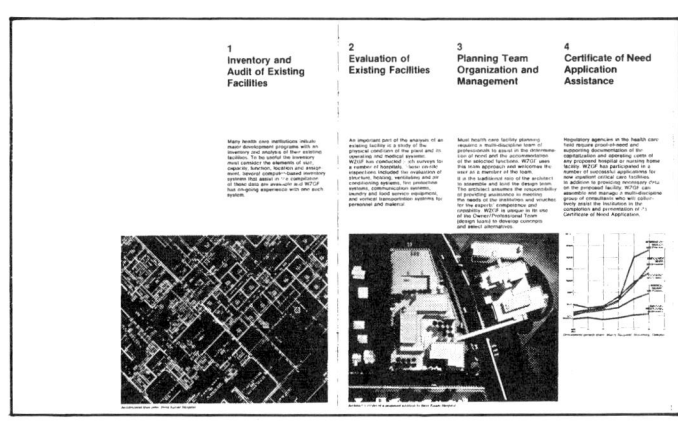

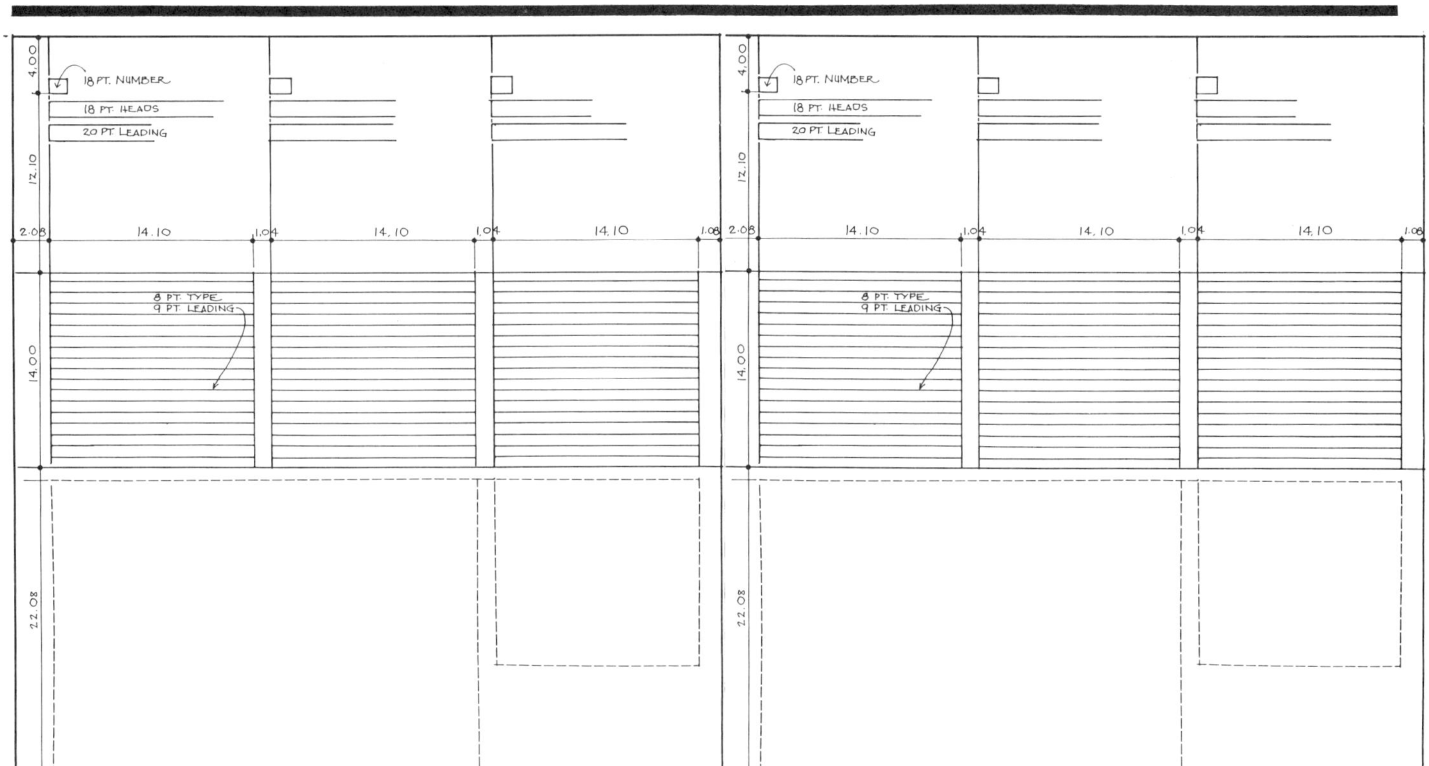

See Appendix

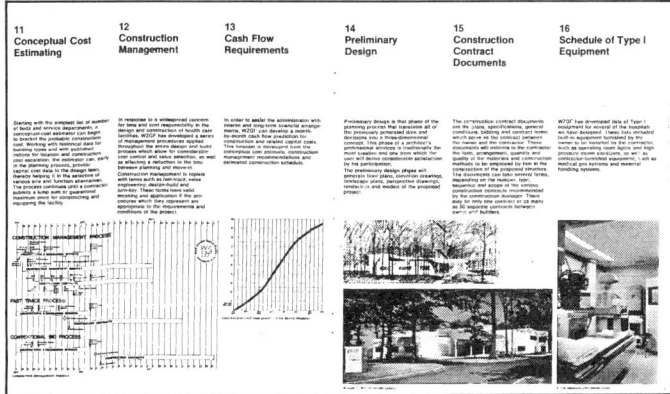

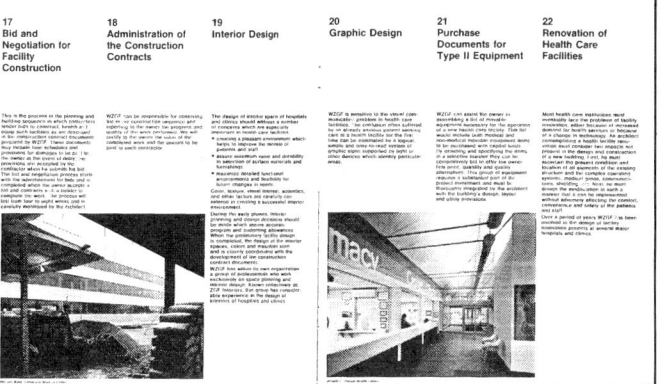

Case Study Formats 151

### Case Study 39

Heavy bars, bold type and dramatic photographs rule this brochure design. The bars first appear on the cover. Inside, the bars have drop-out type as headlines. Bold sub-heads below are separated by rules. The general arrangement of photos is the same for all right-hand pages. Opposite, there is a full-bleed color photo, occasionally with small photo inserts. This is a strong design, with bold strokes, and is meant to represent a strong company with equally strong experience in satisfying its clients.

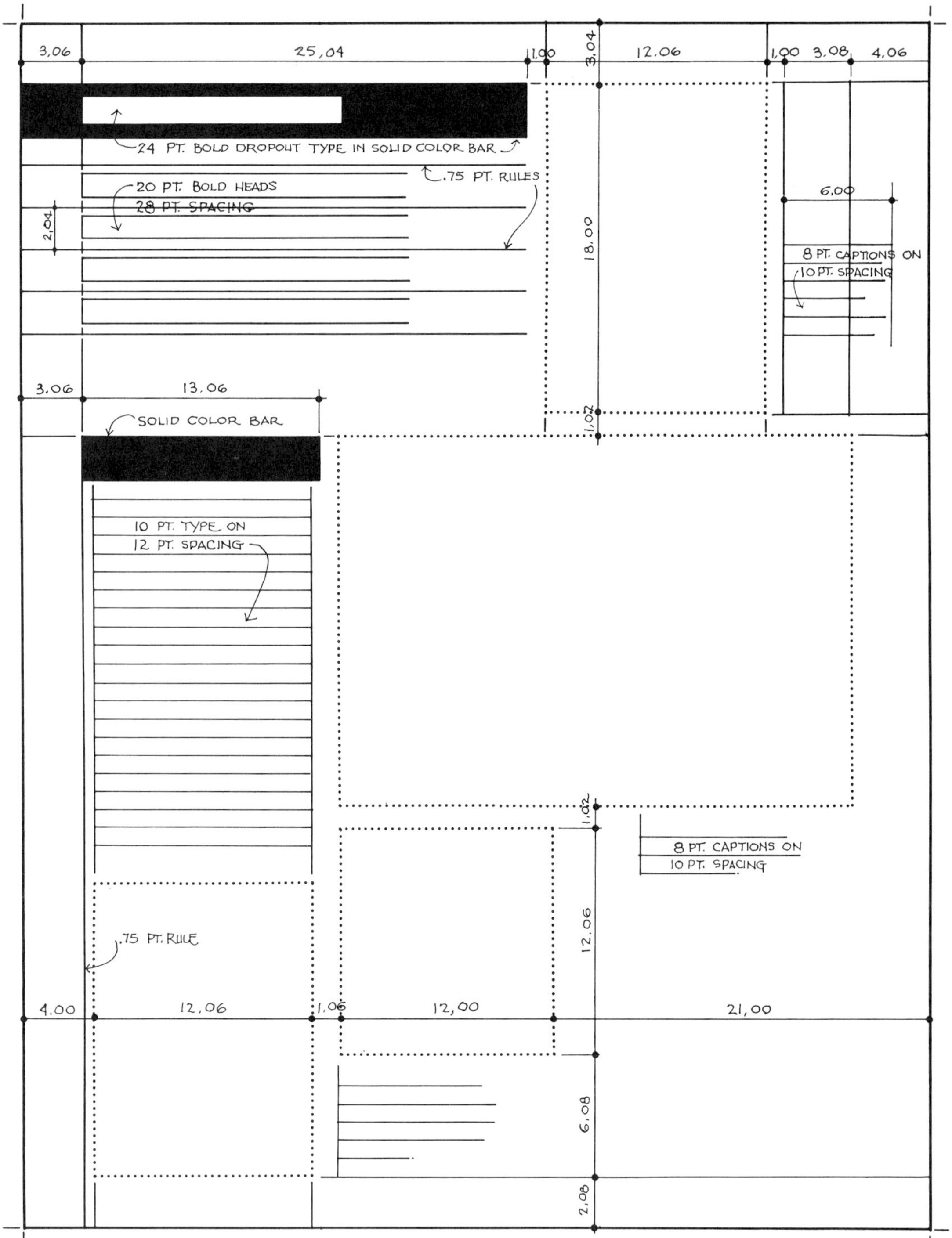

Enlarge to 154% for actual size   Trim size 8 1/2 x 11" (51.3 x 66.4 picas)   See Appendix for full spread grid

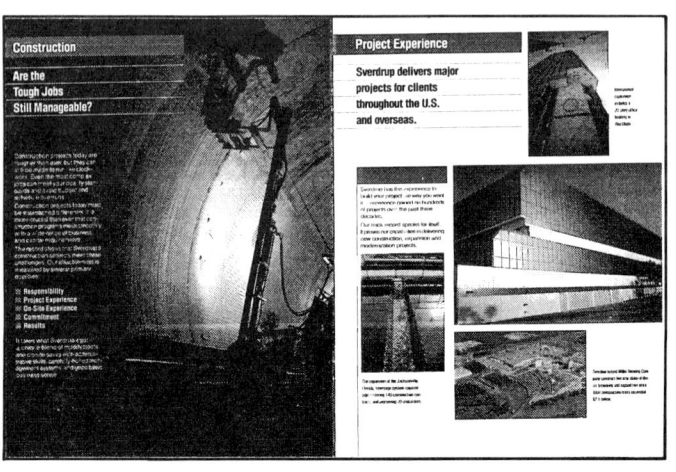

## Case Study 40

The use of photographs determined the layout of these pages. The color bar design and column width on the cover are repeated inside with a similar single column of type. The photo grouping is somewhat more complex since there are essentially only two major columns per page.

Enlarge to 154% for actual size   Trim size 8 1/2 x 11" (51.3 x 66.4 picas)   See Appendix for full spread grid

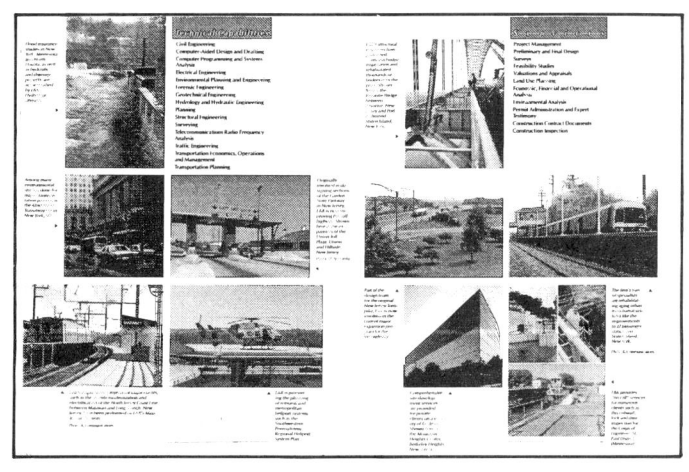
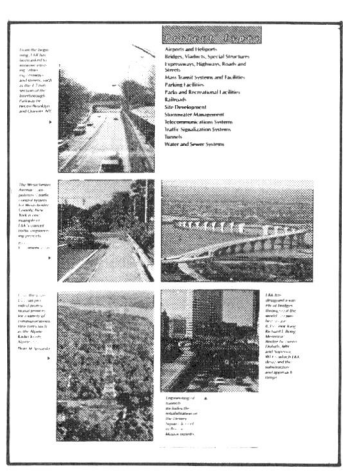

**Case Study 41**

The main theme for this brochure is set by the cover. The firm used photos of the principals set into three squares. Directly above each square is a heading outlining three disciplines. Inside, each page contains only three square photos, and a column of type at the top of each grid. The text on each page is introduced by a large capital letter, set in italics. On a smaller scale, each photograph has a small colored square in the upper left corner which corresponds to the same color square at the top of each column.

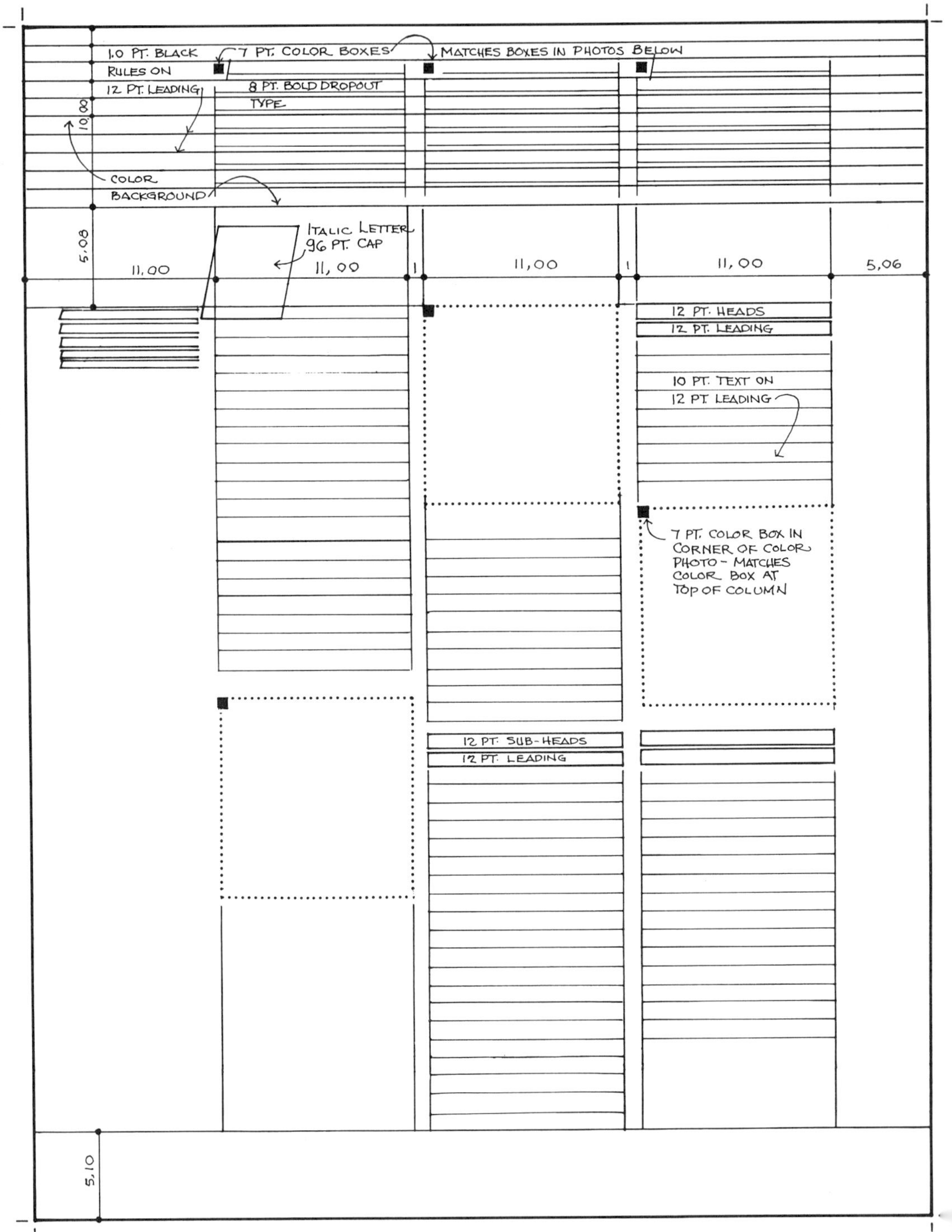

Enlarge to 154% for actual size   Trim size 8 1/2 x 11" (51.3 x 66.4 picas)   See Appendix for full spread grid

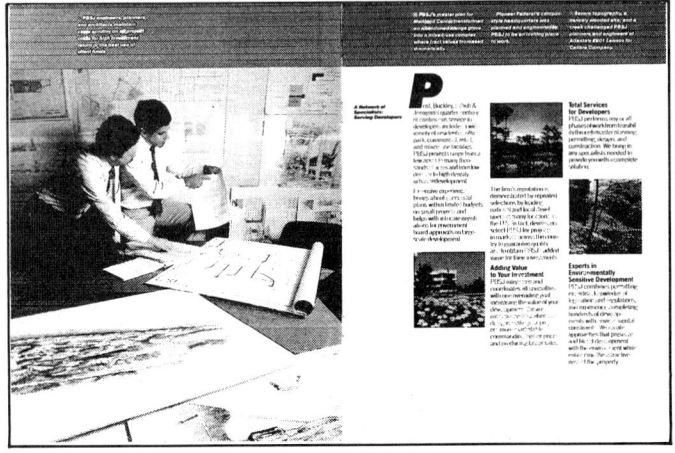

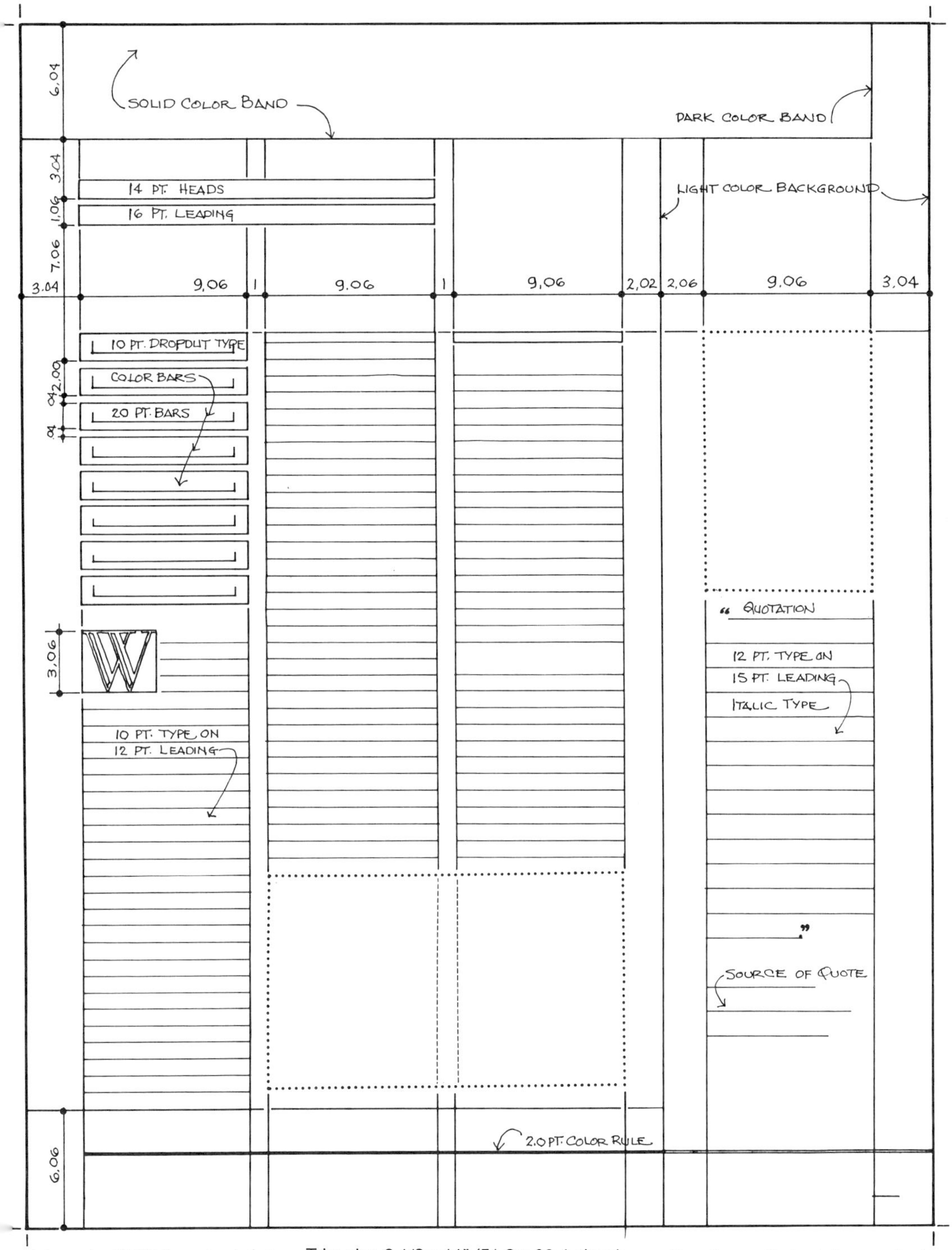

### Case Study 42

Color bars and sensitive layout make this brochure so appealing. It is a three column grid using traditional spacing for the text, plus an auxiliary column with much deeper spacing for quotations from clients. The use of color bars with drop-out type for the opening headlines gives it a very distinctive look. The color bar running across the top of the page echoes the color of the small bars. The colors change from spread to spread.

Enlarge to 154% for actual size    Trim size 8 1/2 x 11" (51.3 x 66.4 picas)    See Appendix for full spread grid

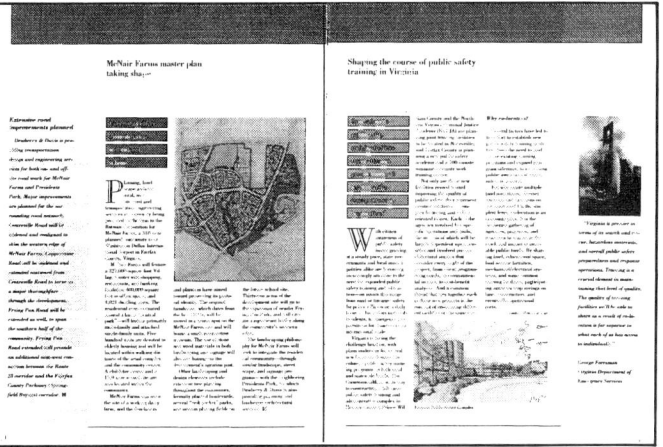

Case Study Formats 155

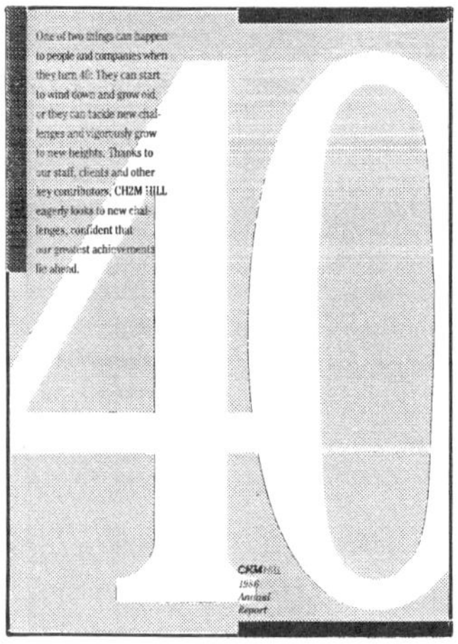

## Case Study 43

The cover of this anniversary piece makes no mistake about the number of years being celebrated. Once inside there is no other indication of the number 40. This design is a "field" grid, in that the entire spread is part of the grid, rather than a single page being either repeated or mirrored. The bars on three sides of the cover are repeated on the inside. Although the dimensions are the same for the main columns of type, the far left column is much wider, leaving the balance open to white space.

See Appendix

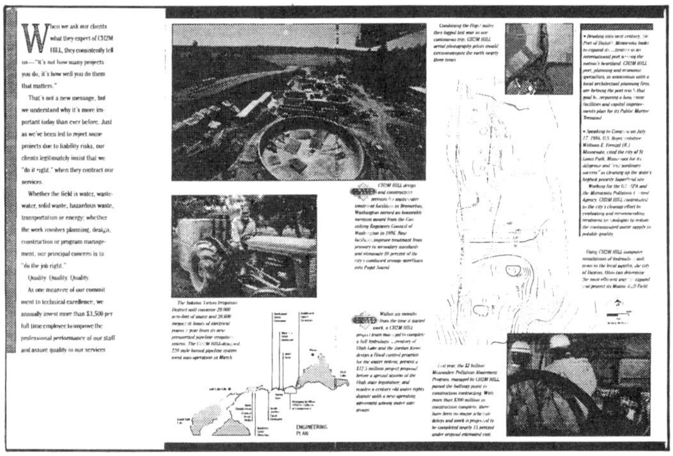

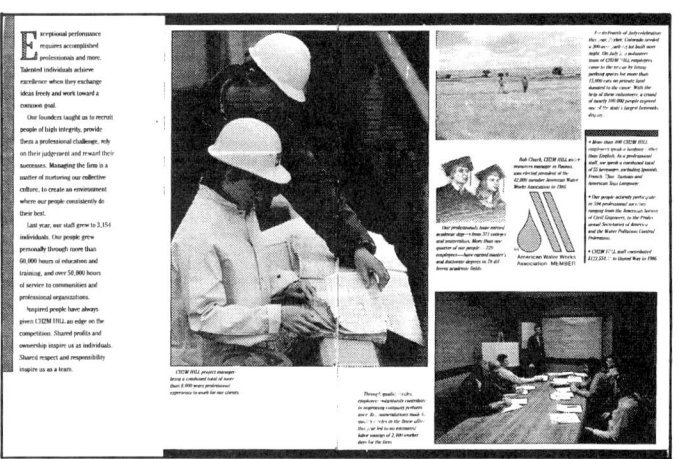

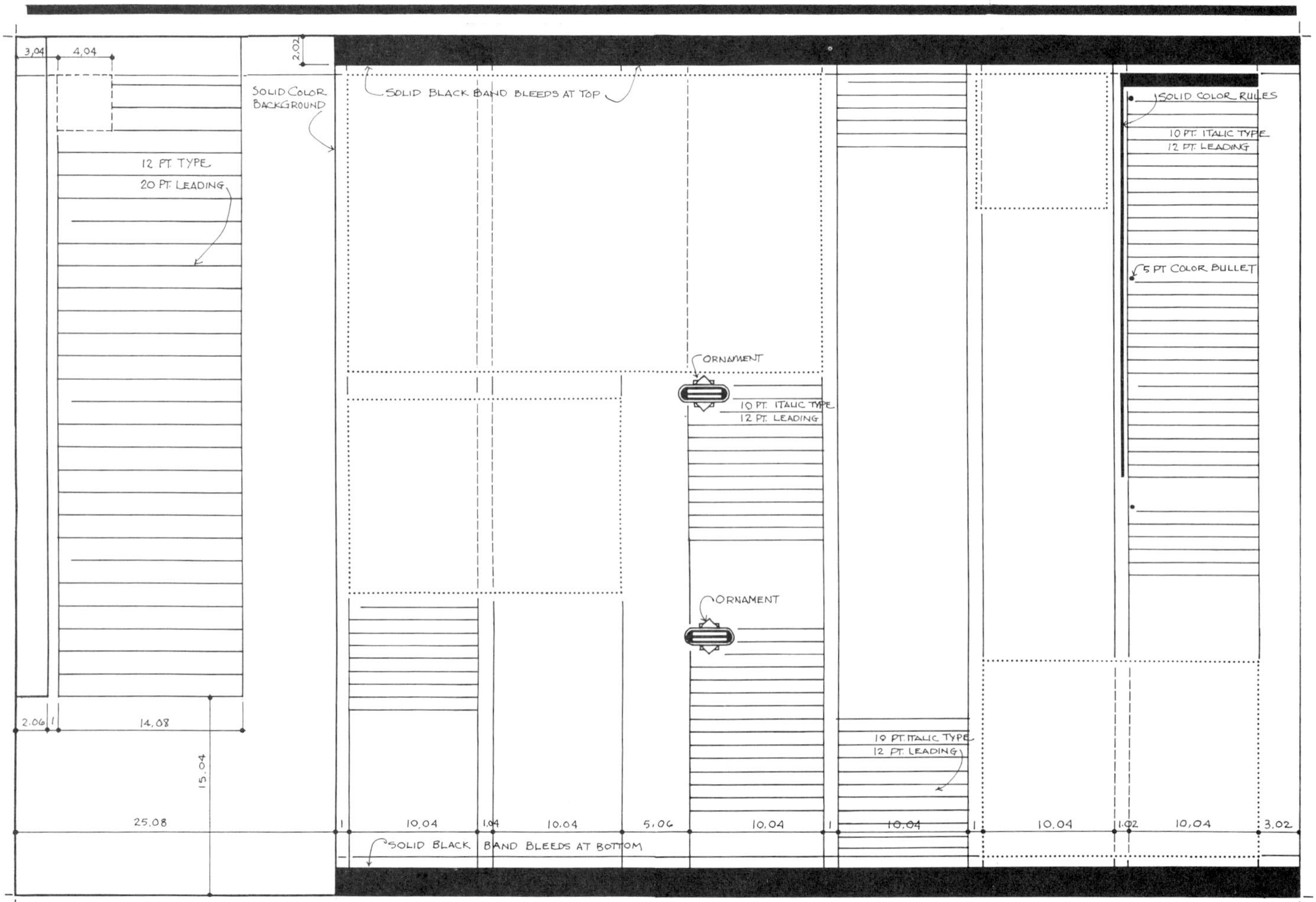

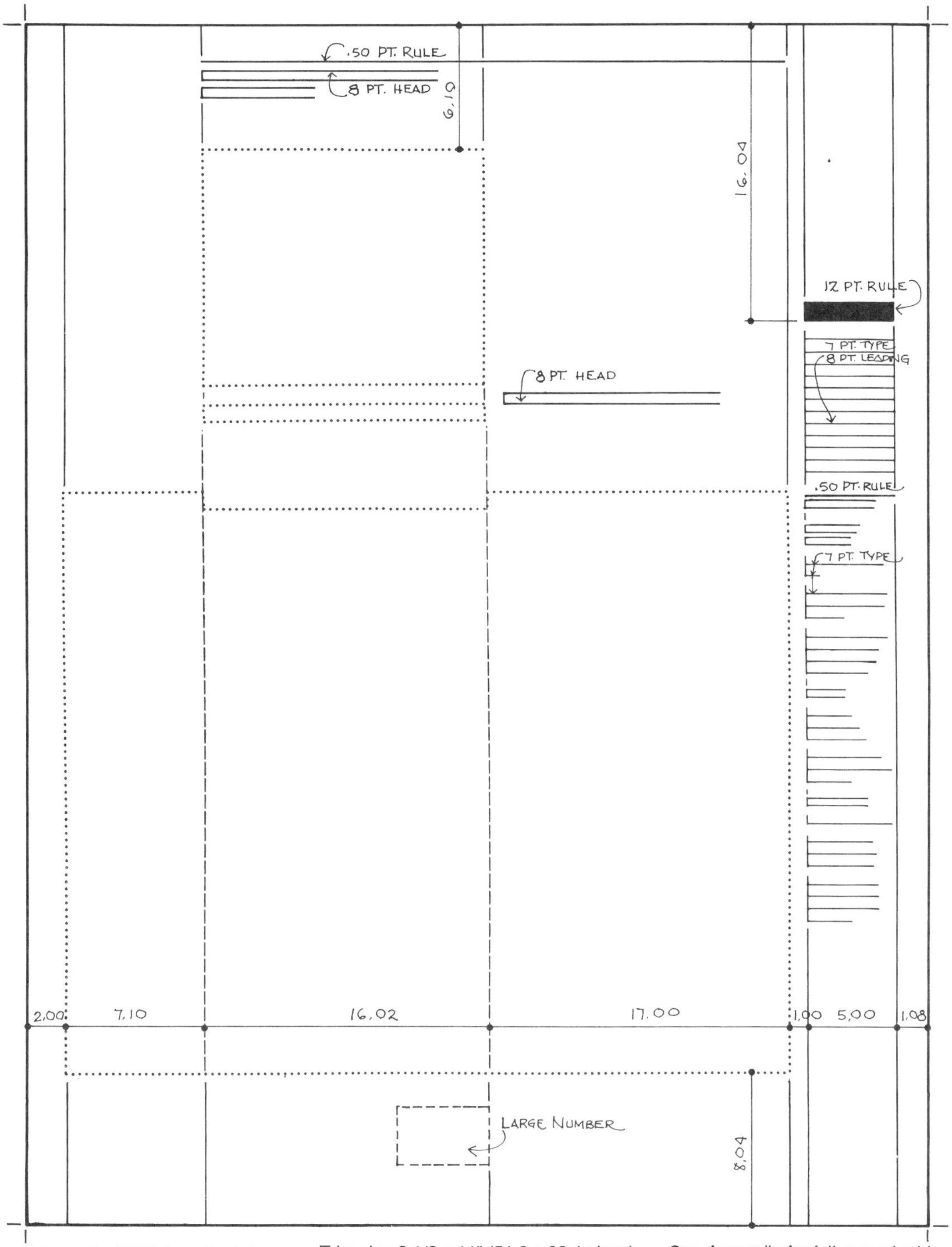

## Case Study 44

There is a different treatment on every page of this architectural firm's fortieth anniversary brochure. The main column of type describing the projects that are illustrated always appears on the outer edge of the page. The thin rule and type at the top of the page remain in that position on every page. The main design element is the removal of parts of each photograph and the distribution of them in an organized manner to other parts of the page. The pieces are not haphazardly lifted out but carefully placed to emphasize the design of the layout. The number "40" appears on every page somewhere, either as a drop-out or solid but always in a different location and in a different type style and different size.

Enlarge to 154% for actual size    Trim size 8 1/2 x 11" (51.3 x 66.4 picas)    See Appendix for full spread grid

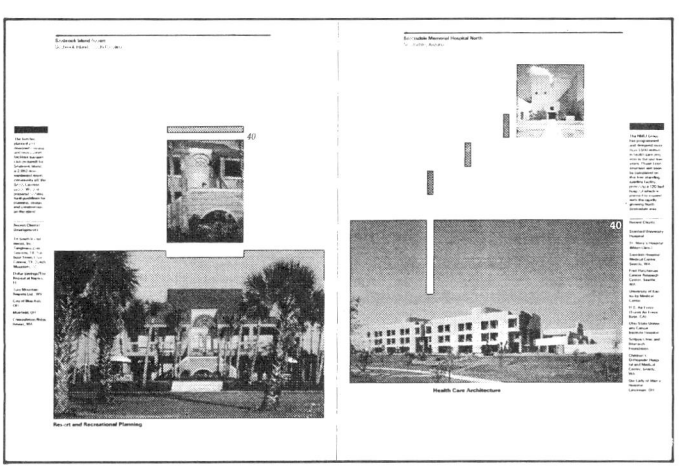
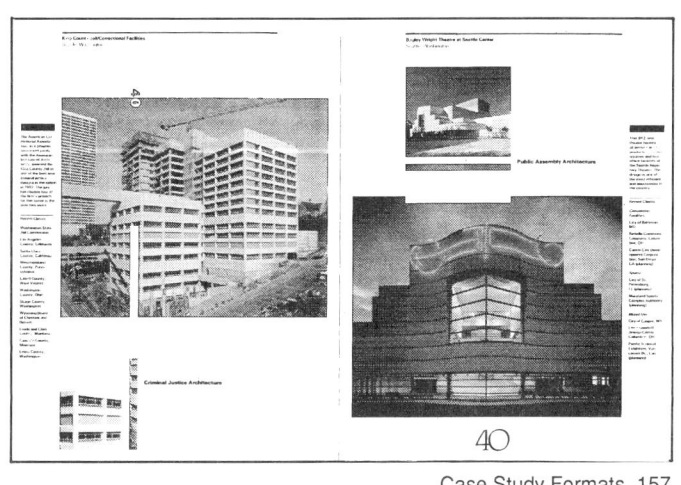

### Case Study 45

Some architectural firms have established internal graphic design studios to produce environmental graphics, signage and promotional material. This firm is at the top of that list, not only in the quality of the work, but in the application. These pages are from their environmental graphics brochure, and feature a four column format, brimming with design ideas, graphic devices, and colorful spot photographs. The strict grid design did not inhibit a lively portrayal of ideas for their brochure, or of their clients' projects.

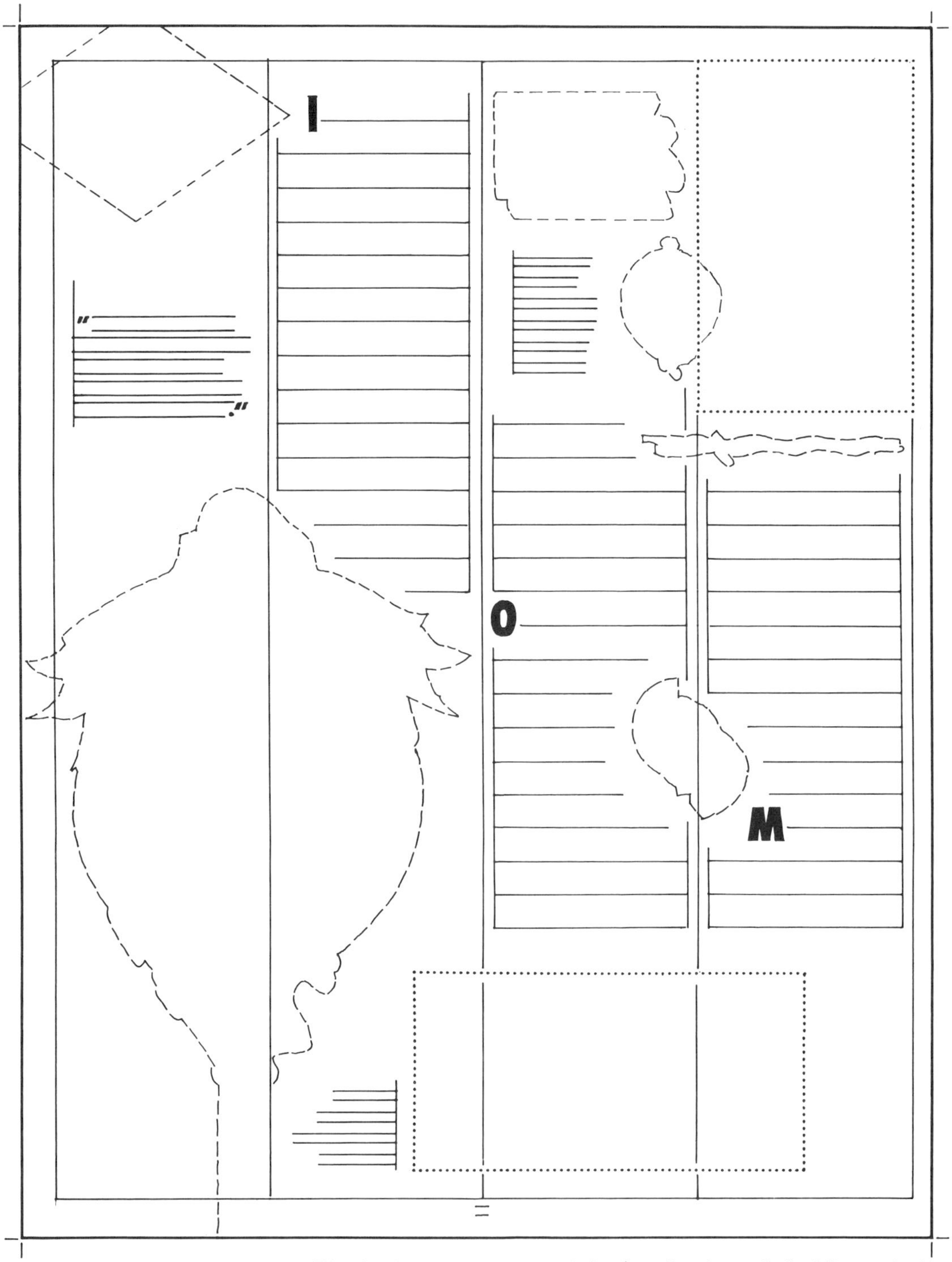

Enlarge to 154% for actual size    Trim size 8 1/2 x 11" (51.3 x 66.4 picas)    See Appendix for full spread grid

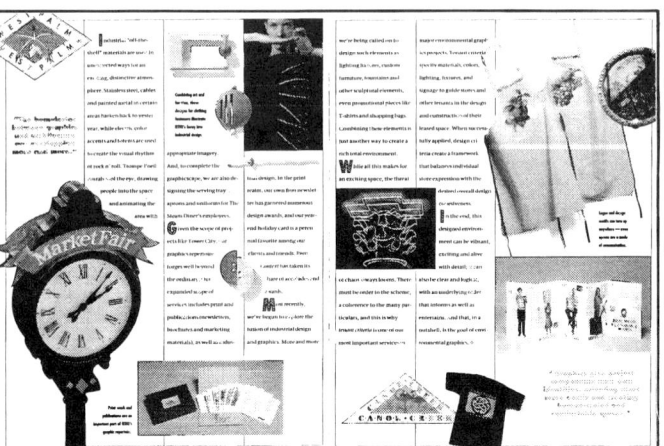

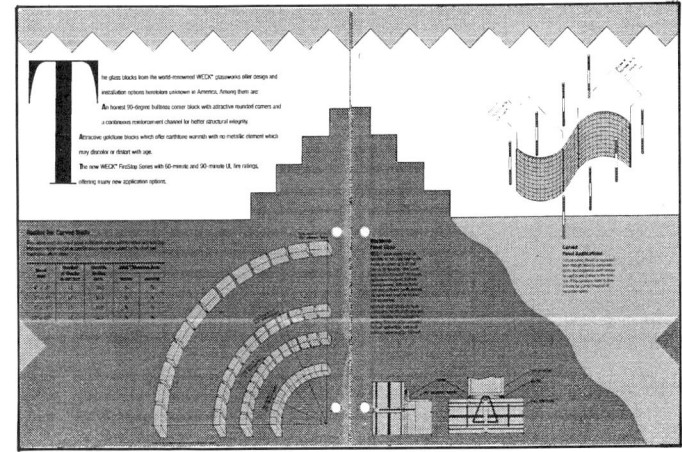

The stepped design in this brochure in indicative of the square modular building product. The steps are composed in several scales and in a variety of arrangements. The steps are turned on the diagonal across the top of the page. A large serif-face letter presides over each page of text. The stepped design is carried throughout the entire brochure.

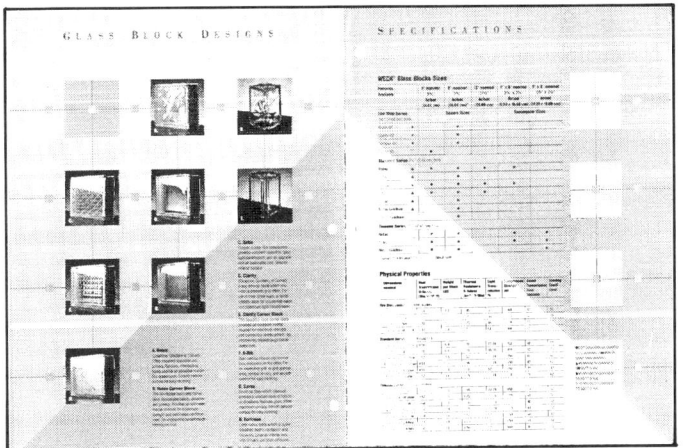

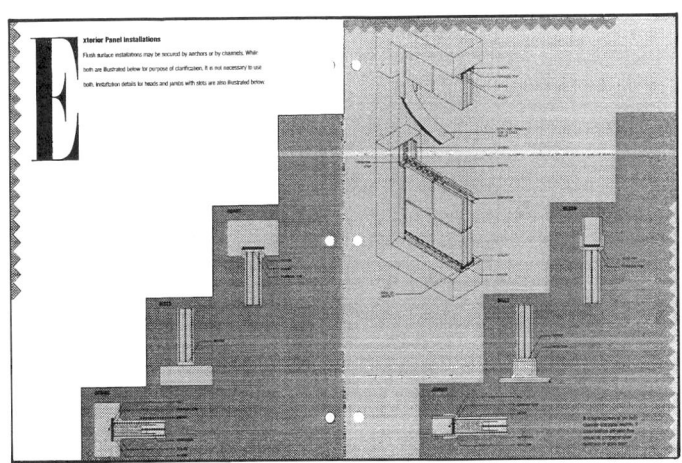

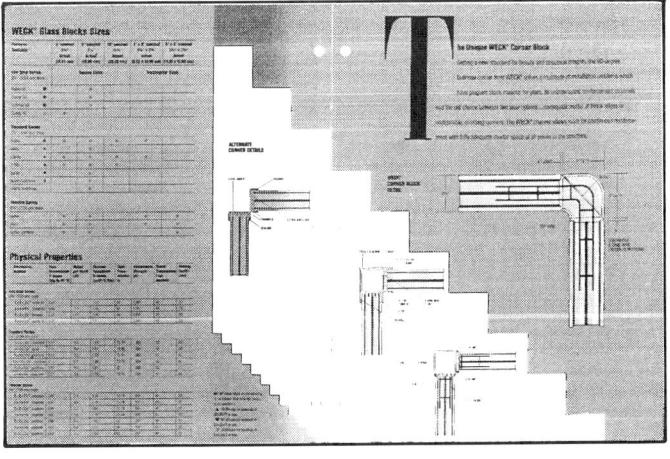

### Case Study 46

A series of overlapping elements characterize this grid. The main columns of type are precisely located throughout this publication. However, the positioning of the photographs overlap these grids and adds a variety to the layouts. Once past the first page, all left pages are full bleed photos, which add drama to the spreads. A solid bar appears as a constant in the upper corner of each right page, reflecting the solid color bar on the right edge of the cover.

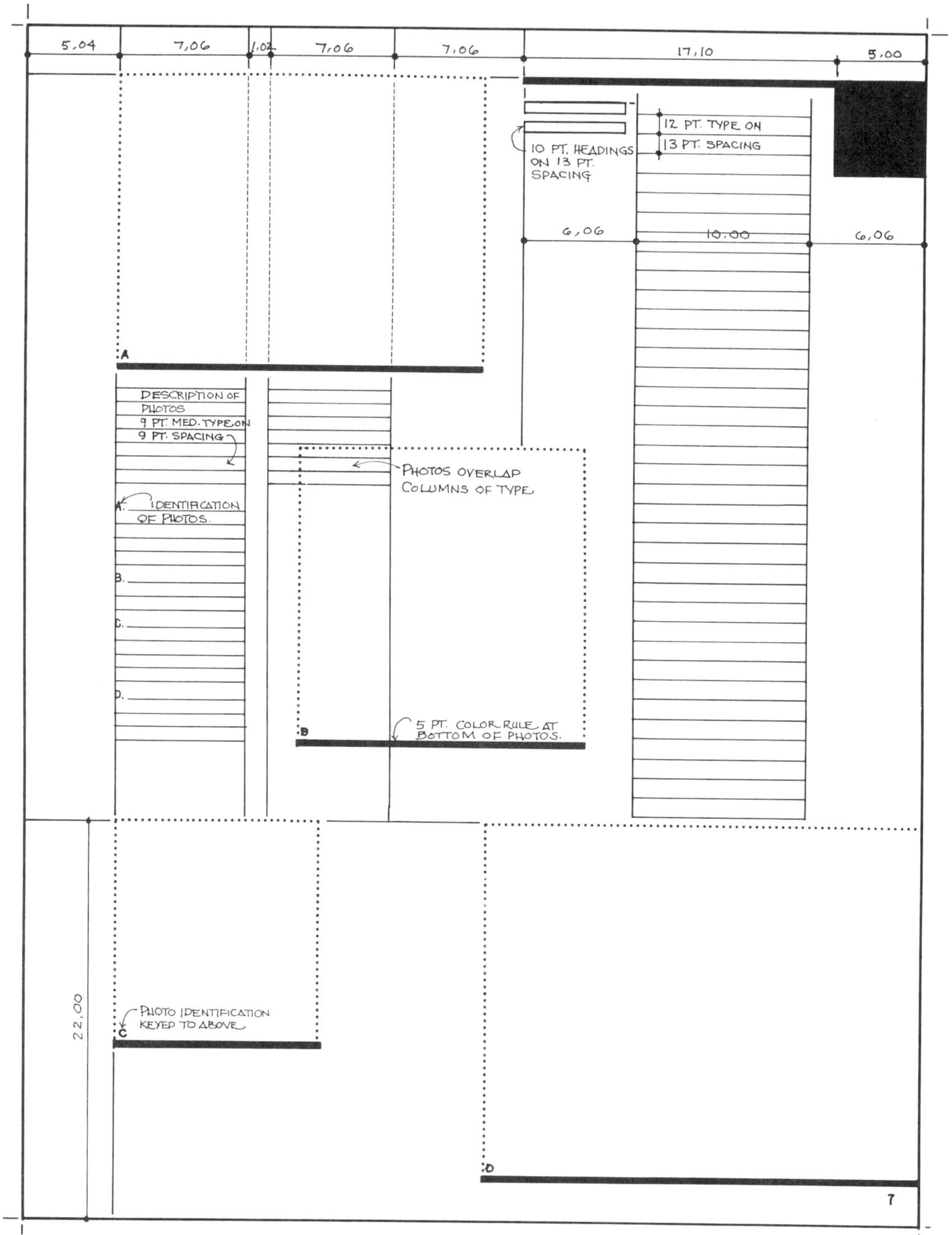

Enlarge to 154% for actual size   Trim size 8 1/2 x 11" (51.3 x 66.4 picas)   See Appendix for full spread grid

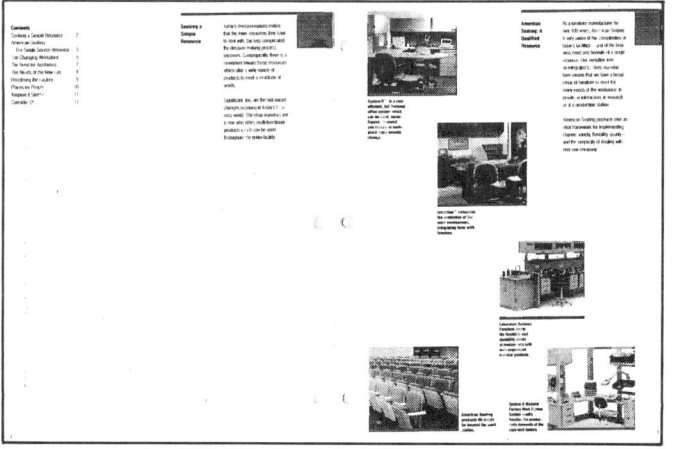

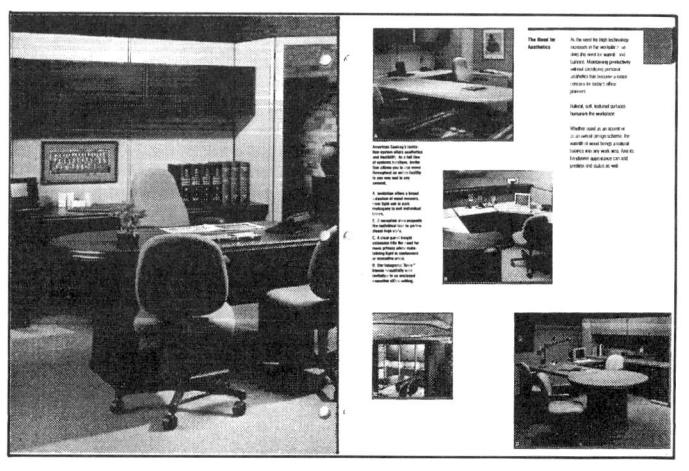

160

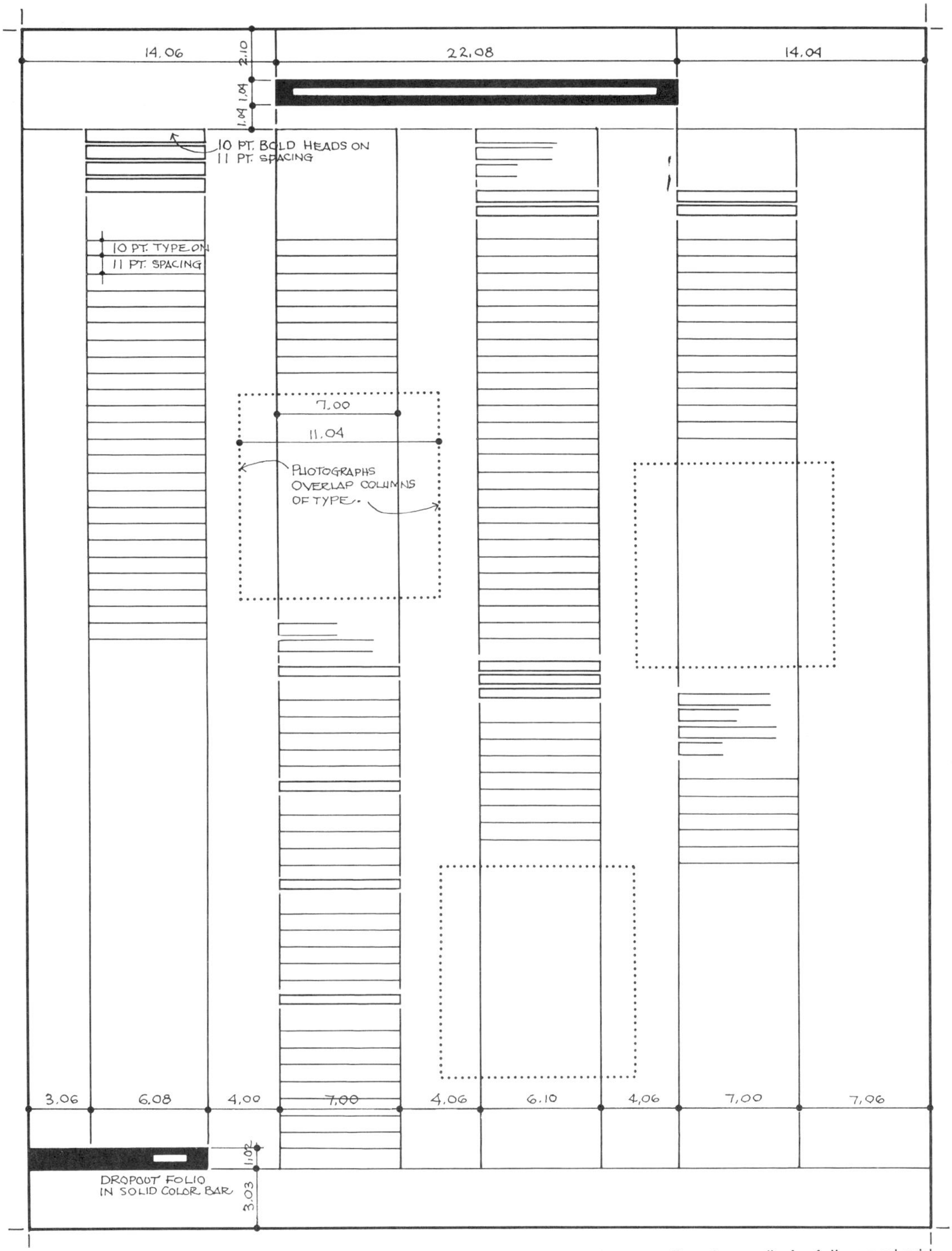

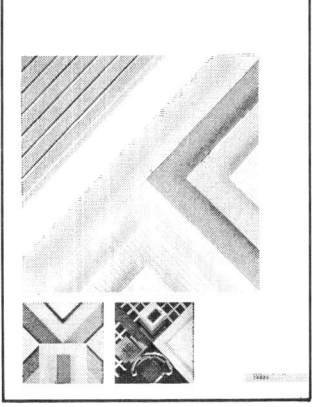

## Case Study 47

Four narrow columns with wide spacing between them characterize this grid. The reason for this wide separation is that the photographs, which are mostly square in format, overlap into this margin area. This creates an openness to the page while allowing a picture of ample size to be shown. Other design elements include the solid color bar at the head margins and the solid color bar for the folios. Some pages are dominated by a large square photo, comprised mainly of installations showing the product.

Enlarge to 154% for actual size    Trim size 8 1/2 x 11" (51.3 x 66.4 picas)    See Appendix for full spread grid

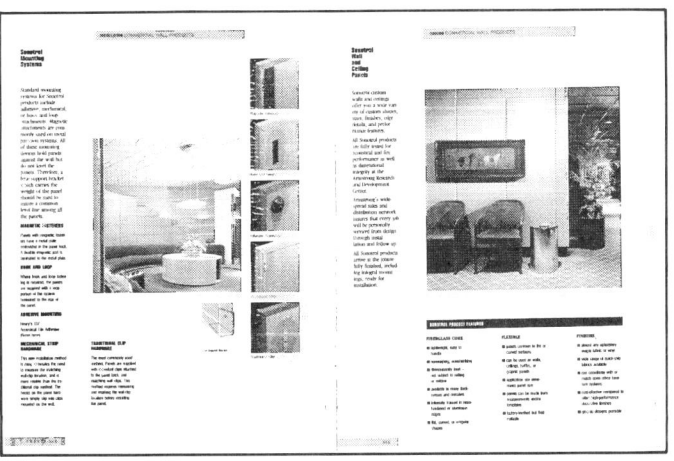

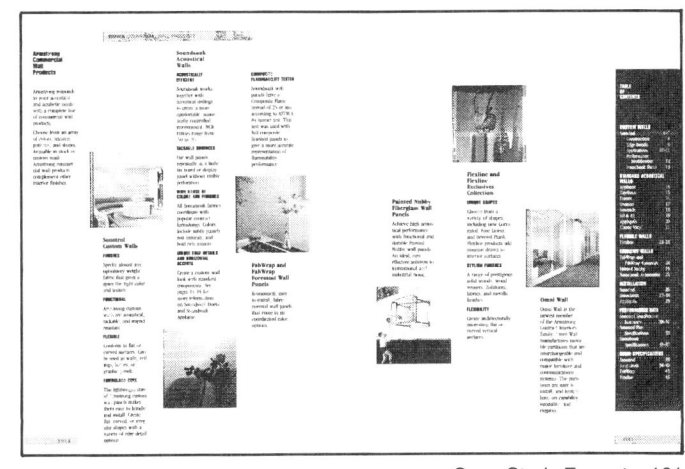

## Case Study 48

Standard grids are very easy to work with, perhaps too easy, as this very traditional layout shows. The four column grid appears in reverse outline on the cover. Some of the grid squares are filled in with photographs. The name of the firm appears in the blank space below the grid modules. The inside photographs are precisely cropped according to the modules. There are no descriptive captions except for bold headings calling out each section. The brochure opens up into 3 panels.

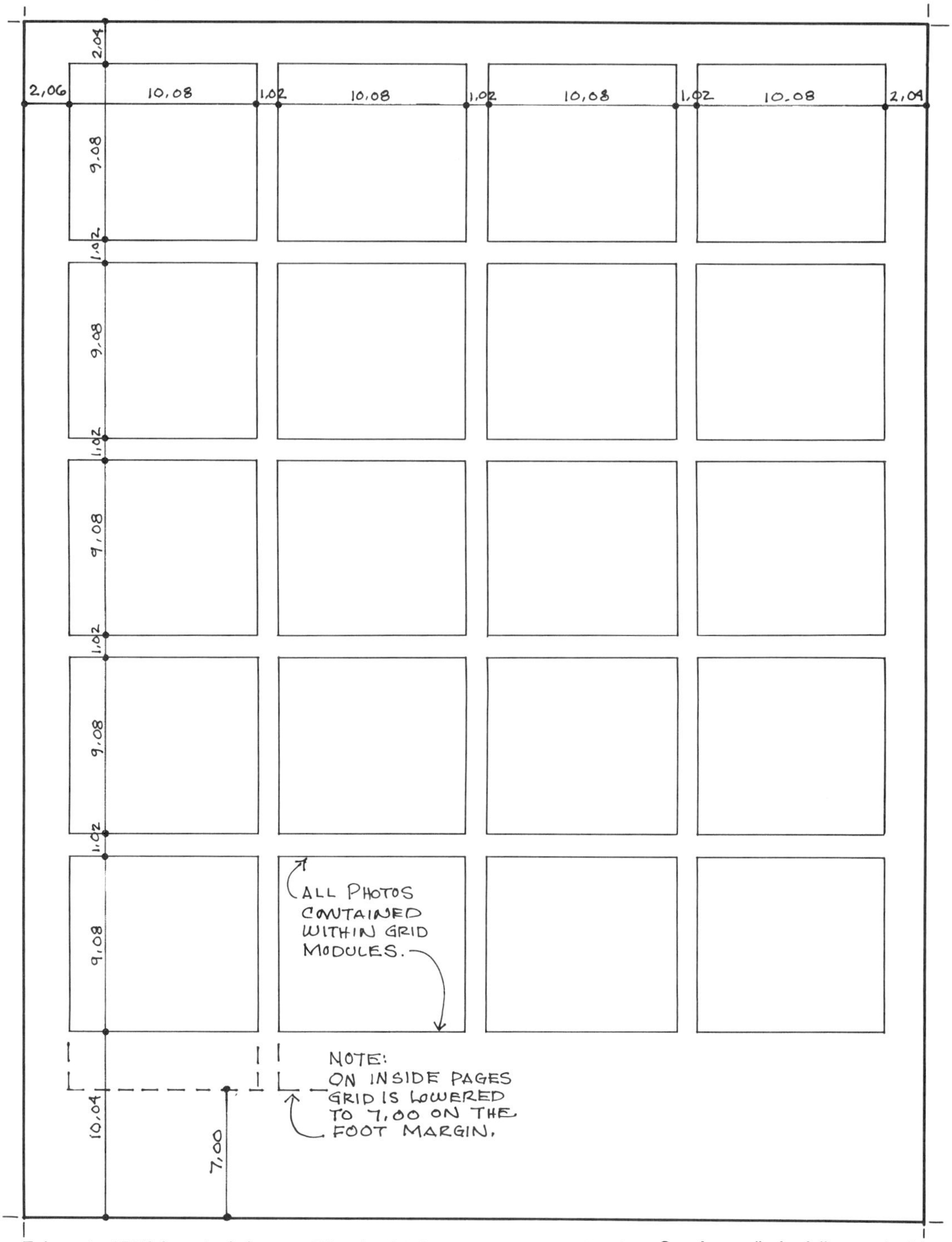

ALL PHOTOS CONTAINED WITHIN GRID MODULES.

NOTE: ON INSIDE PAGES GRID IS LOWERED TO 7.00 ON THE FOOT MARGIN.

Enlarge to 154% for actual size   Trim size 8 1/2 x 11" (51.3 x 66.4 picas)   See Appendix for full spread grid

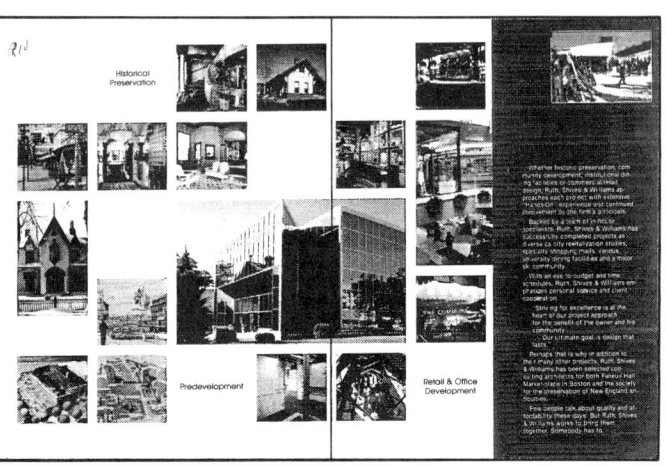

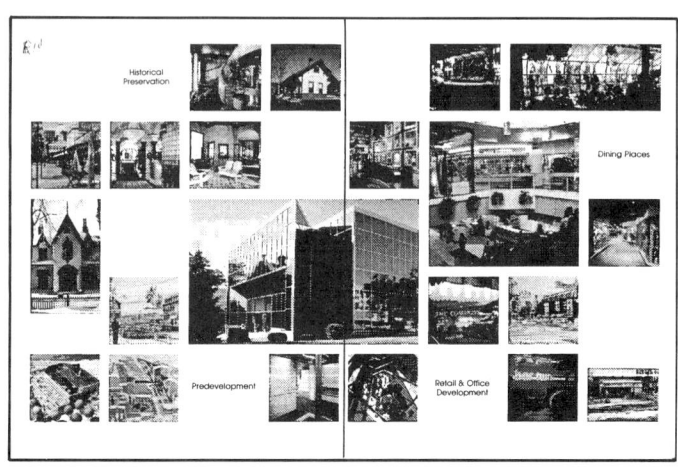

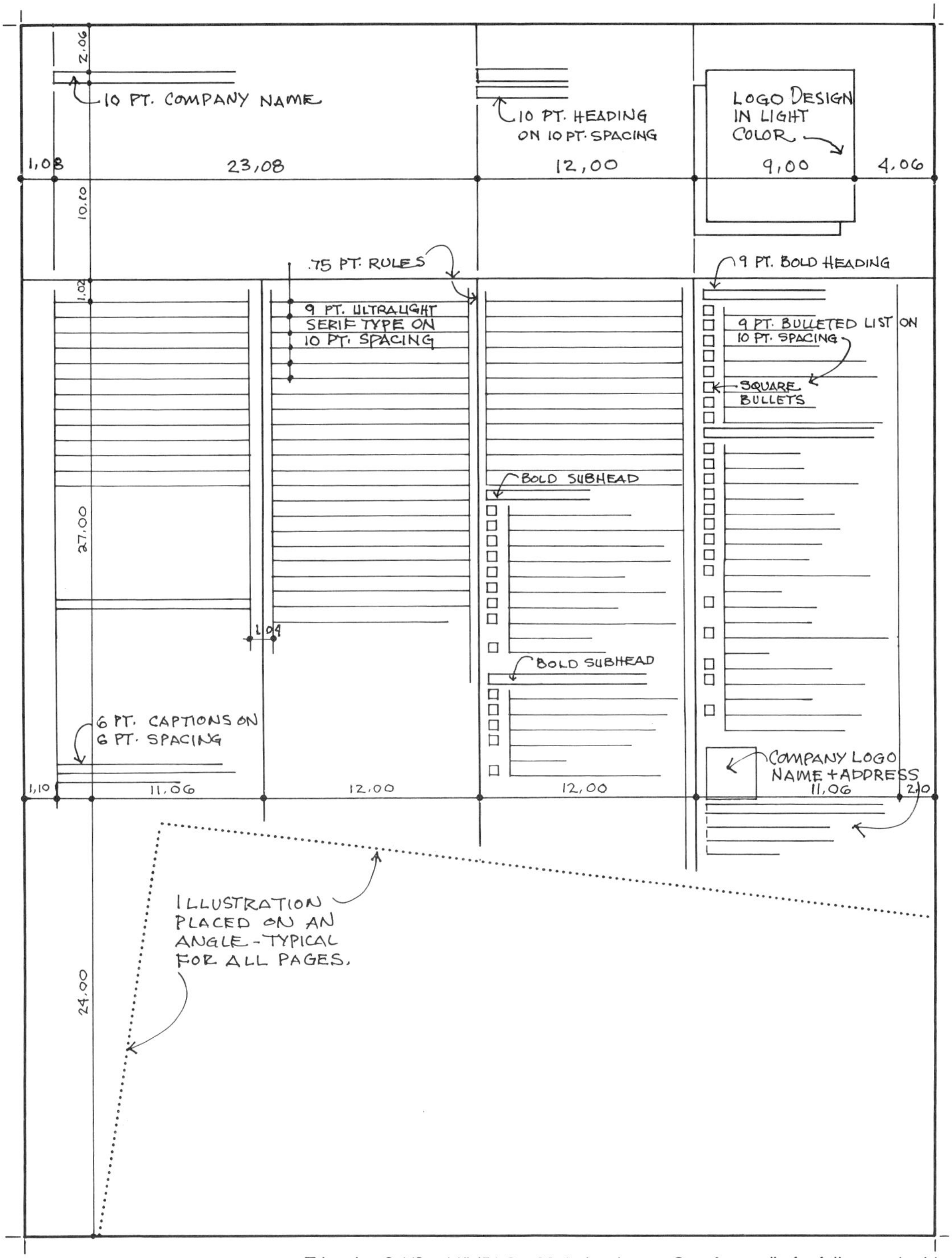

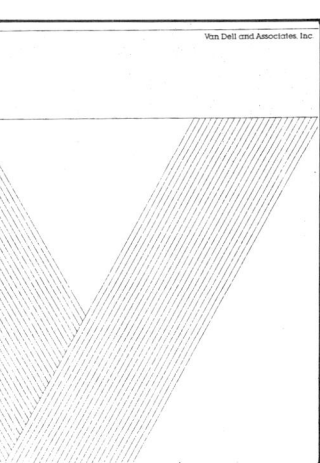

## Case Study 49

The most distinctive aspect of this environmental engineering brochure is the use of graphic designs to depict the specialized services offered. These appear in the upper corner of each page. Under the top margin is a four column grid separated by thin rules. Bulleted lists elaborate on the services and are highlighted by square shapes which echo the square graphics. At the bottom of each page is a detail from a drawing that relates to the featured service. For emphasis it is placed askew across the bottom.

Enlarge to 154% for actual size   Trim size 8 1/2 x 11" (51.3 x 66.4 picas)   See Appendix for full spread grid

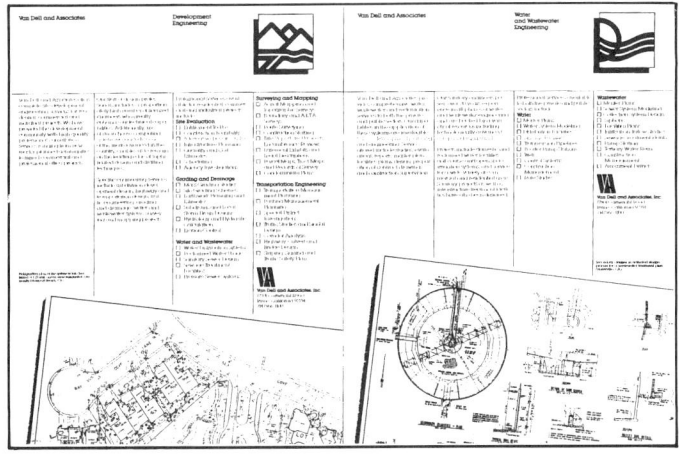
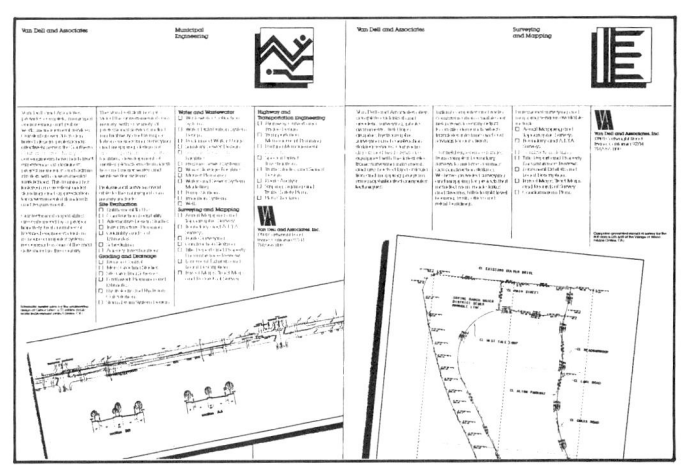

### Case Study 50

A very precise grid governs this manufacturer's brochure, which relates to modular wall panel units. Thin rules outline the grid format except for the areas where type or photographs appear. The photographs range from small to very large, and are all carried within the grid lines. This grid is composed of a four column wide by three column high format. At the top of each right-hand page is a solid color bar (yellow), with the initials of the company printed inside it with black ink.

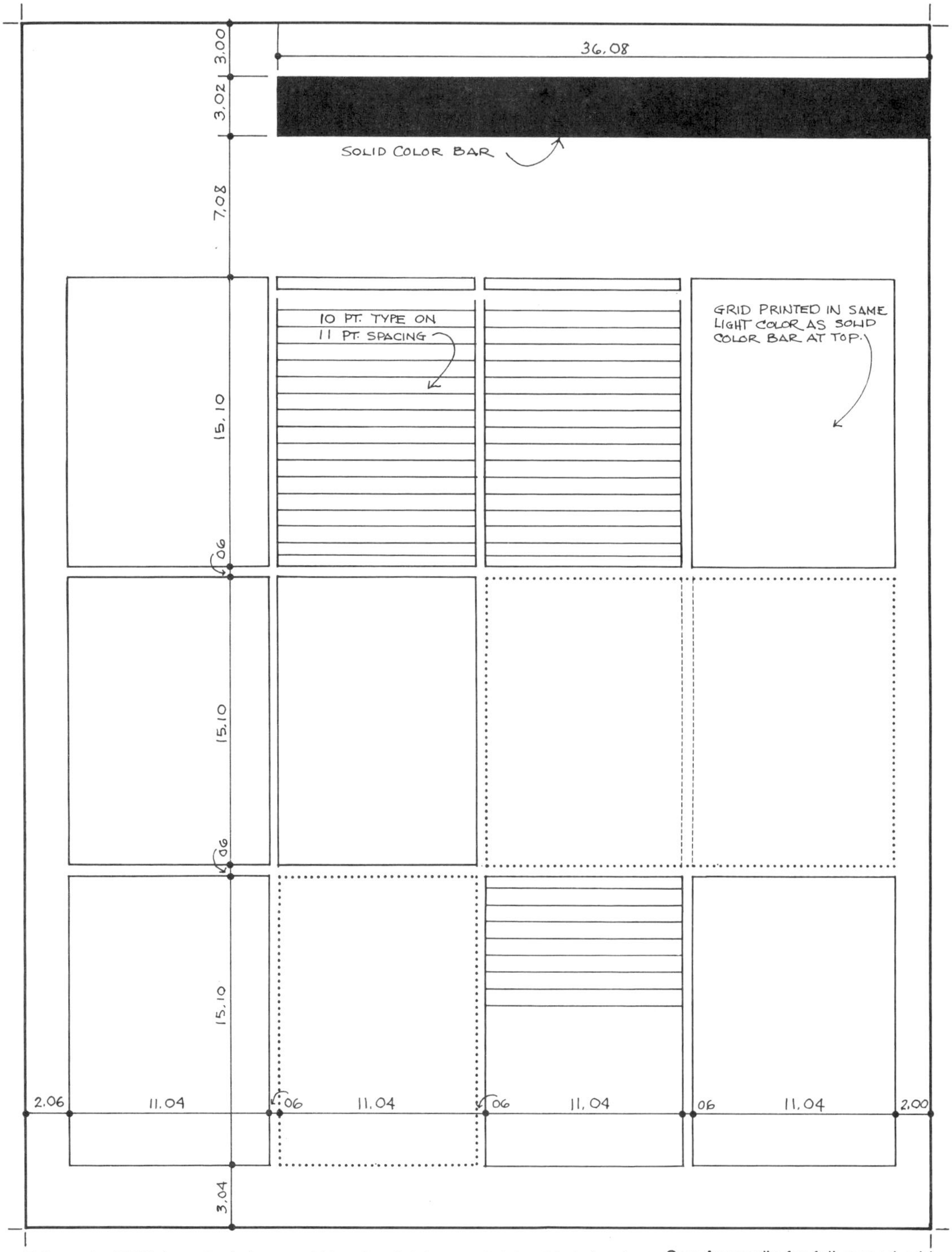

Enlarge to 154% for actual size   Trim size 8 1/2 x 11" (51.3 x 66.4 picas)   See Appendix for full spread grid

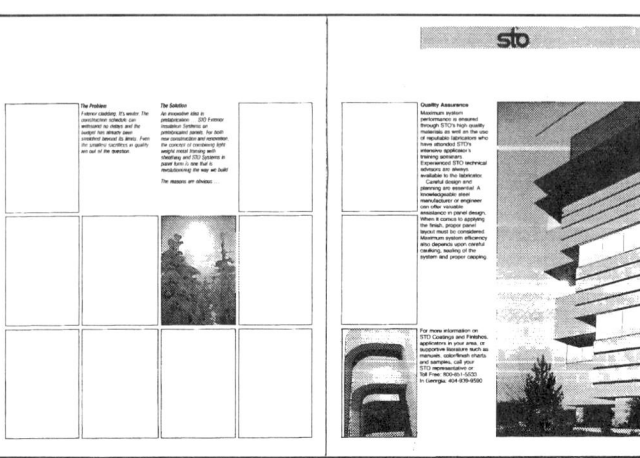

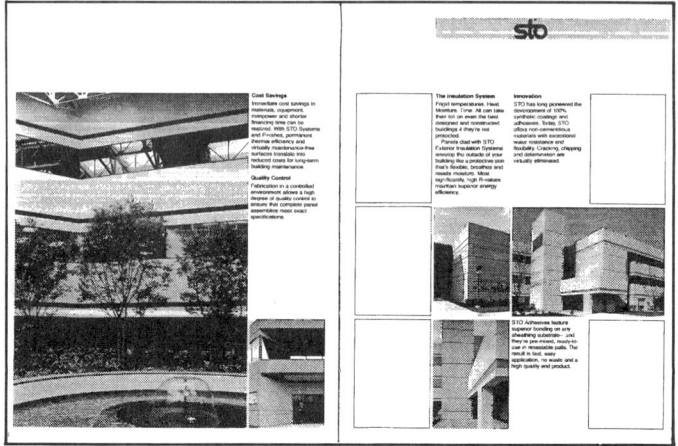

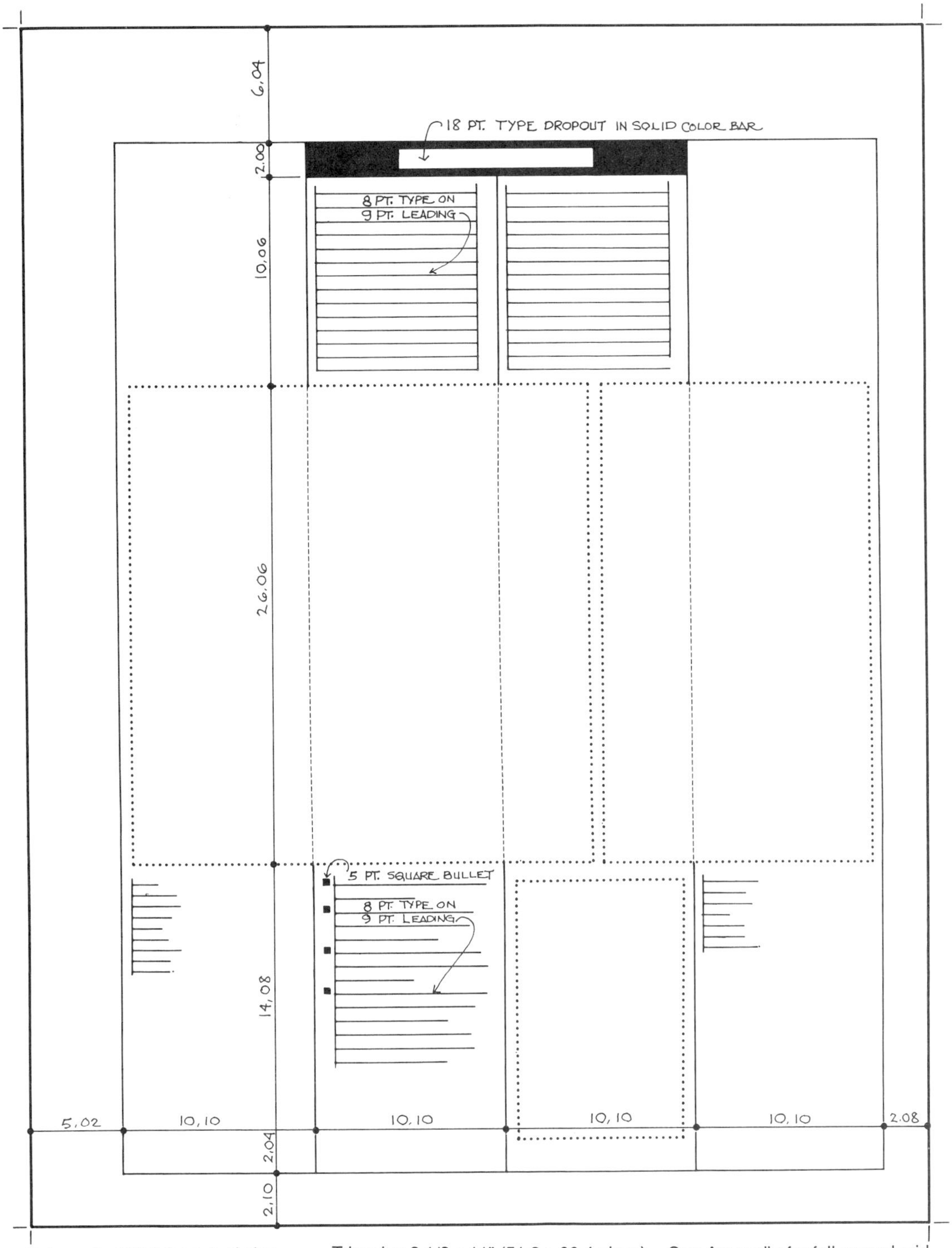

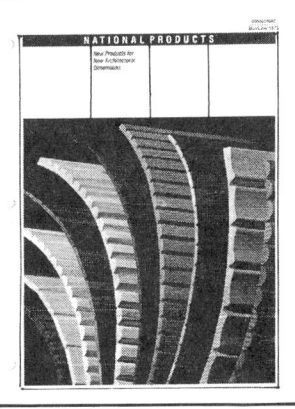

### Case Study 51

The cover for this catalog sets up the grid design that is used throughout, and establishes the four column grid separated by thin black rules. A solid black bar with drop-out type appears at the top of each page. The dominant element on each page is a large horizontal panel of photographs with smaller ones below. The two center columns carry the type, and the solid bar relates to this type.

Enlarge to 154% for actual size   Trim size 8 1/2 x 11" (51.3 x 66.4 picas)   See Appendix for full spread grid

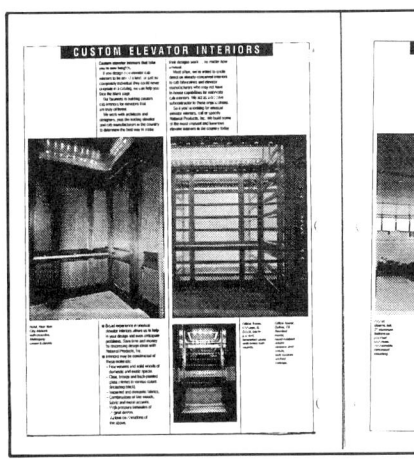

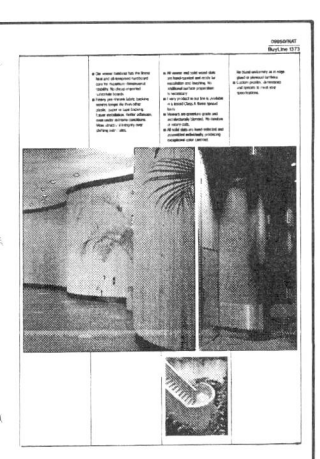

Case Study Formats

**Case Study 52**

The cover of this brochure on water services establishes the theme of the four column grid. Thin white rules are dropped out of the photograph along the vertical grid lines. Inside these appear as black rules between the columns of type. The cascading water on the cover is echoed in the stepped design of the main title blocks, under which appears the text. This same stepped design is repeated on all left hand pages. The main headlines are set in blue type to further emphasize the clean water aspect of the firm's services.

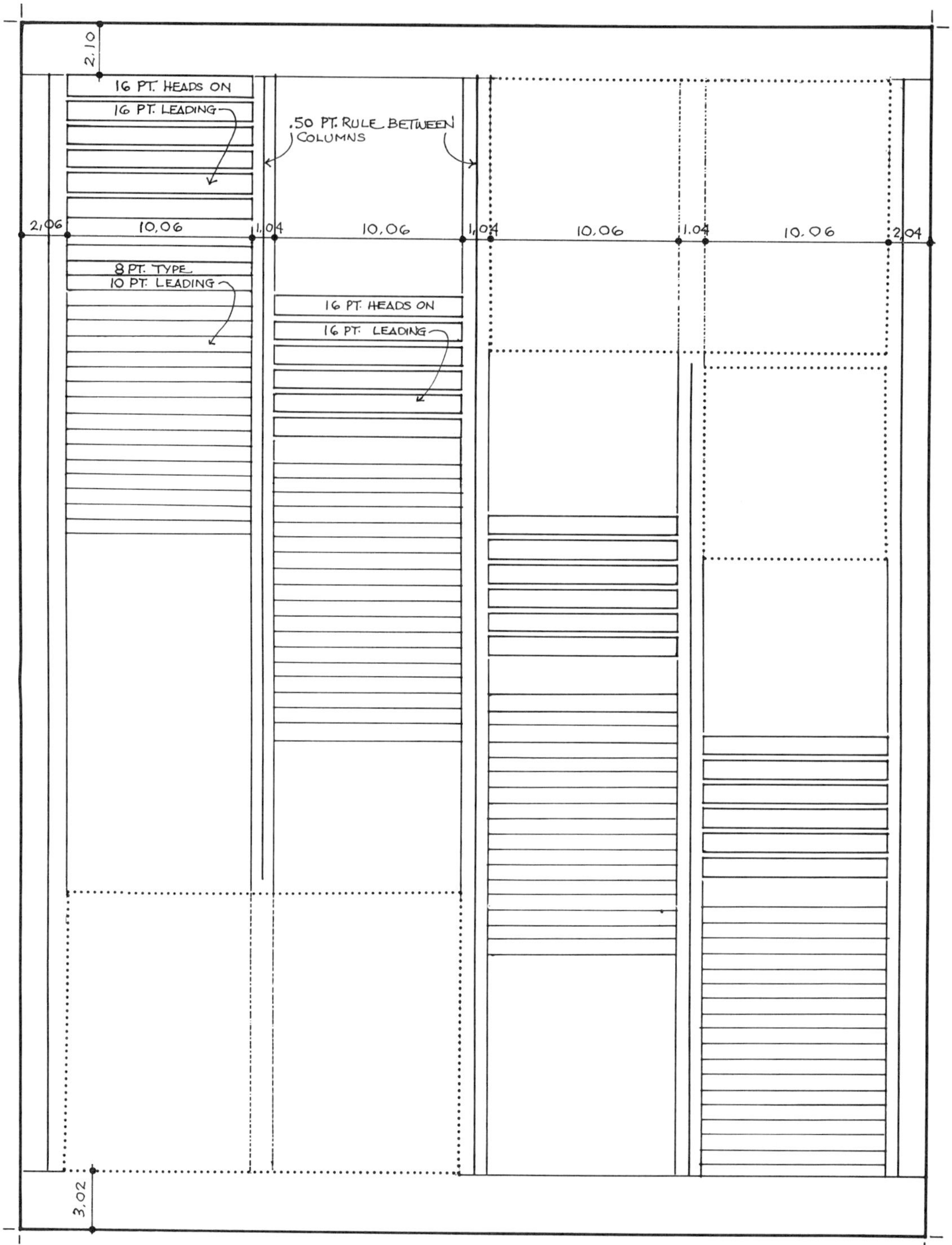

Enlarge to 154% for actual size      Trim size 8 1/2 x 11" (51.3 x 66.4 picas)      See Appendix for full spread grid

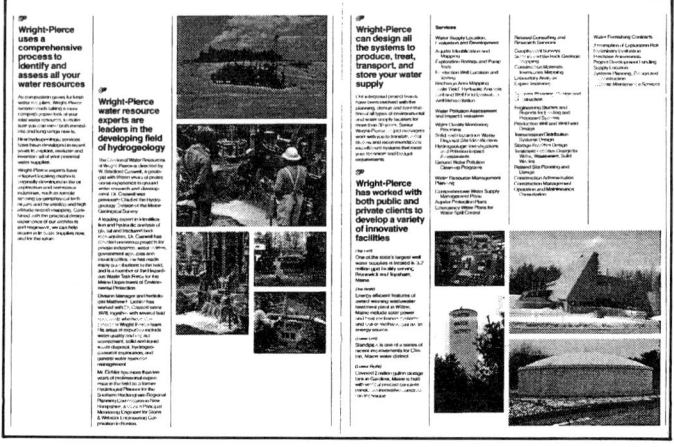

166

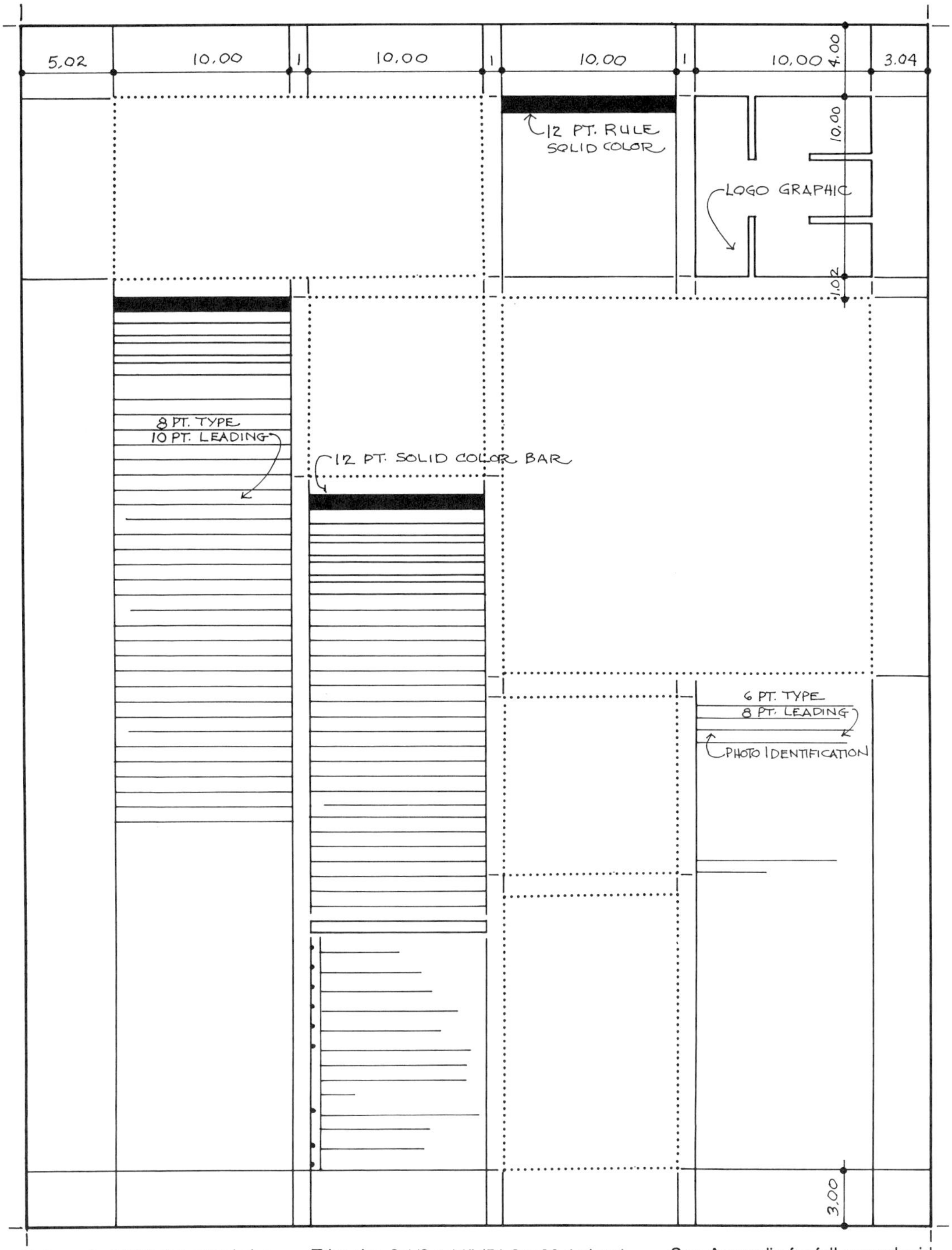

## Case Study 53

This brochure also used a four column grid (designed by the same graphic artist). The major difference is the use of heavy rules to establish the tops of columns and the looser arrangement of photographs. The firm's logo appears first on the cover where it holds a color photograph within its outline, and again on the outside corner of each right-hand page. Many of the color photographs are cropped into a square format reflecting the design of the firm's logo.

Enlarge to 154% for actual size   Trim size 8 1/2 x 11" (51.3 x 66.4 picas)   See Appendix for full spread grid

Case Study Formats 167

## Case Study 54

The conceptual development of this grid was shown previously, and related how the grid was set up to accommodate the format of the photographer's camera. The photo project pages that resulted show the variety of images that could be obtained for each project, keeping an overall consistency to the entire program. The rules at the top were reminiscent of those found in interior design magazines. The text was kept in a single column on the left of the page. The firm's logotype was in line with the text above, but was placed beneath a rule at the bottom. Although many of the negatives were the same size, there was a great deal of variety within the layouts while avoiding excessive cropping of the images.

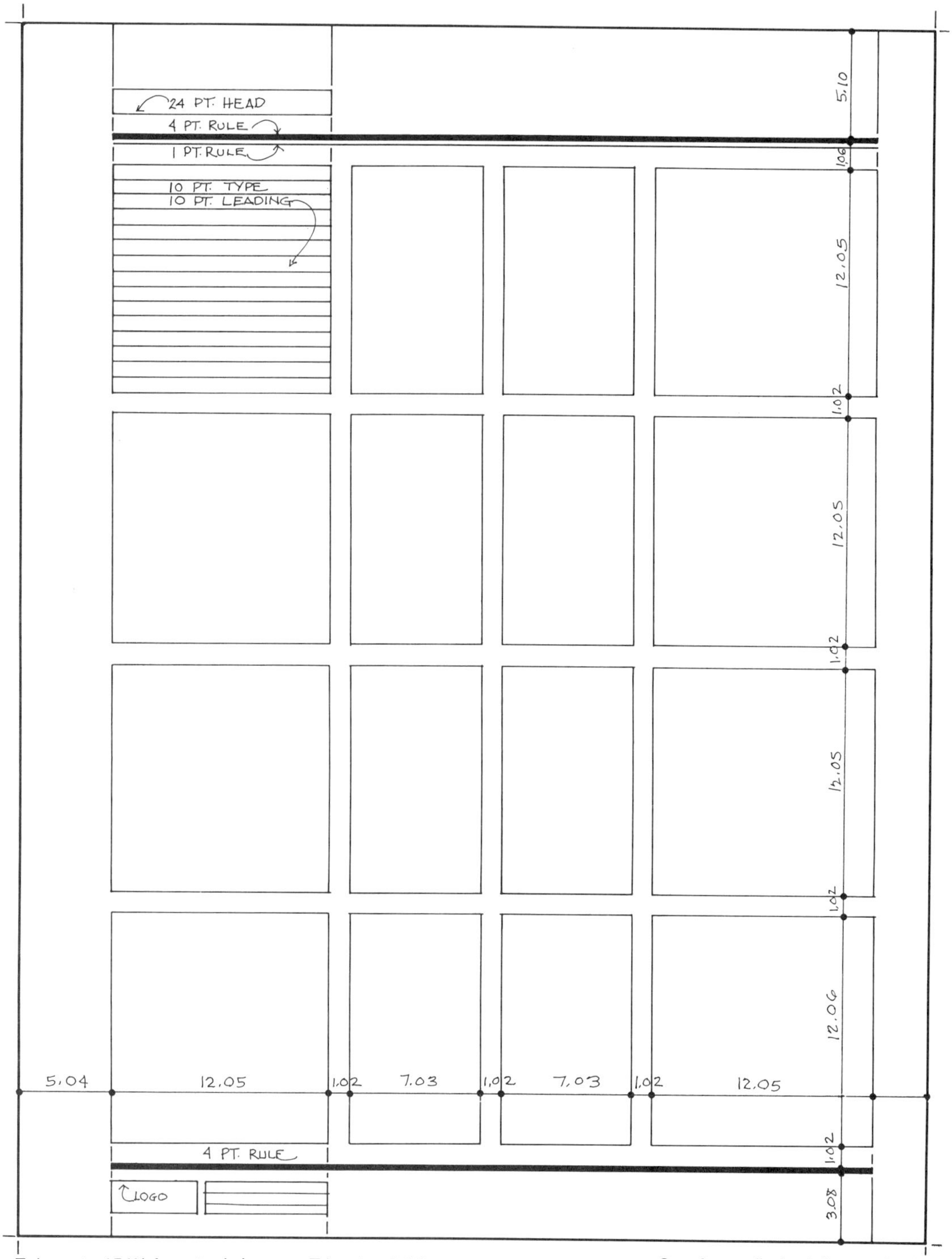

Enlarge to 154% for actual size        Trim size 8 1/2 x 11" (51.3 x 66.4 picas)        See Appendix for full spread grid

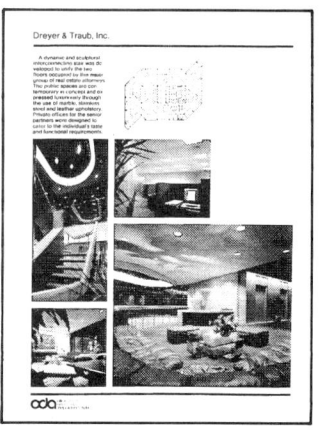

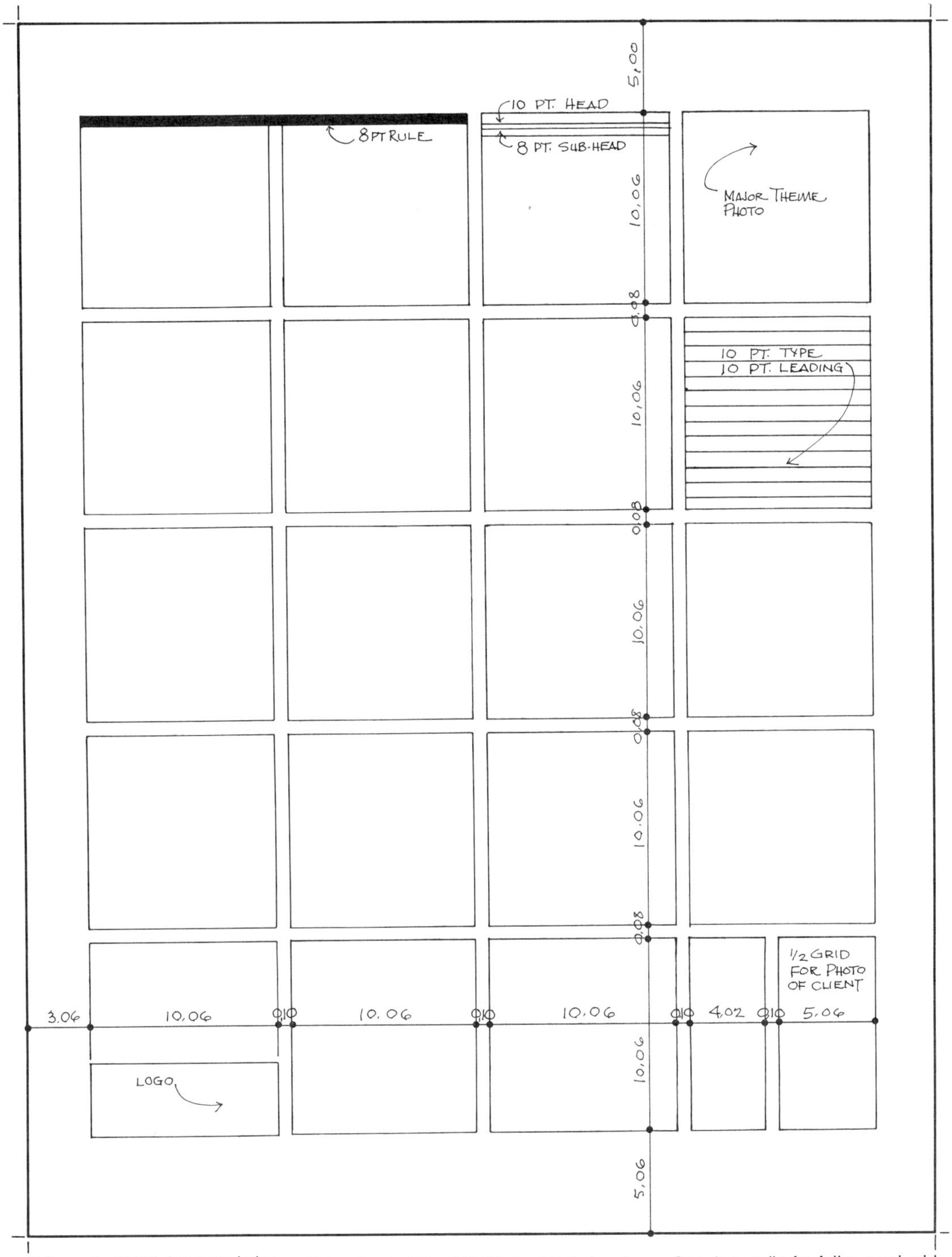

## Case Study 55

The conceptual development of this grid was shown previously, and demonstrated how the grid originated from the firm's logo. It also showed other pieces that were developed using this grid, such as client lists and other pieces. The key element in each of these designs was the square theme picture at the top of each page. Below this appeared a column of type. At the bottom grid module a small photo of the client appeared, along with a quotation and company identification. The photos were arranged within the grid modules, adding a floor plan in most cases. The firm's logo appeared at the bottom of each page.

Enlarge to 154% for actual size    Trim size 8 1/2 x 11" (51.3 x 66.4 picas)    See Appendix for full spread grid

 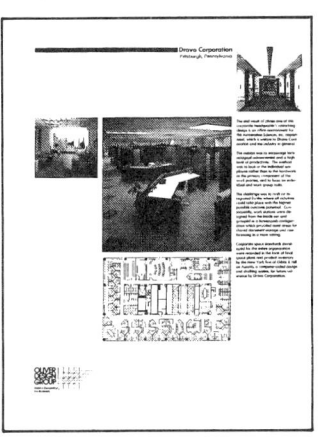  

Case Study Formats 169

### Case Study 56

One would hardly imagine the number of combinations that are possible using a four column grid. But this brochure shows many of them. The grid is four square units wide and six units high. There is a wide array of variations, from a one-quarter by one-quarter module, all the way to larger photo images covering six to eight modules. The grid is expressed horizontally, vertically, in parts, and in wide views. The larger modules are used to illustrate installation plans and drawings.

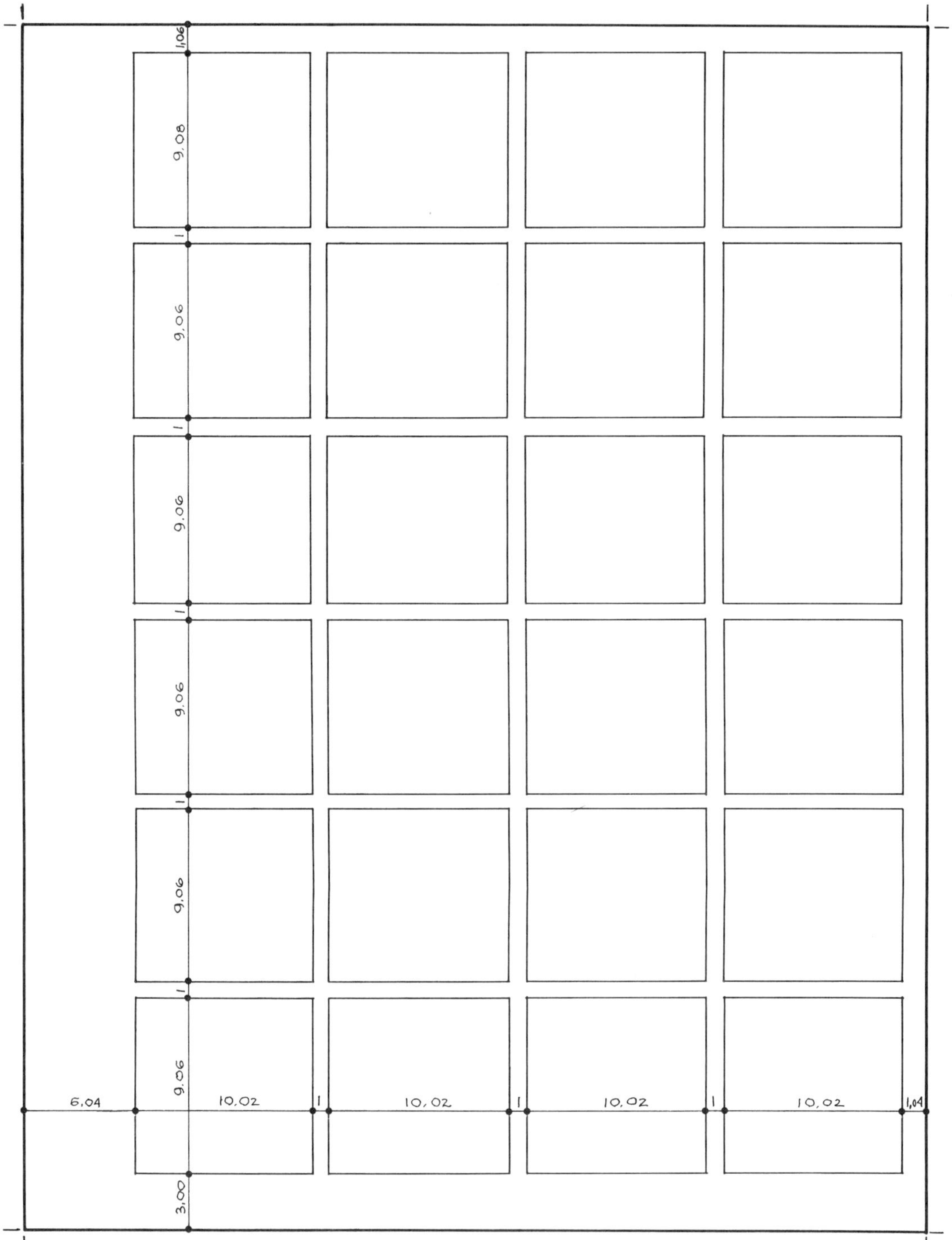

Enlarge to 154% for actual size    Trim size 8 1/2 x 11" (51.3 x 66.4 picas)    See Appendix for full spread grid

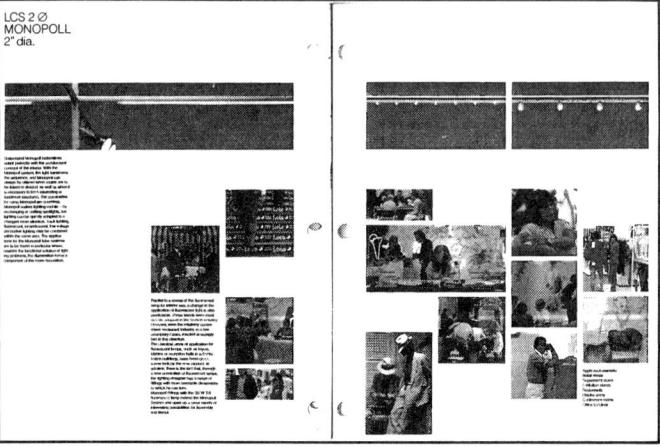

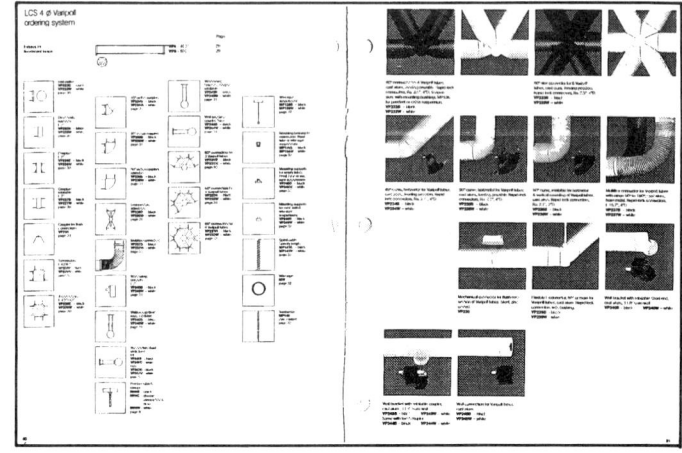

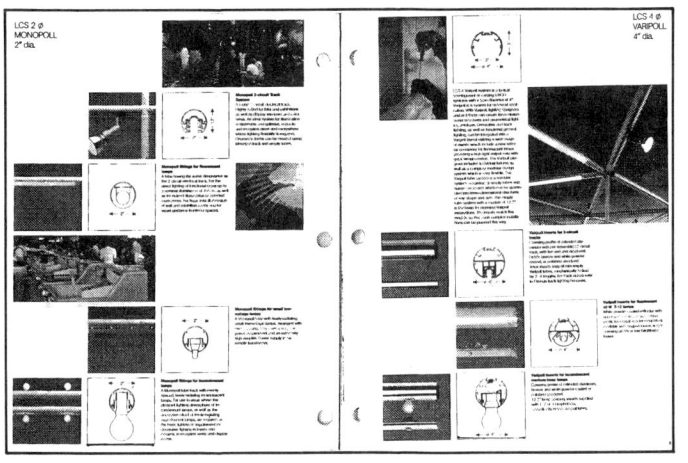
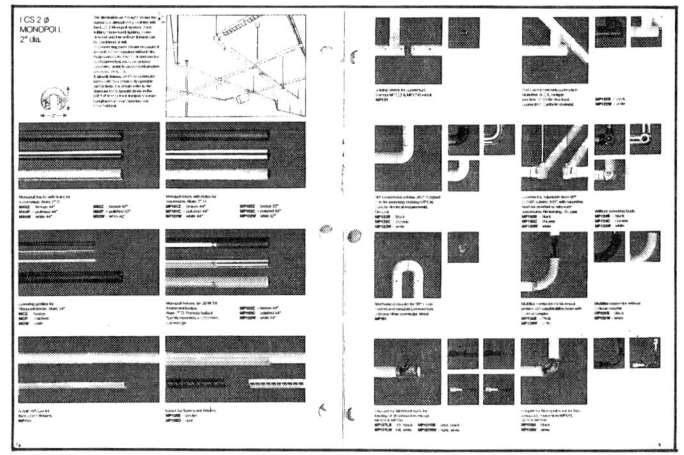
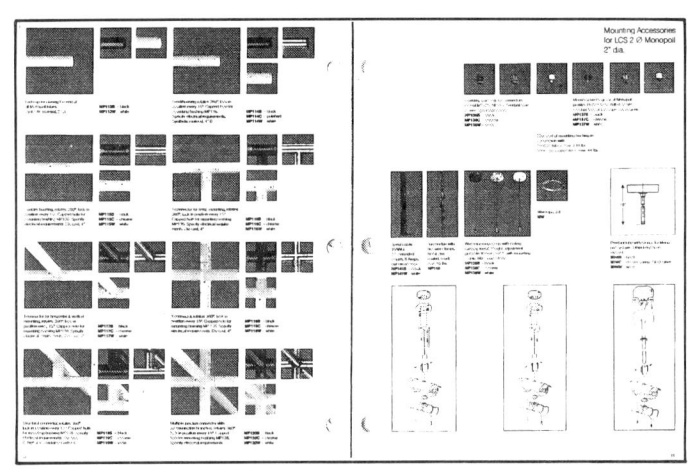
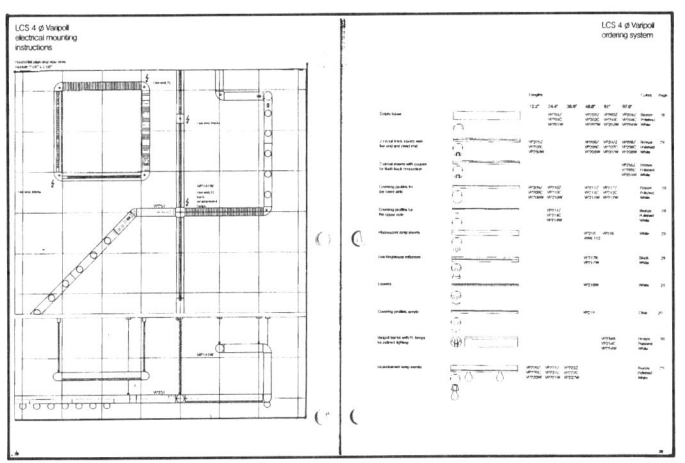
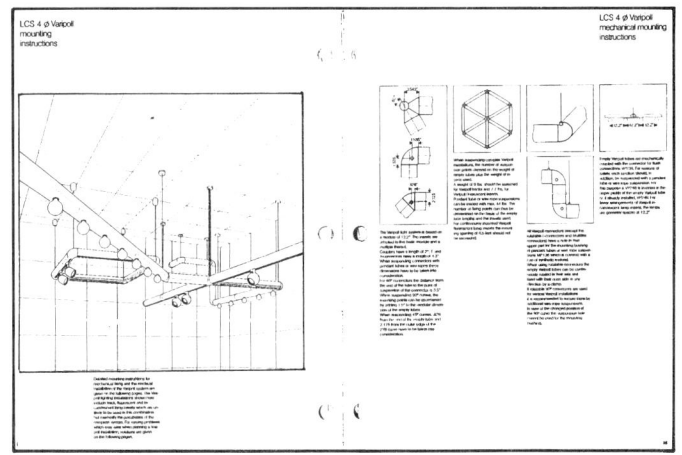
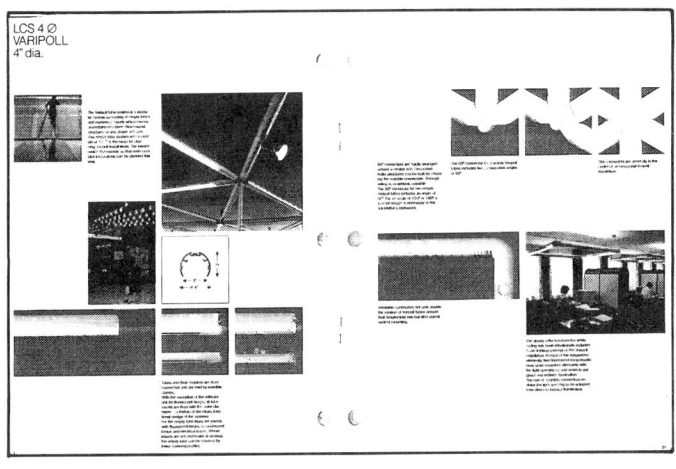
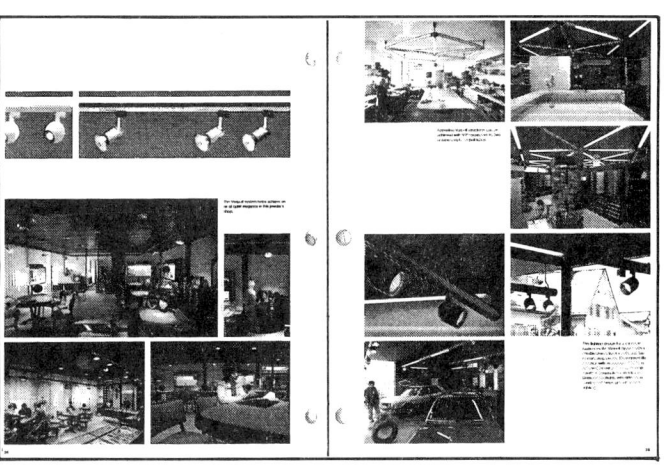

Case Study Formats 171

## Case Study 57

This firm's angular logo design was the starting point in the development of this grid. The logo appears on the cover in a reverse shadow outline. The grid is most apparent from the angular display of photographs and arrangement of type. Each column of type has a solid color bar over it to help separate it from the color photos. This is a creative grid pattern which may have far more applications on a white background, where the text would be easier to read.

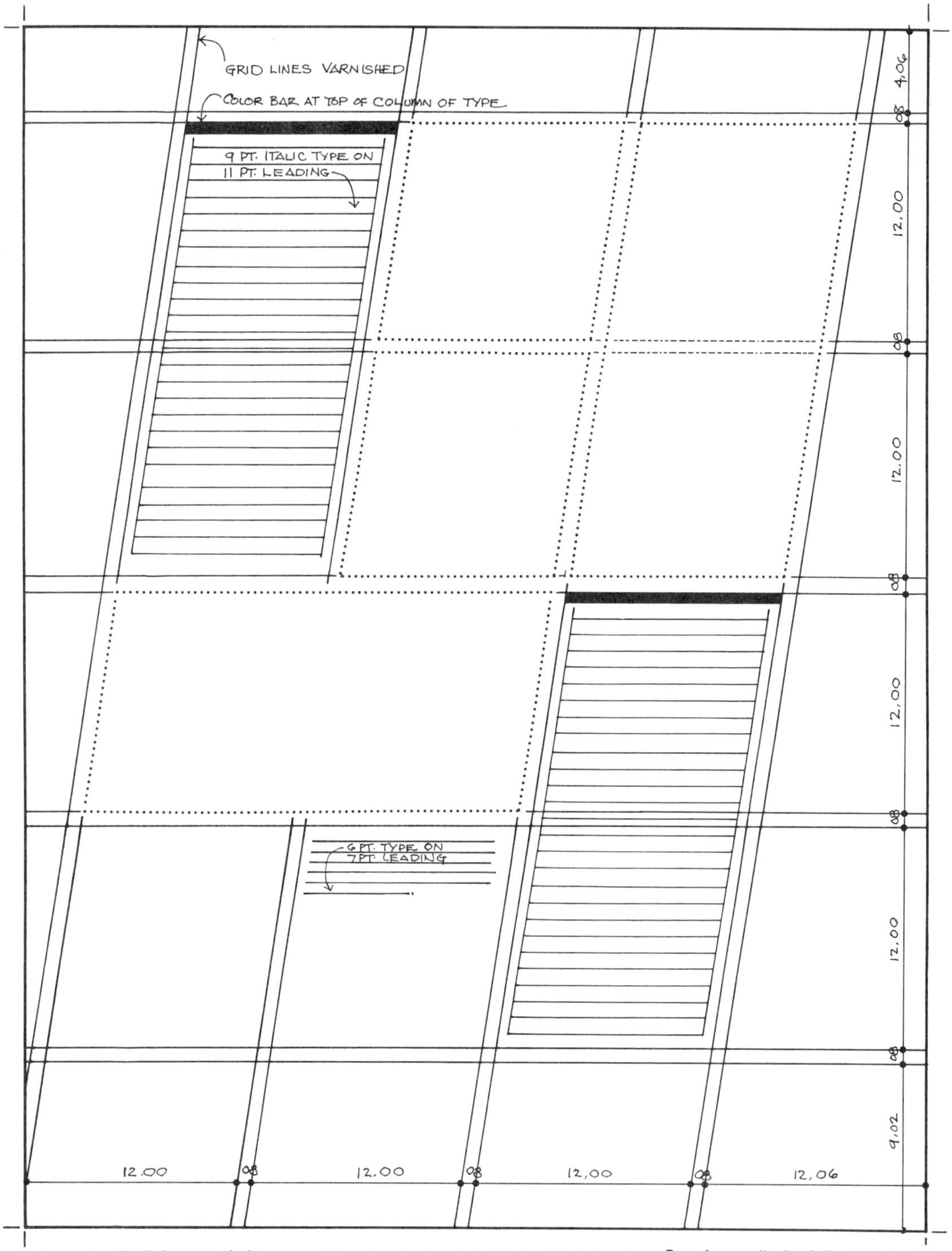

Enlarge to 154% for actual size    Trim size 8 1/2 x 11" (51.3 x 66.4 picas)    See Appendix for full spread grid

## Case Study 58

This grid is designed on an angle of 78 degrees. It is a simple three column grid using two of the main columns of text to describe the product, and one column for other headline information. The angular feature is derived from the firm's logo, which is a lowercase italic typeface. The pages shown are single sheet advertisements, and the reverse side has a standard three column grid using traditional vertical columns.

Enlarge to 154% for actual size   Trim size 8 1/2 x 11" (51.3 x 66.4 picas)   See Appendix for full spread grid

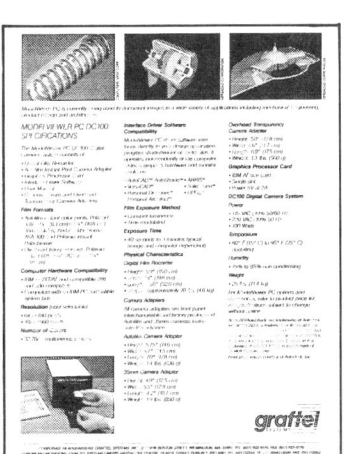
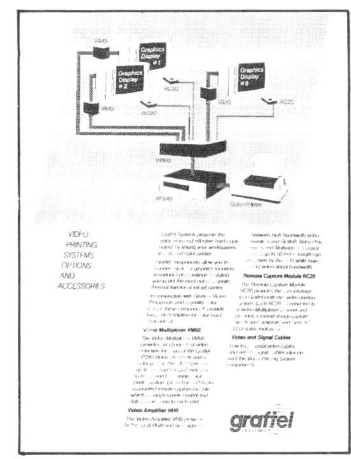

Case Study Formats 173

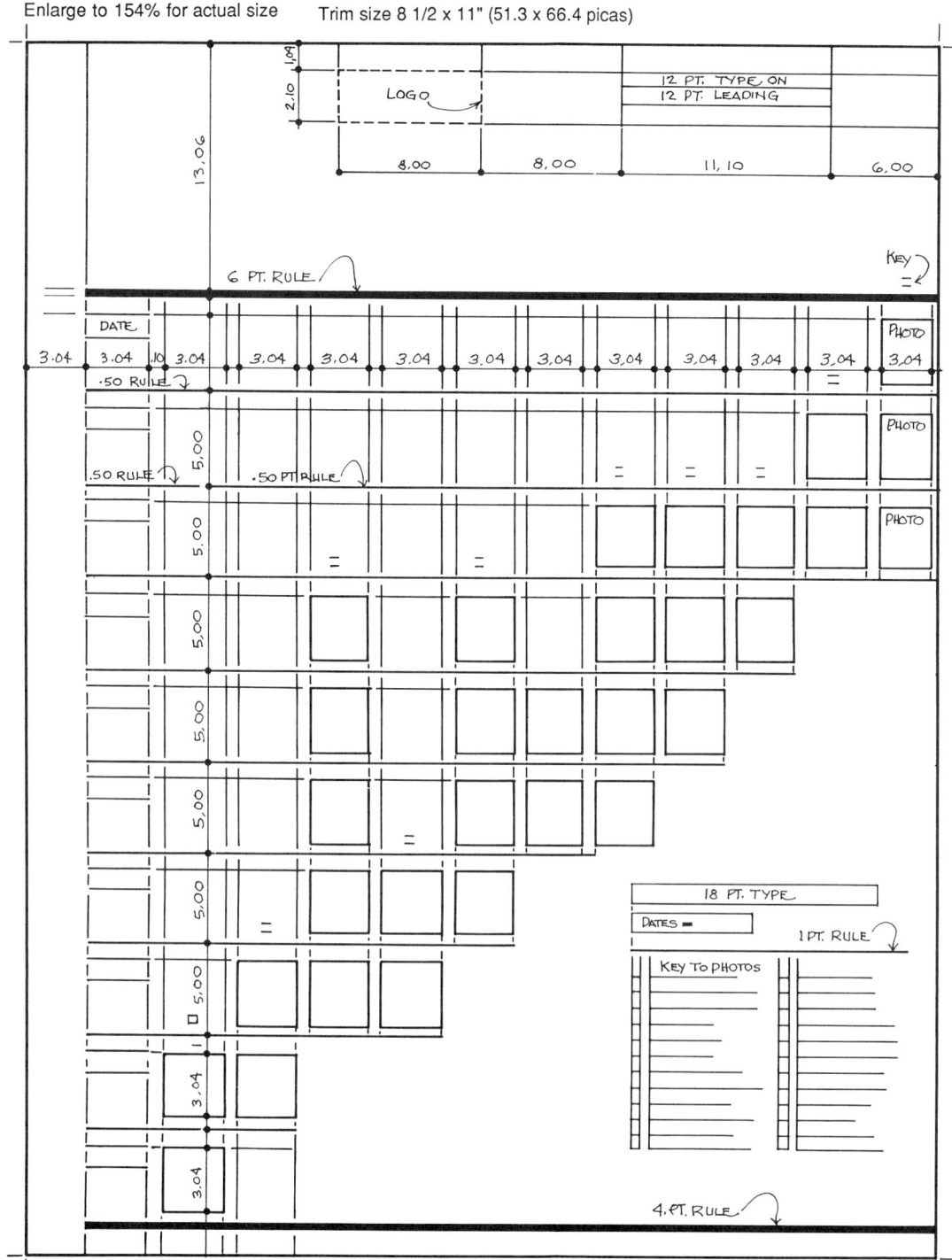

## Case Study 59

The design of this piece is an ingenious way to show the lineage of a design firm. It does so both graphically, statistically and photographically. By repeating the small head shot of each partner throughout the yearly divisions, one can clearly see how the firm evolved over ninety years to become a multidiscipline firm with many partners. Each person is keyed to the grid by numbers appearing above each photo. The numbers are then keyed to the list at the lower right of the chart. When folded, the piece measures 81/2" x 11".

## Case Study 60

This piece was designed to celebrate a fifteenth anniversary, as portrayed by the cover graphics. It is a four panel gatefold design wherein each panel has a four column grid for photographs separated by wide margins. Across the top of the page is a chronological dateline and listing of projects. Across the bottom is a series of square photographs in vertical columns. These step up as the number of projects increased over the years.

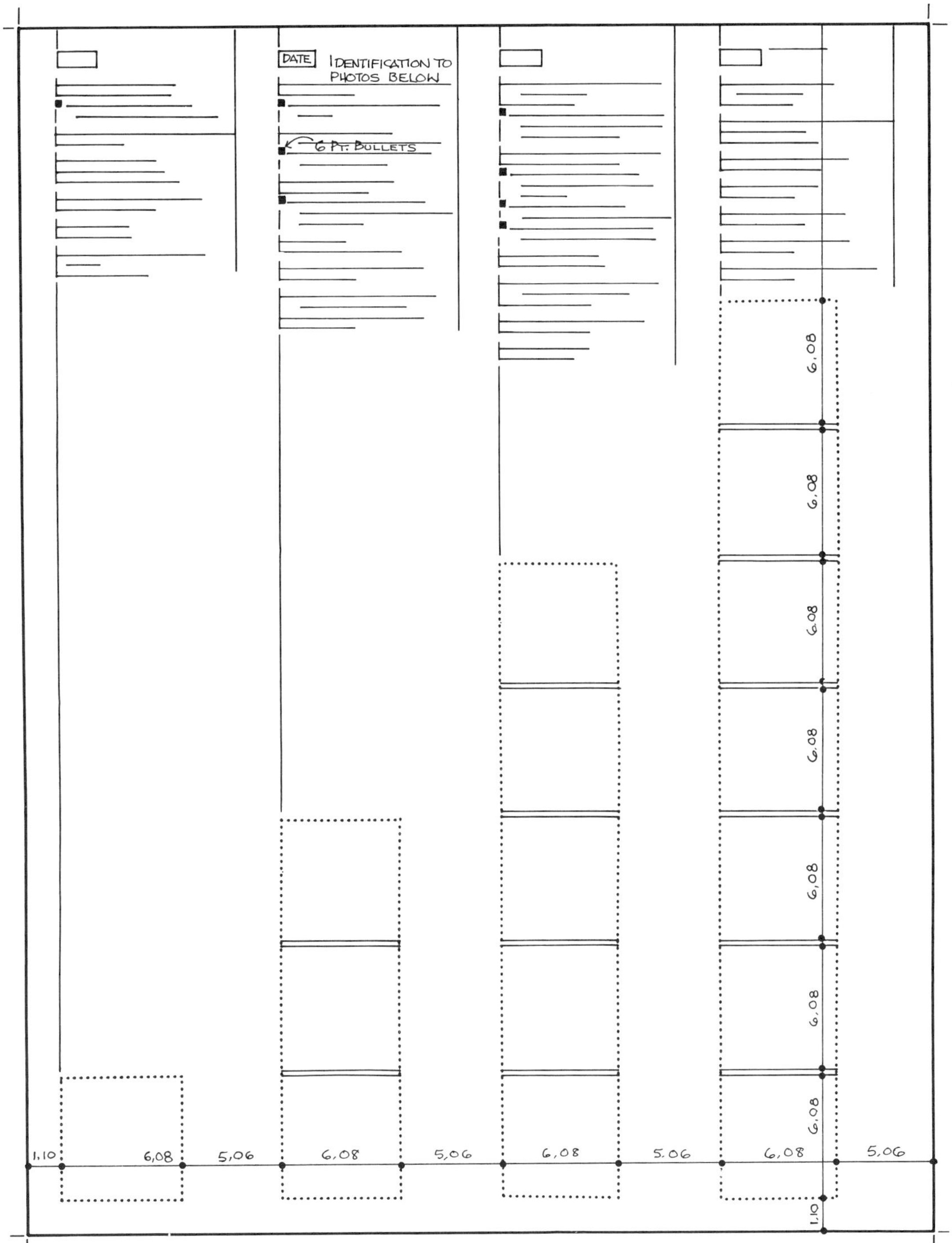

Enlarge to 154% for actual size   Trim size 8 1/2 x 11" (51.3 x 66.4 picas)   See Appendix for full spread grid

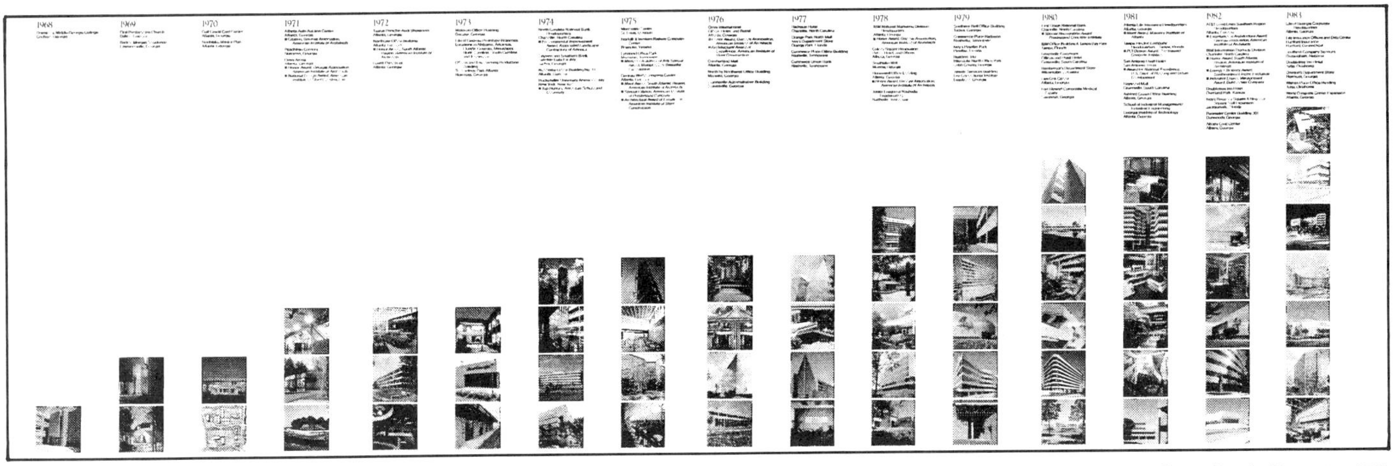

Case Study Formats 175

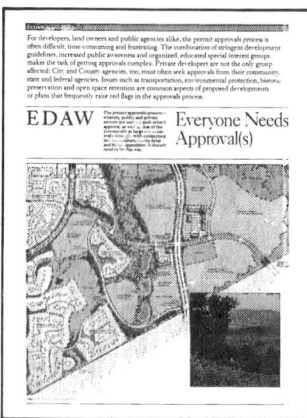

### Case Study 61

The concept for this grid is quite simple, yet complex and varied in application. It is a combination of different grids, tiered vertically. On the cover it is used as a one column grid. The single column is split into two equal columns in the second tier, which is split again into four columns in the third tier. A thin rule is used to separate these tiers. Under this rule is the fourth tier, which is a three column grid. Each of these columns is split in half, making a six column grid at the bottom.

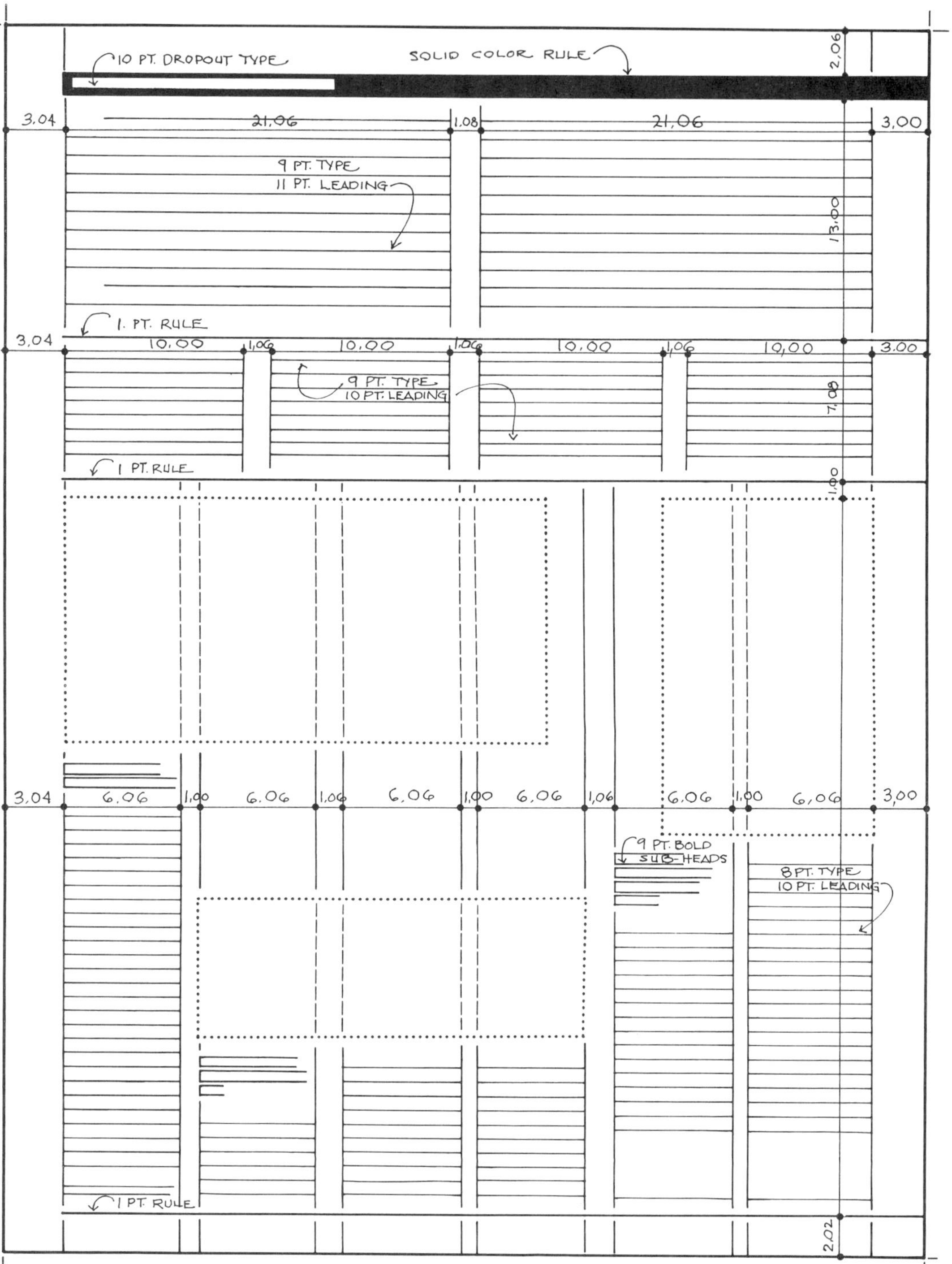

Enlarge to 154% for actual size   Trim size 8 1/2 x 11" (51.3 x 66.4 picas)   See Appendix for full spread grid

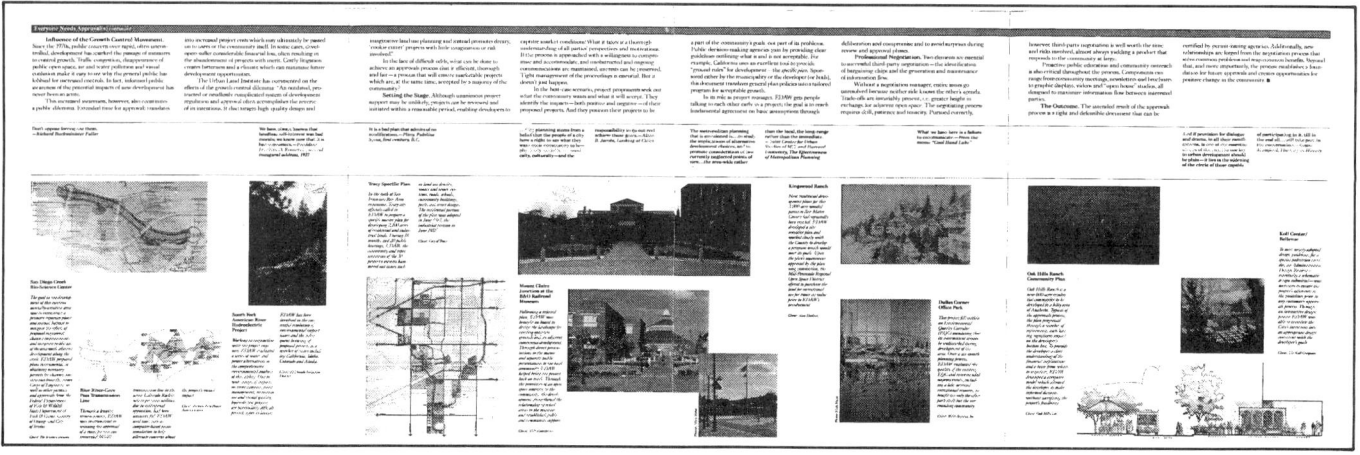

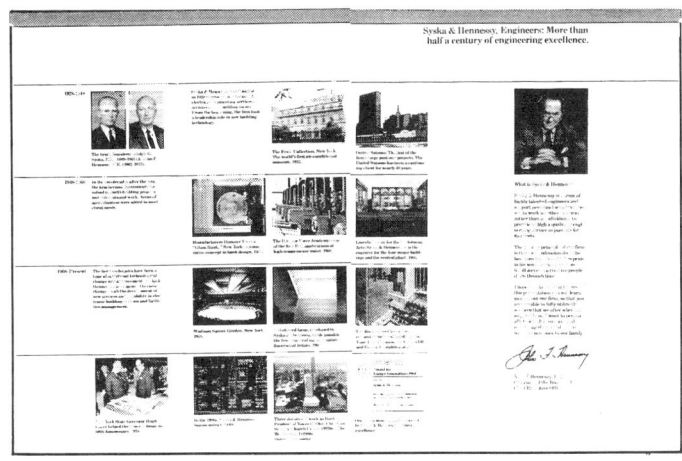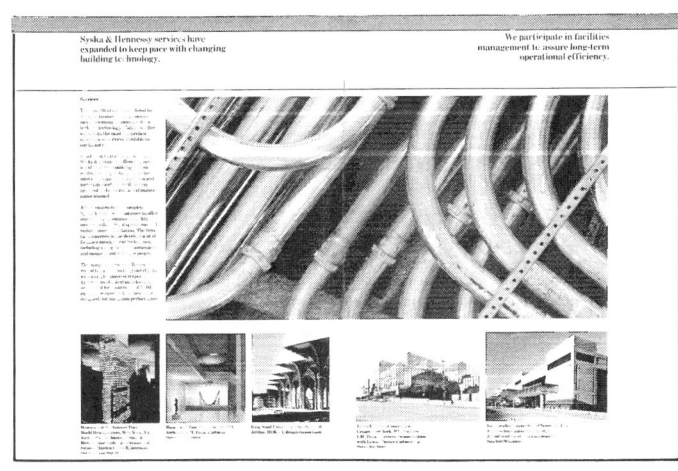

Enlarge to 154% for actual size    Trim size 8 1/2 x 11" (51.3 x 66.4 picas)    See Appendix for full spread grid

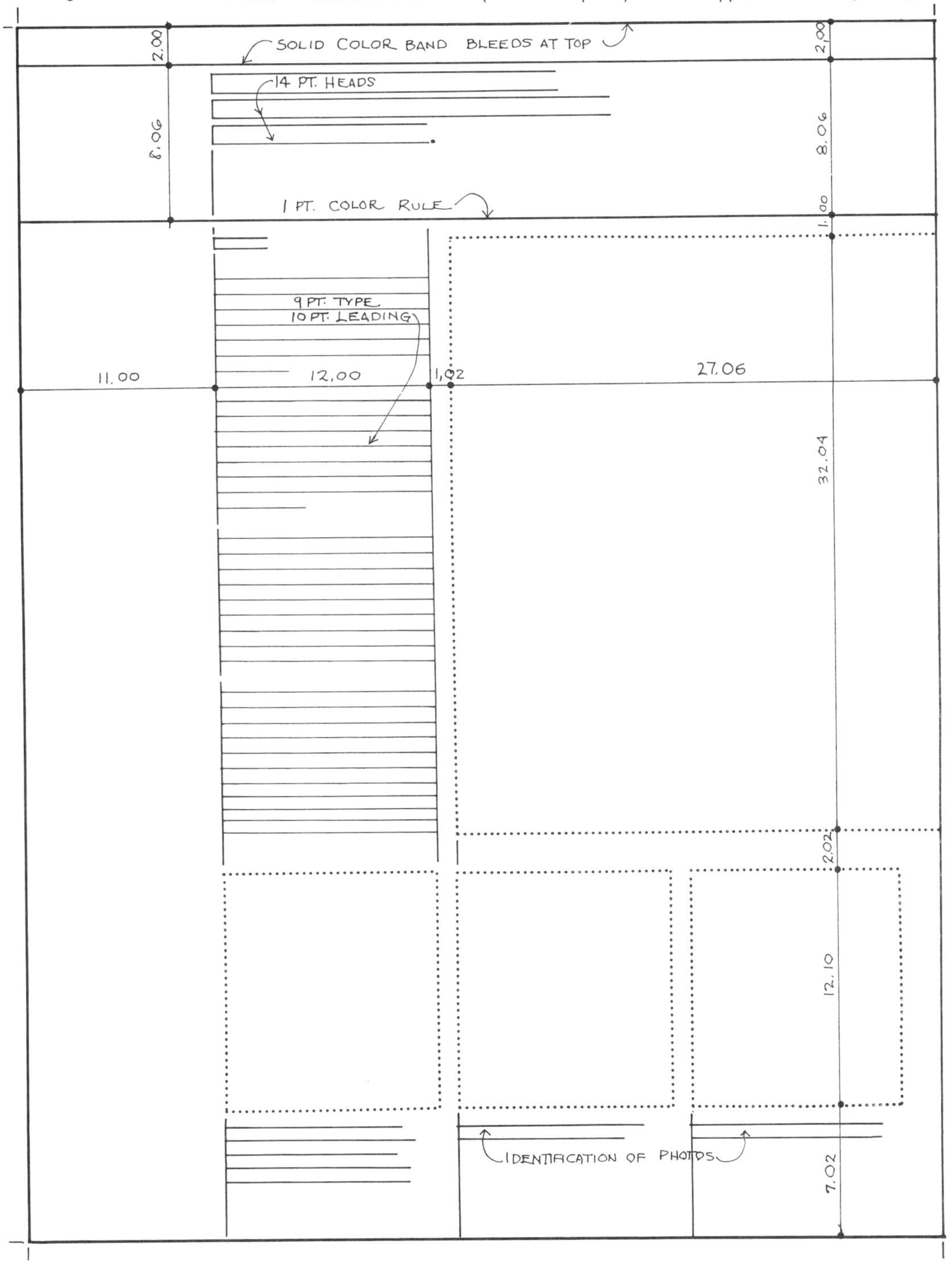

## Case Study 62

A five column grid can be very useful in situations where you want to show a lot of small photos. Here the application was for a series of photos of key people in the firm and for the list of clients. Larger photos can always be included. In fact, it helps to offset the monotony of a lot of small pictures. Other photos can be added at any place in the grid.

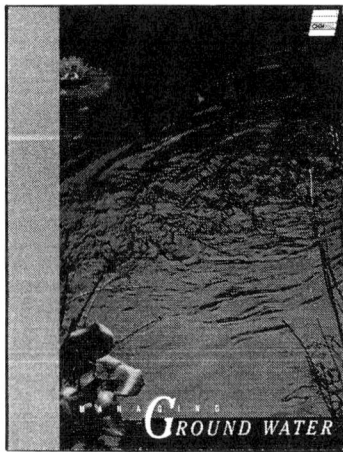

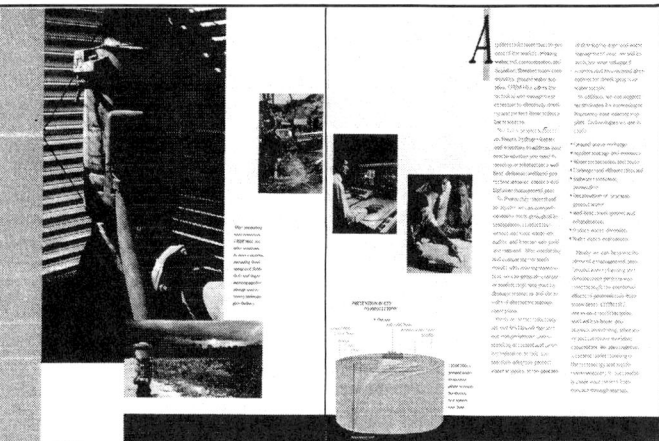

## Case Study 63

There is little on the cover of this specialty brochure to indicate the imaginative layout within, except for the color bars.The wide side bar of solid color and the wide bar at the bottom of the page are included in the page layout. The composition inside is perfectly balanced around the illustration and small photographs. They offset the main large photo on the left page. A large decorative letter begins the two columns of text. Decorative triangular bullets call out items in the illustration.

See Appendix

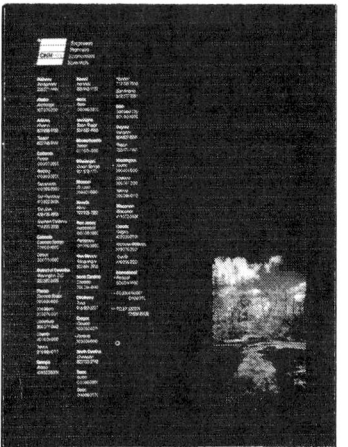

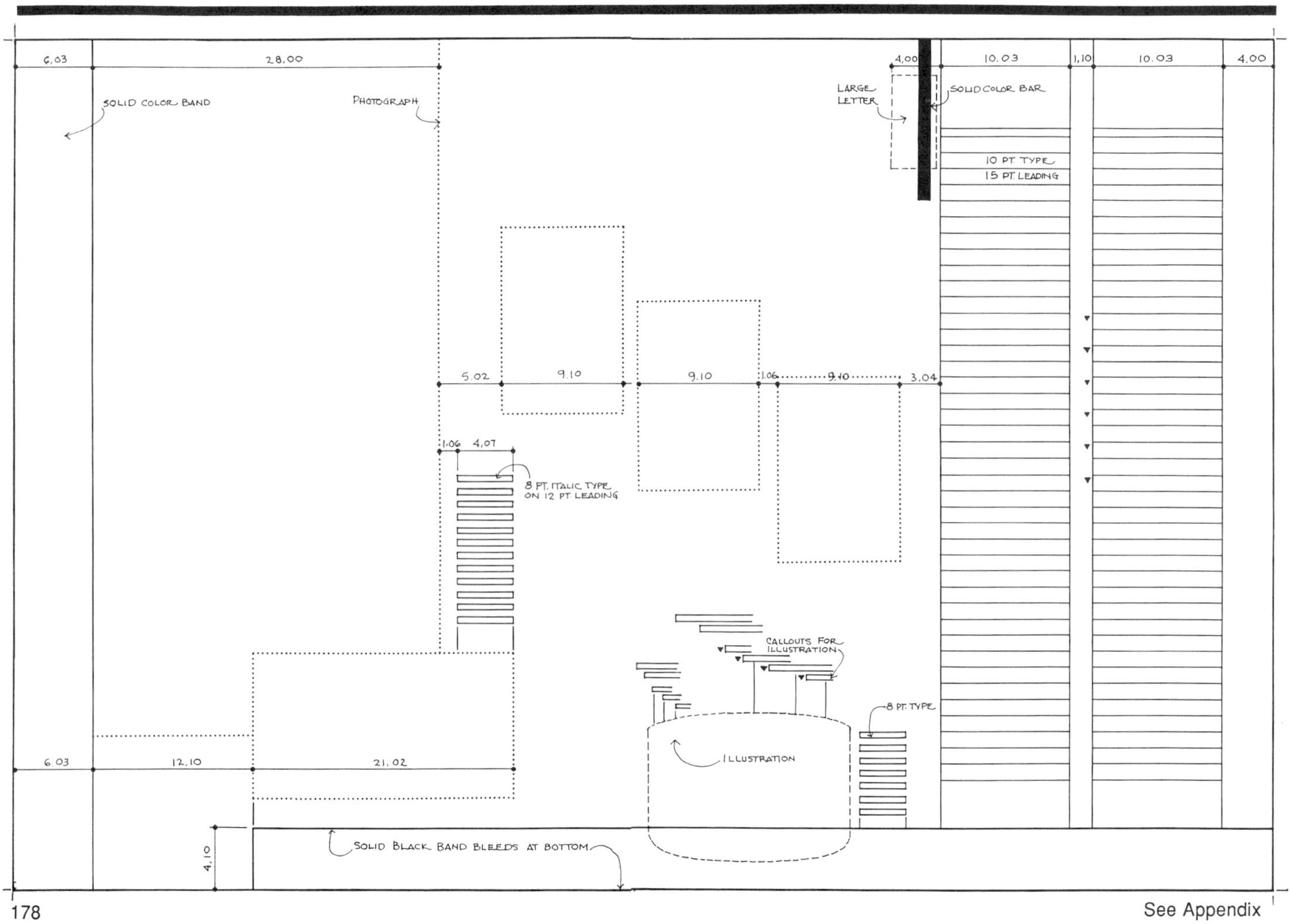

## Case Study 64

The cover of this specialty services brochure strongly influenced the layout and design of the inside spread. The stepped arrangement of core samples on the cover photo is picked up in the stepped layout of photos and text. Inside, the single column of type is headed by a large decorative letter, positioned in front of a solid color bar. The text is set in a drop-out type against a black background, and consists mainly of a bulleted list of specialty services.

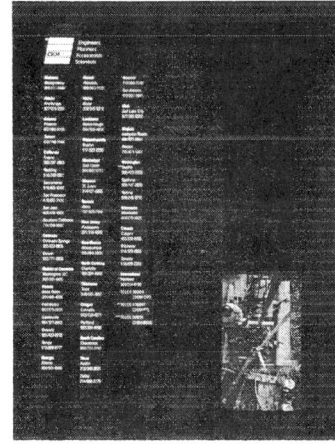

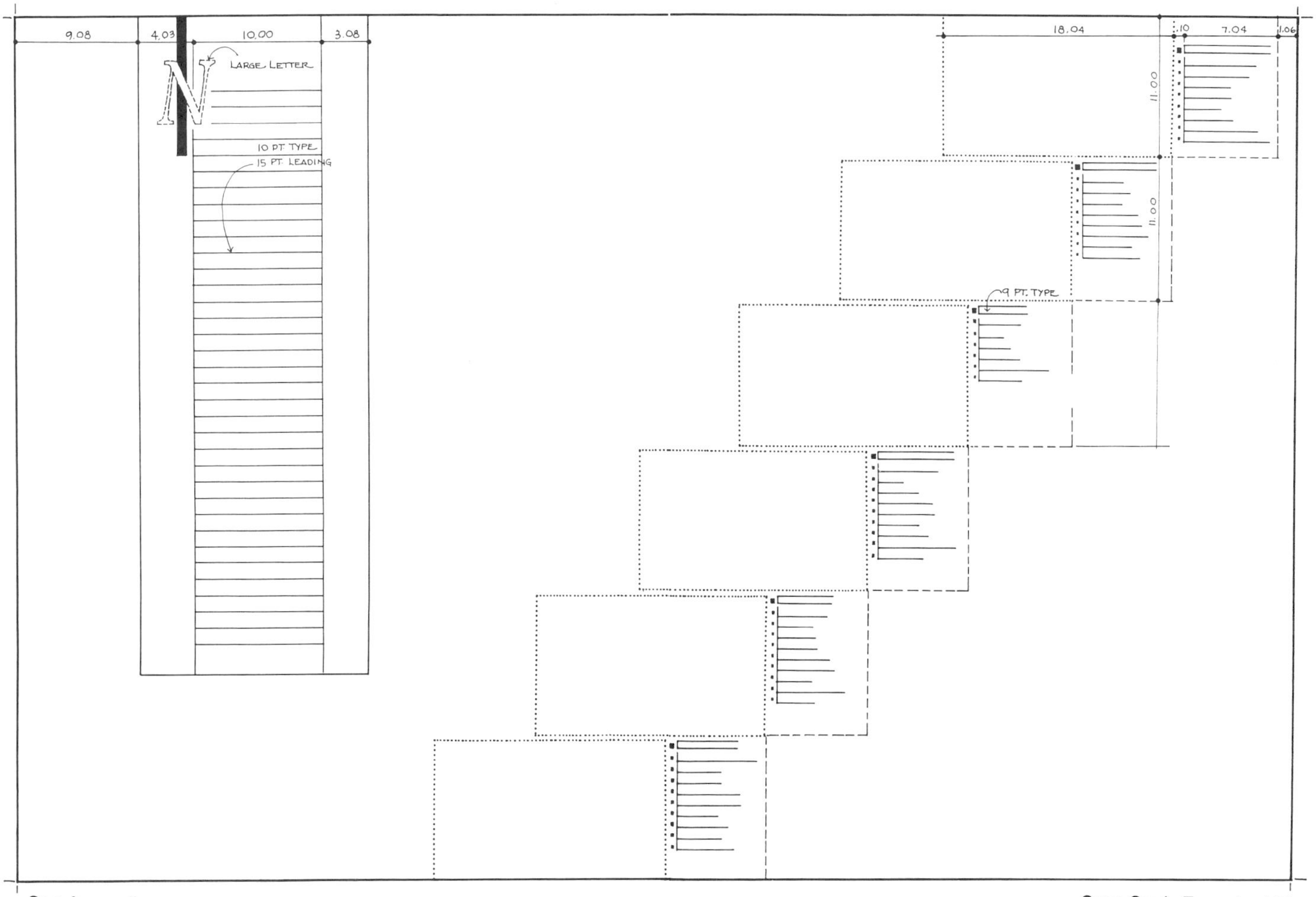

See Appendix

**Case Study 65**

Two elements from the cover are used throughout this striking brochure for an architectural firm. The first is a bright color bar set into the black background, and the second is the use of a theme photo placed on a diagonal bias. The inside of the cover has a wide column of text, but the remaining pages show photos within this grid margin. This looks like a single column format, but in reality it is a five column grid with wide margins between the columns.

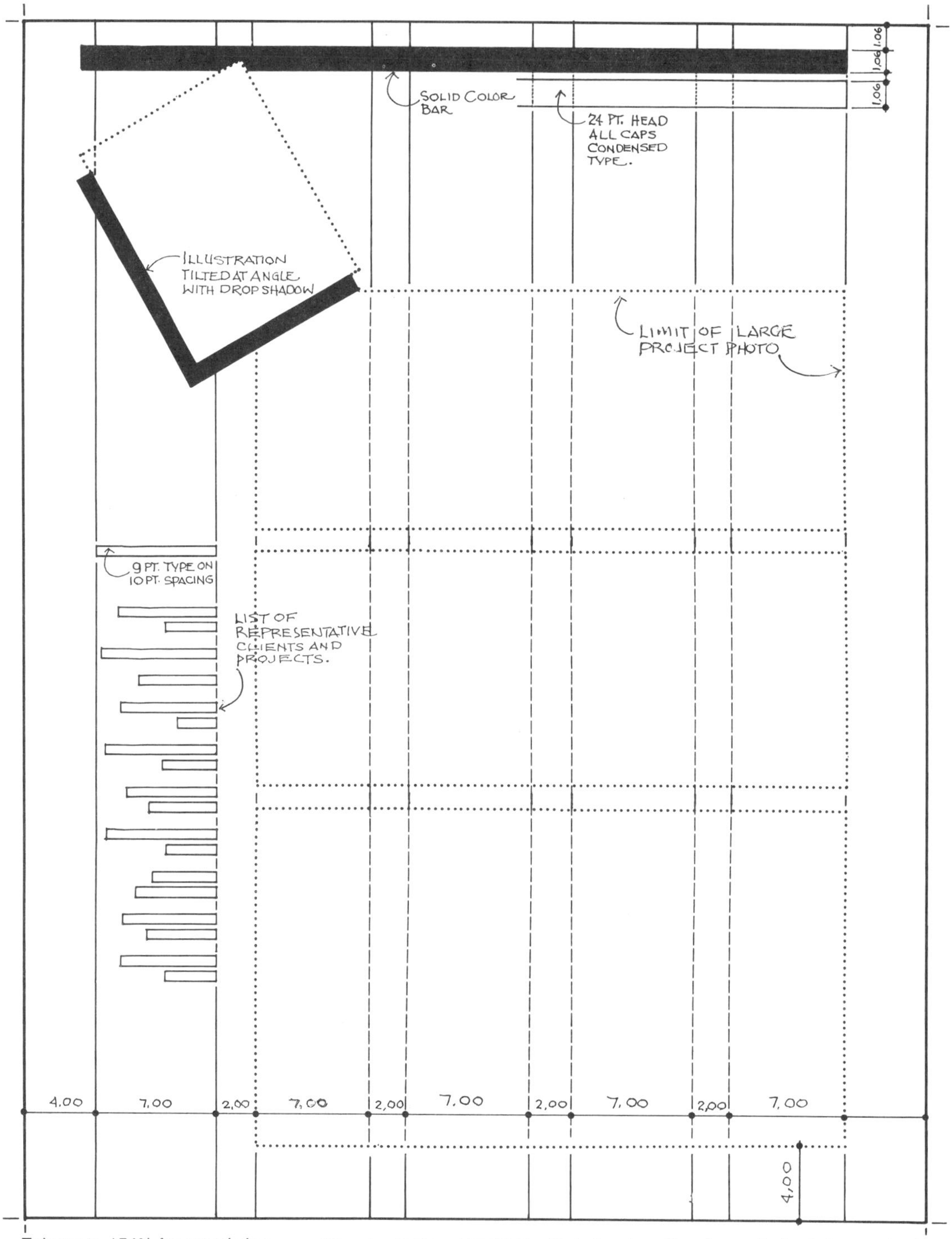

Enlarge to 154% for actual size    Trim size 8 1/2 x 11" (51.3 x 66.4 picas)    See Appendix for full spread grid

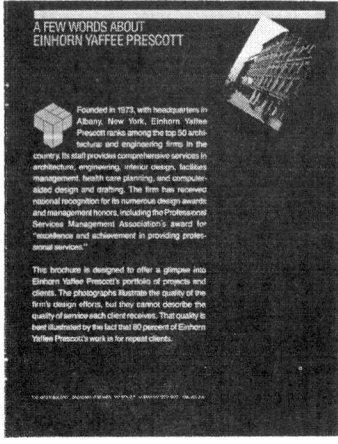

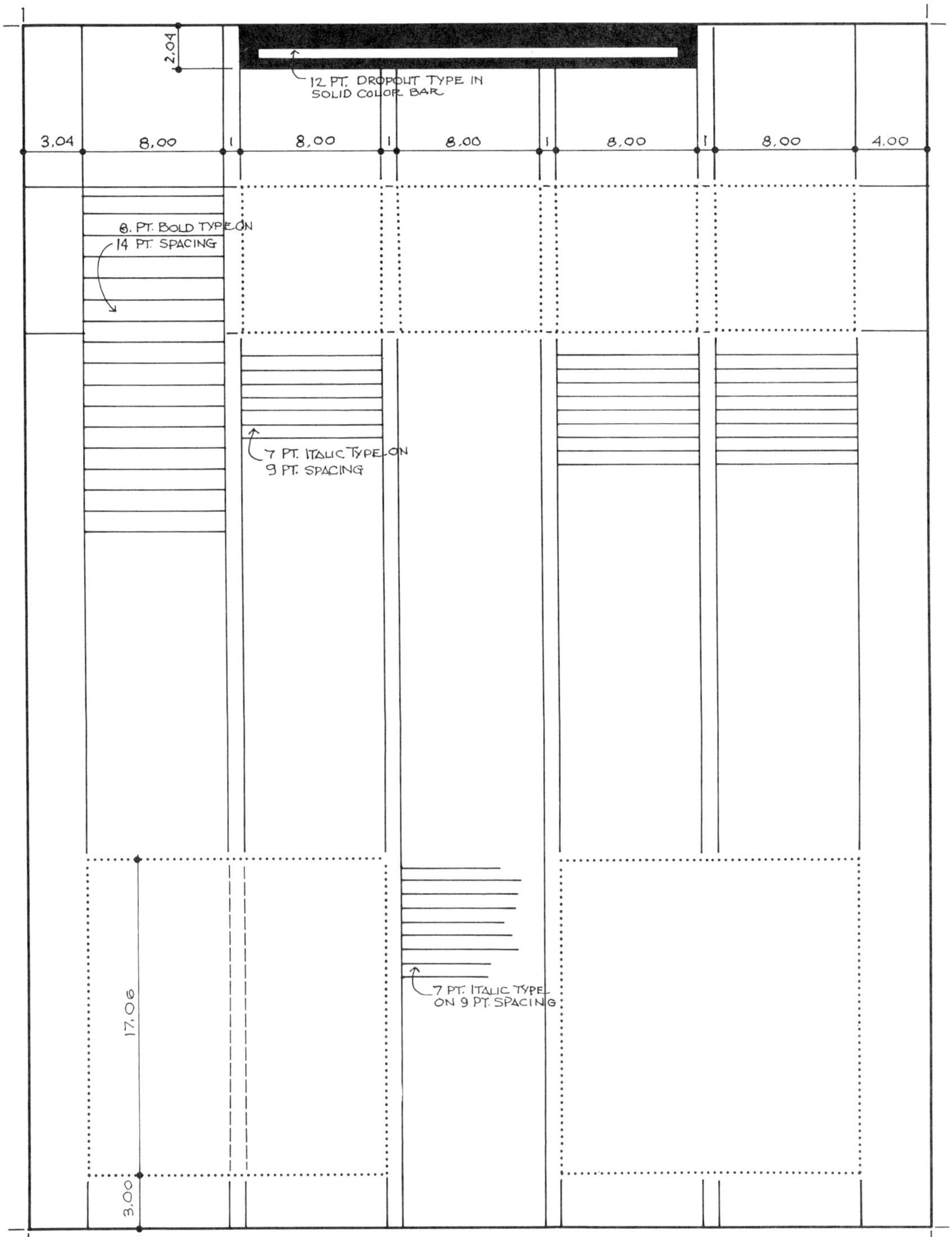

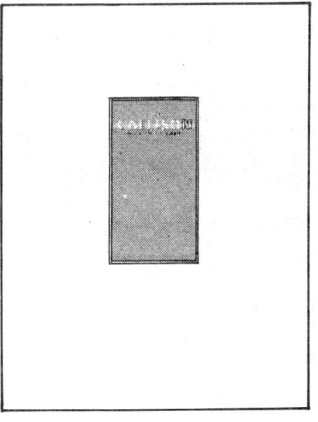

## Case Study 66

The flexible layouts of this brochure stem from the creative use of a five column grid. It is easy to recognize in the illustrated list of clients and projects, but it is harder to define on most of the interior layouts. The columns of text are narrow, and text is limited to minimal project descriptions. These pages have a lively look which stems from the highly creative layout, one which seems to defy its rigid geometry while at the same time respecting it.

Enlarge to 154% for actual size    Trim size 8 1/2 x 11" (51.3 x 66.4 picas)    See Appendix for full spread grid

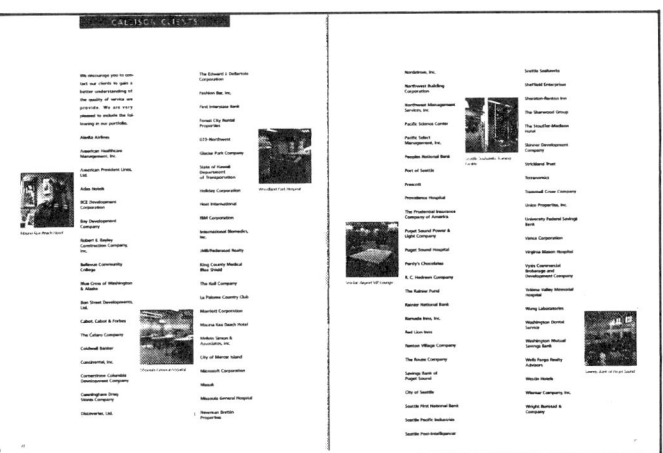

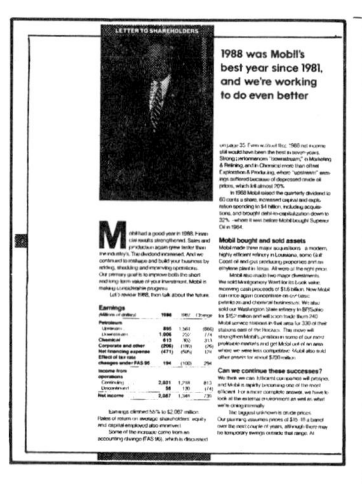

## Case Study 67

The possible combinations of a five column grid are limited only by the imagination. This is expressed in this layout for an annual report. At the top it used a five column grid to highlight different aspects of the company's services. Below, the two outer columns were combined, leaving a small center column for bold sub-heads, small photos or just blank space. The solid color bar at the top of the page is in line with the other color bars on the page, one in the middle and one at the bottom.

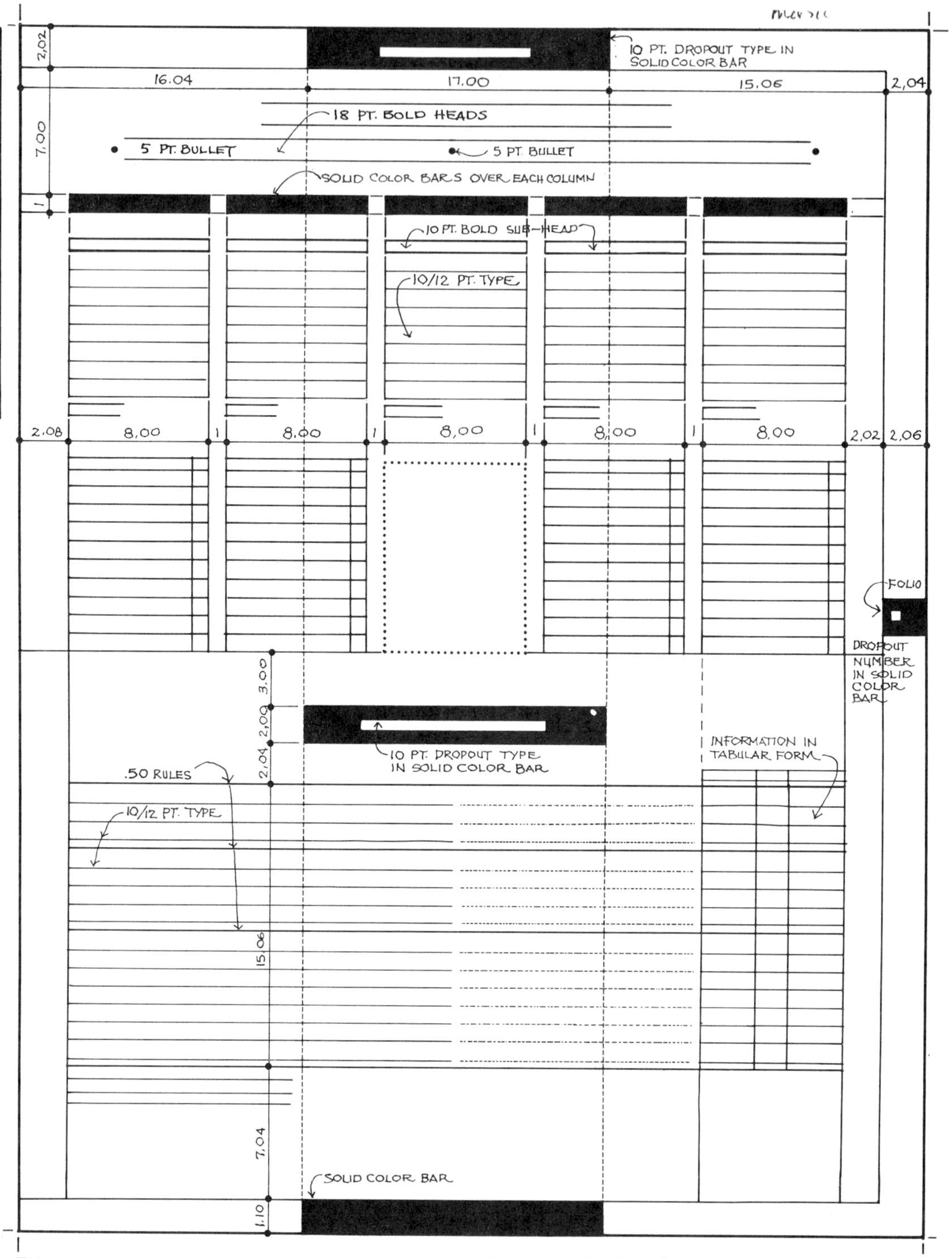

Enlarge to 154% for actual size   Trim size 8 1/2 x 11" (51.3 x 66.4 picas)   See Appendix for full spread grid

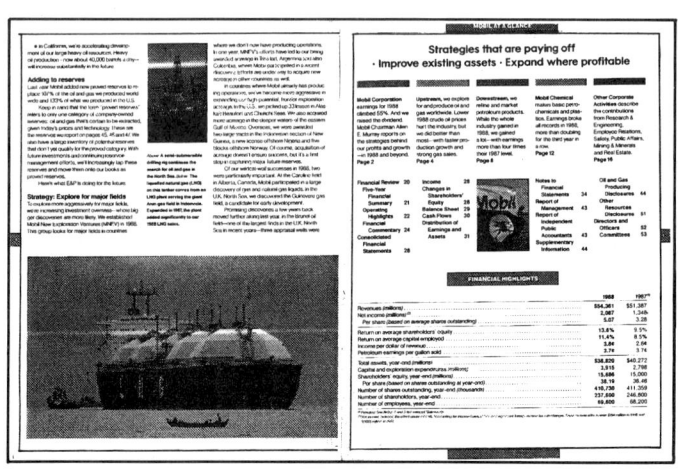

182

## Case Study 68

Each spread of this brochure follows the identical format. It is based on a three column grid for the left page and a two column grid for the right page. There is a single column of text on the left page and a large photo covering the majority of the spread. This photo is always a closeup detail of some aspect of the firm's engineering services. Under this large photo are three smaller vertical ones on the left page and two smaller oblong ones on the right page.

Enlarge to 154% for actual size   Trim size 8 1/2 x 11" (51.3 x 66.4 picas)   See Appendix for full spread grid

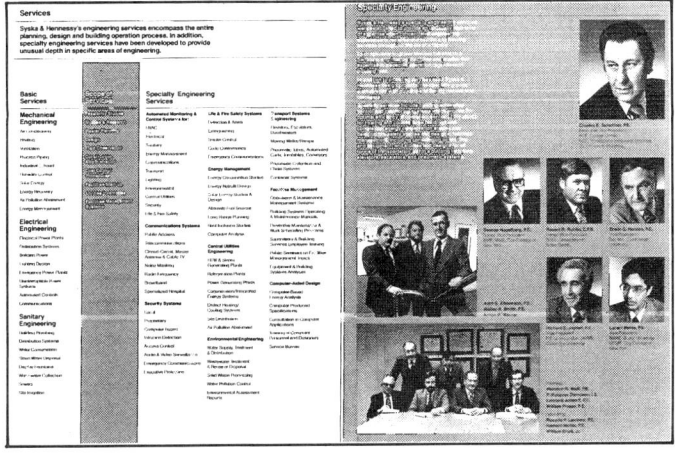

Case Study Formats 183

### Case Study 69

The elements that make up this grid are initiated on the cover. The most noticeable one is the solid black bar. The thinner rules below delineate the horizontal position of the grid lines inside. Few brochures have the horizontal grid expressed as clearly as this one. It shows up as a 6 point rule of gloss varnish printed over the white background. The photos are cropped to fit within this horizontal grid. There is also a corresponding five column vertical grid.

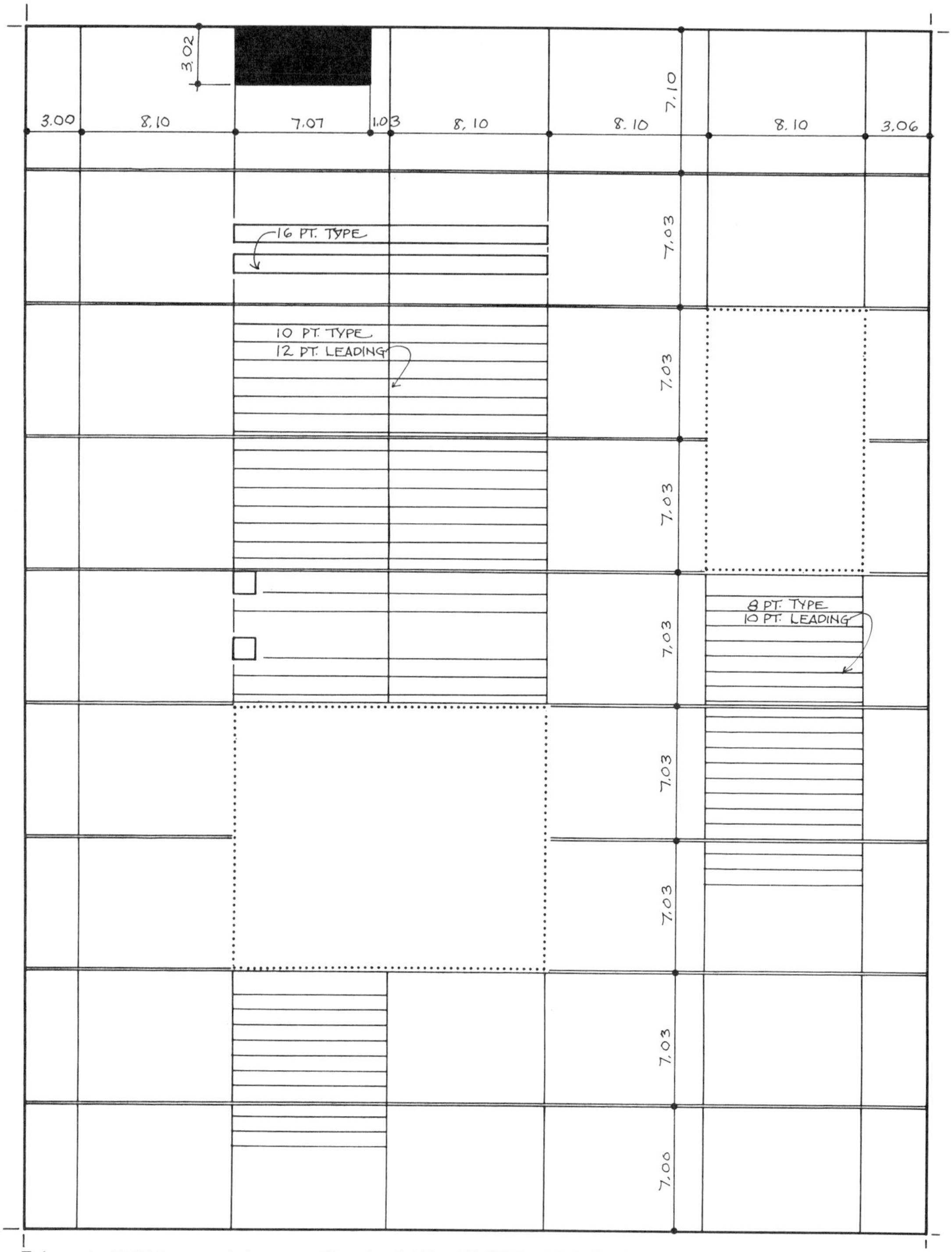

Enlarge to 154% for actual size   Trim size 8 1/2 x 11" (51.3 x 66.4 picas)   See Appendix for full spread grid

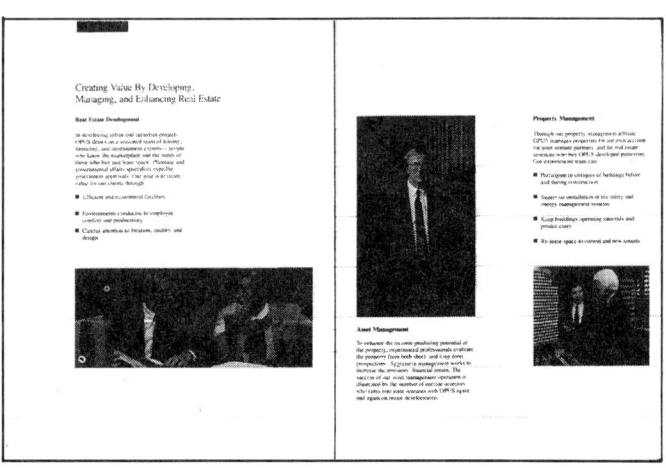

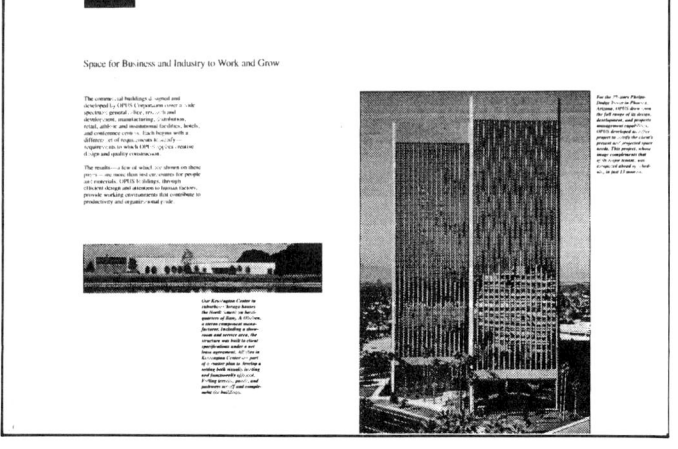

184

## Case Study 70

A designer must have a well designed brochure to showcase his or her services. That is the case with this brochure, which not only highlights the firm's own office but other projects as well. The page is divided into square grids, 3 across and 11 down, and the elements relate to these grid lines in a very deliberate manner. What may appear as a carefree page layout is actually one that has been thoroughly studied. The end result is a casual deliberateness that is visually rewarding.

Enlarge to 154% for actual size    Trim size 8 1/2 x 11" (51.3 x 66.4 picas)    See Appendix for full spread grid

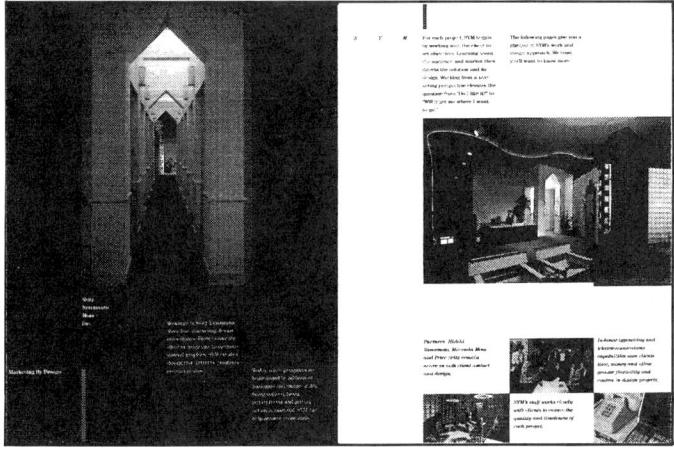

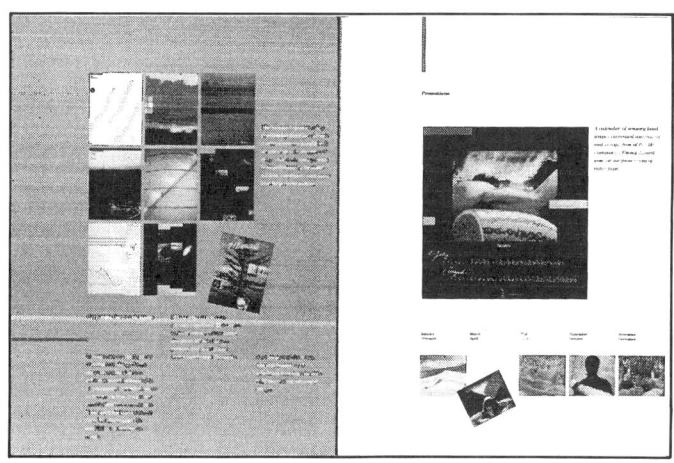

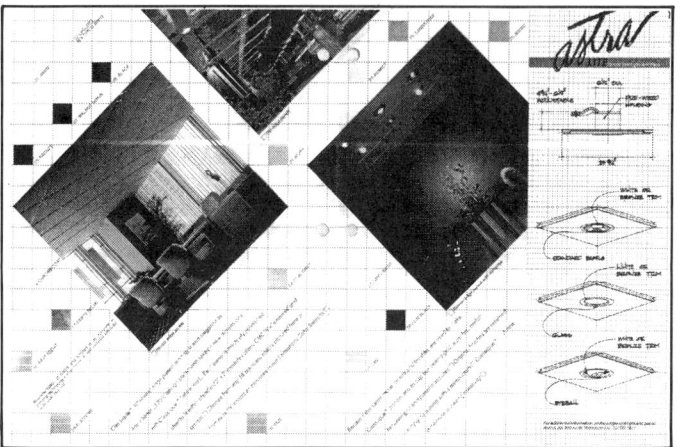
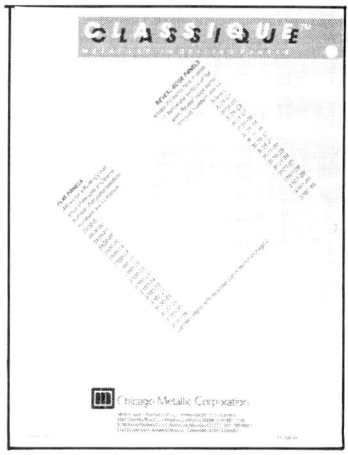

### Case Study 71

The cover portrays the theme of this catalog by the use of a square photograph positioned at a 45 degree angle to the edges of the page. This same design continues on the back cover. Inside, all photos within the rectilinear grid are tipped at the same 45 degree angle. All the descriptions of the products follow this pattern. There is a smaller rectilinear grid in a light color which contains color samples of the product.

Trim size 8 1/2 x 11" (51.3 x 66.4 picas)     Enlarge to 154% for actual size

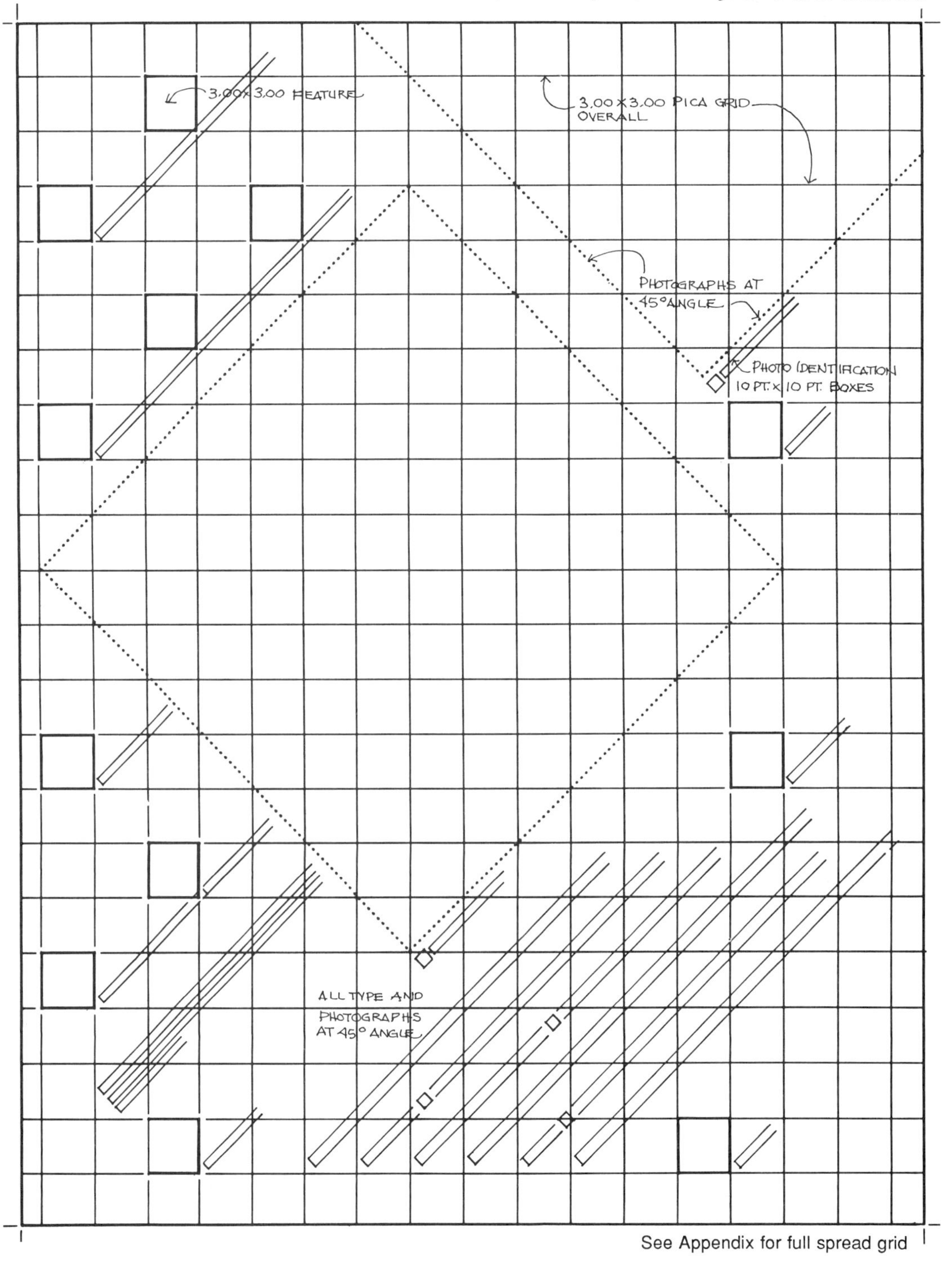

See Appendix for full spread grid

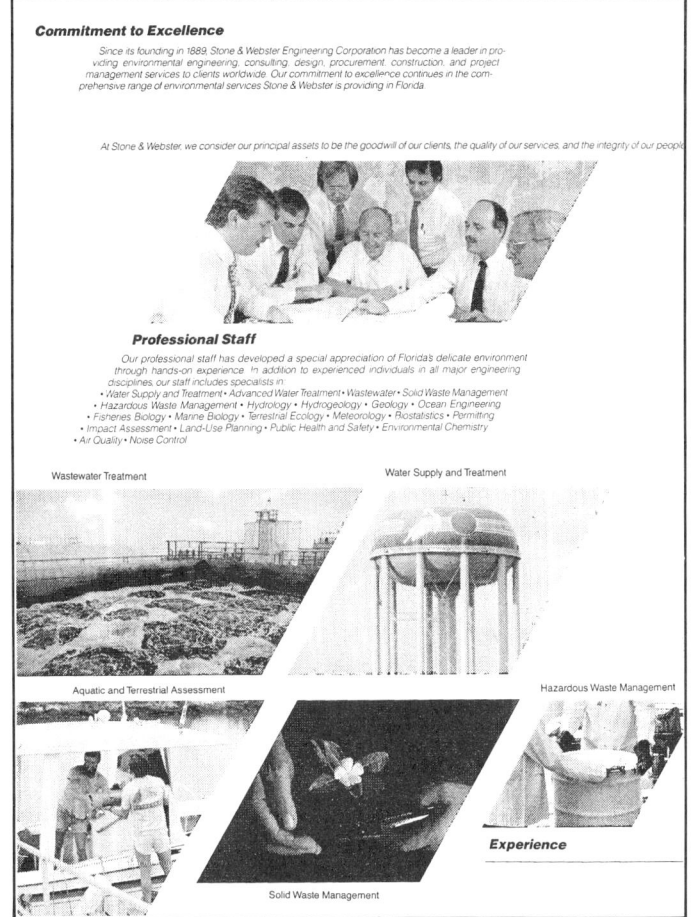
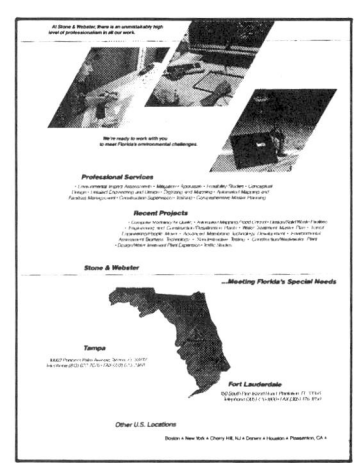

This engineering firm designed one of its services brochure on an angular pattern. This pattern was derived from the firm's logo, which consists of an equilateral triangle. There is not an angular grid that governs the position of photographs and text; there is only an angular layout. All the elements conform to this angle whether they line up with other elements or not. Each element is aligned visually within the angular framework. The back cover uses the same angular pattern.

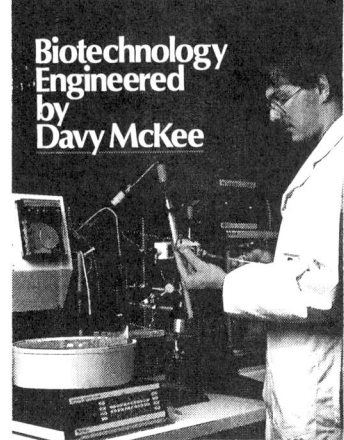
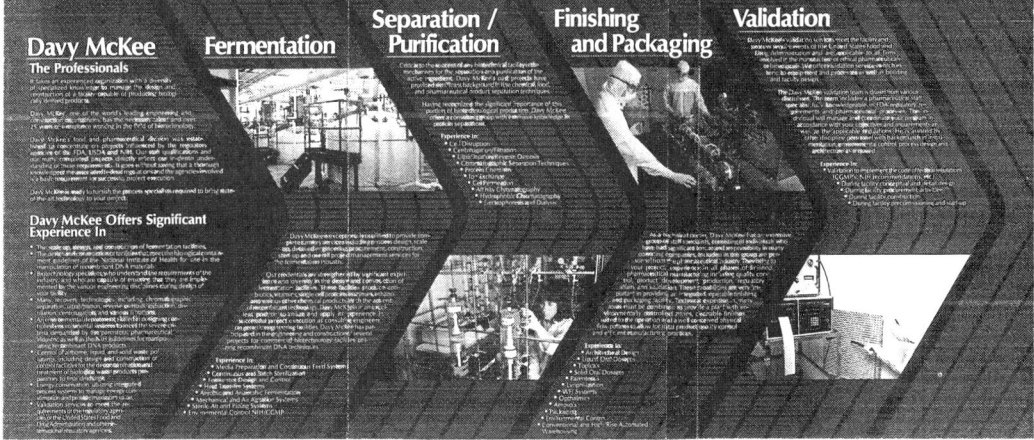
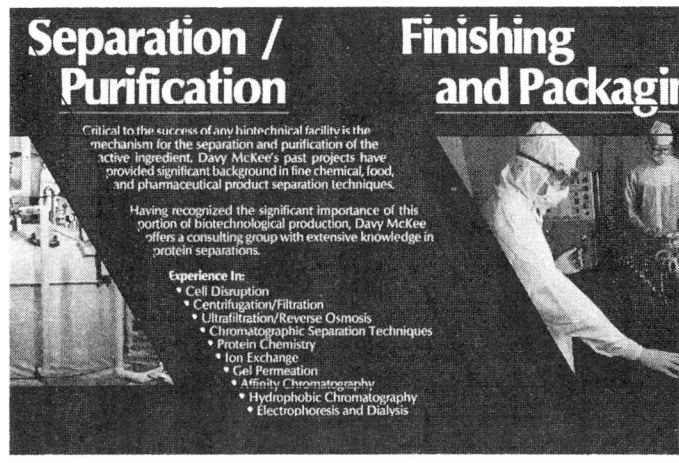

This grid design consists of an acute angular pattern above the center of the page, and a mirror image obtuse angle on the bottom, joined in the center by a rounded curve. The angular pattern of text, plus the reverse type over the complicated patterns, makes it extremely difficult to read. It is also very complex visually, particularly when all three panels are unfolded.

Case Study Formats 187

# 5 Appendix

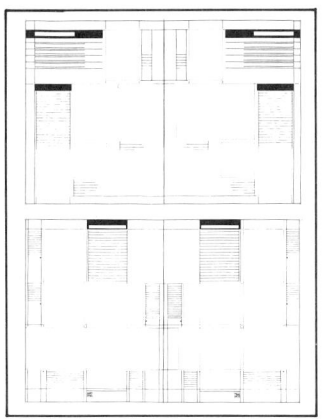

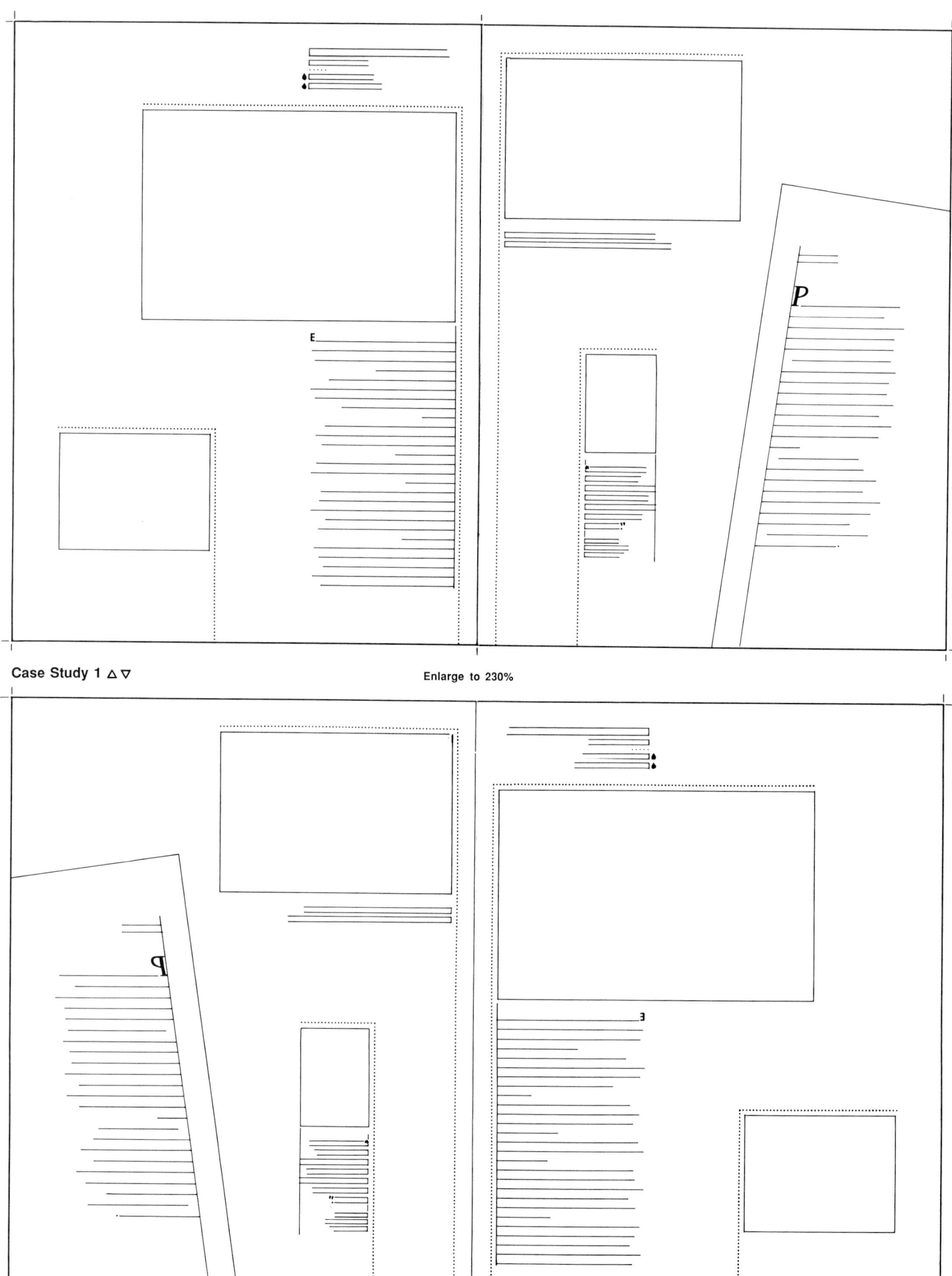

Case Study 1 △▽   Enlarge to 230%

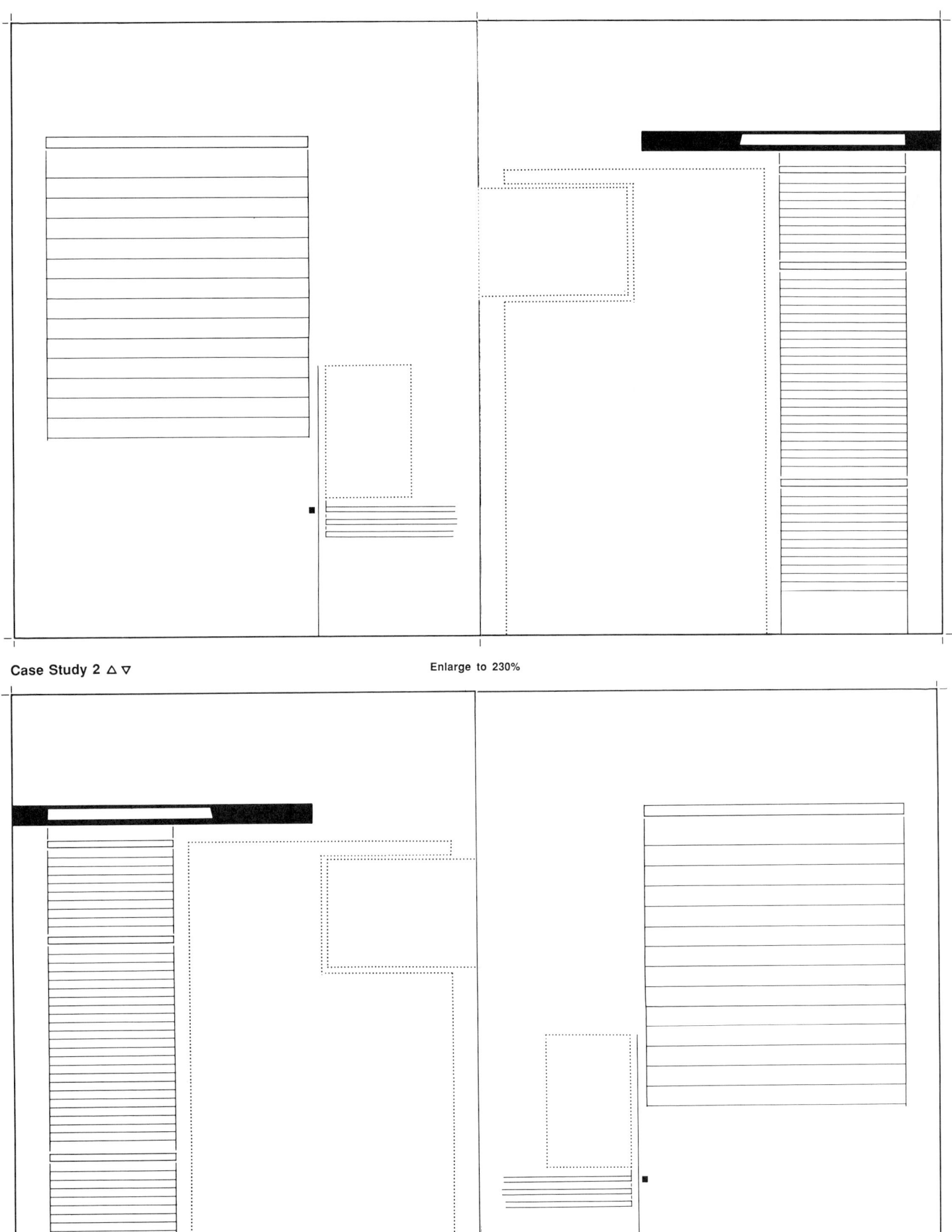

**Case Study 2** △ ▽     Enlarge to 230%

Appendix 191

Case Study 3 △ ▽     Enlarge to 230%

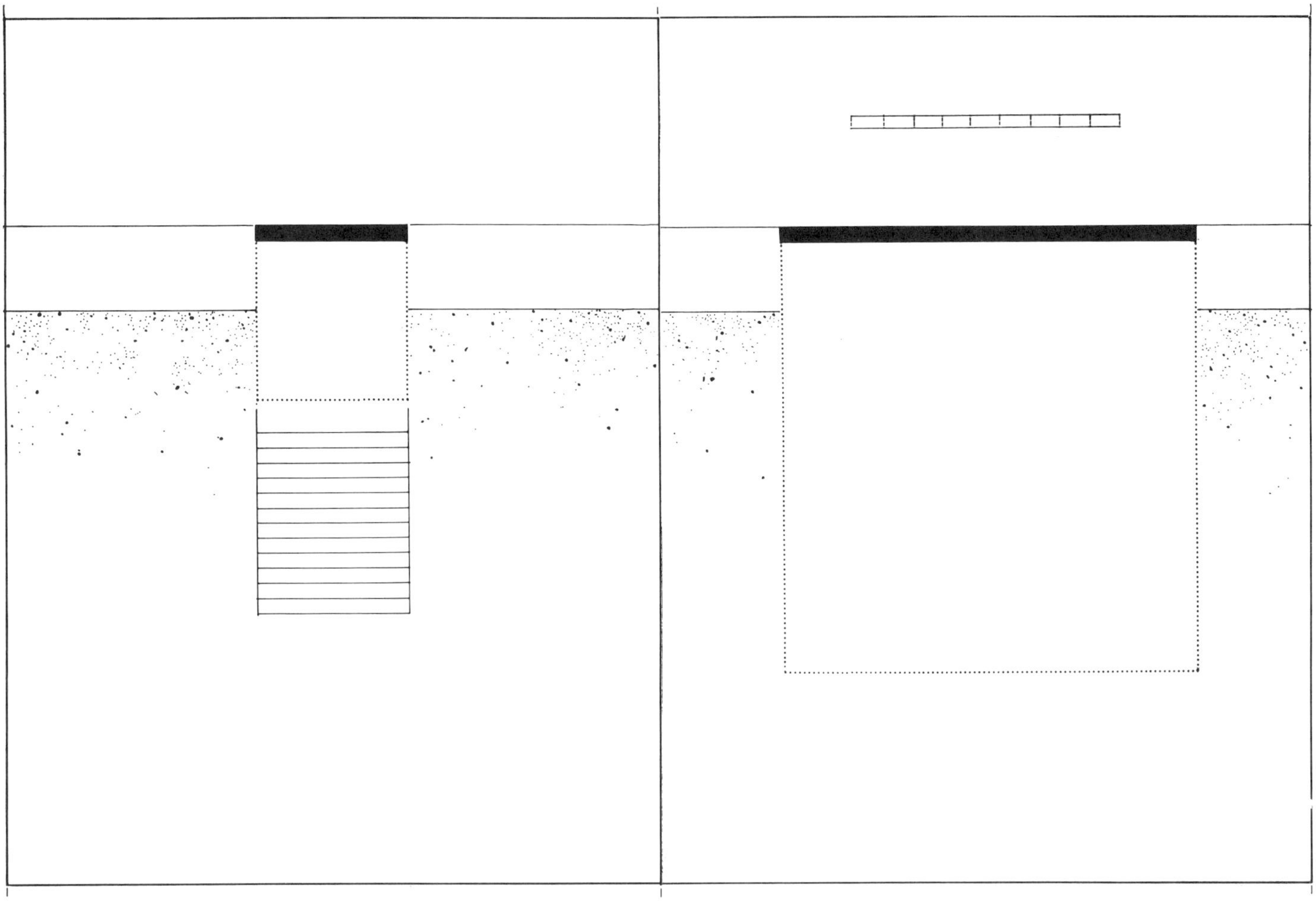

**Case Study 4** △▽

Enlarge to 230%

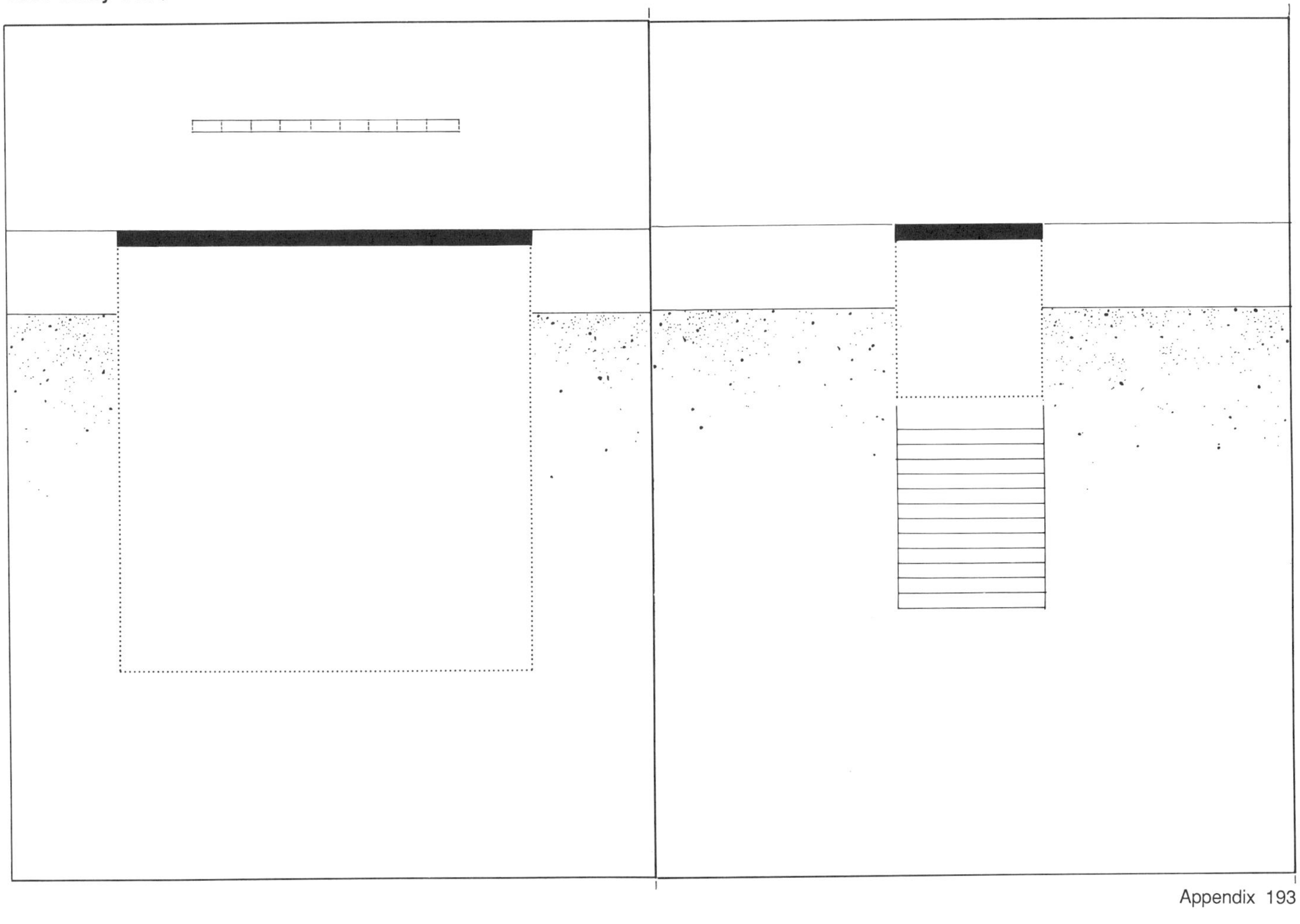

Appendix 193

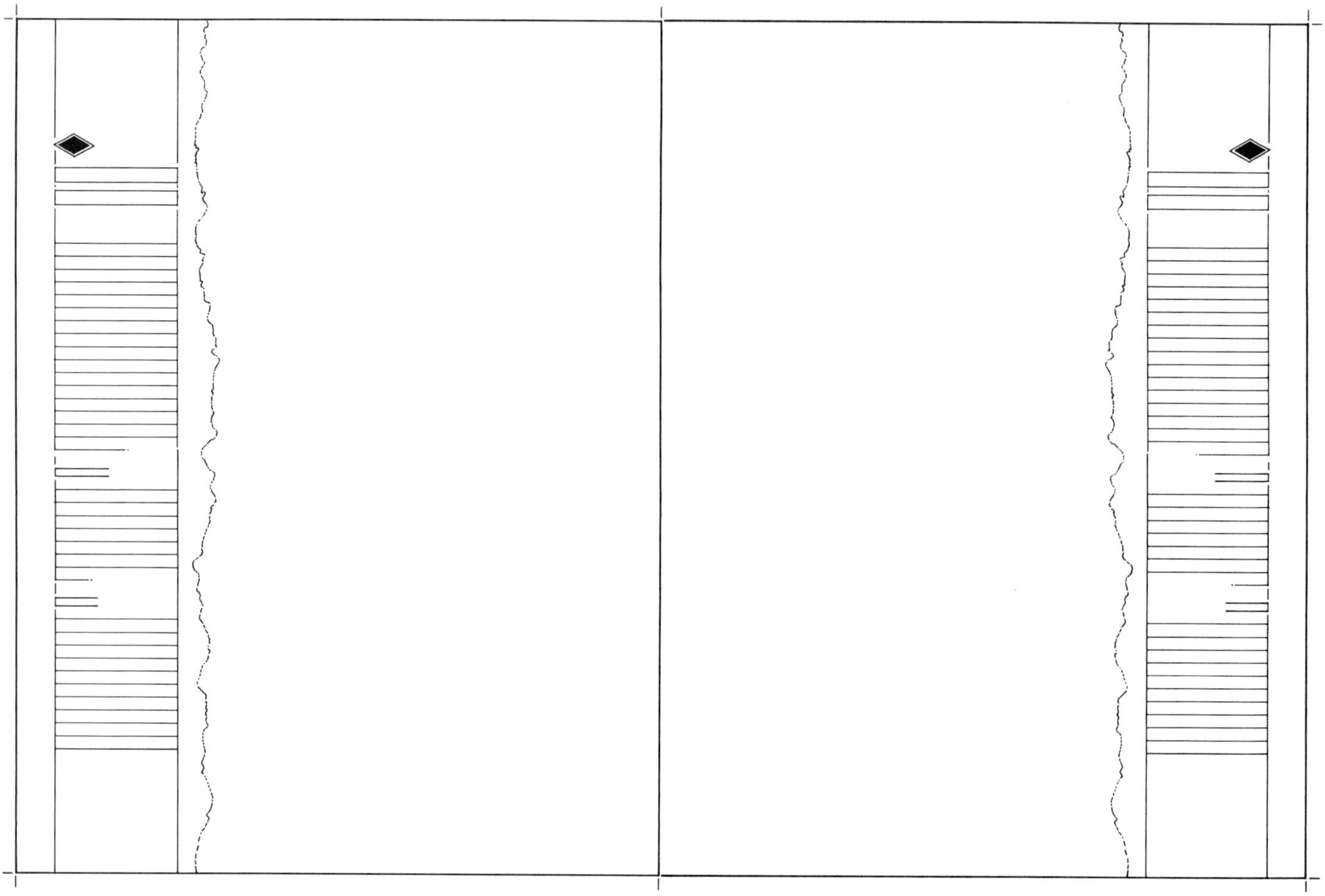

Case Study 5 △    Enlarge to 230%    ▽ Case Study 6

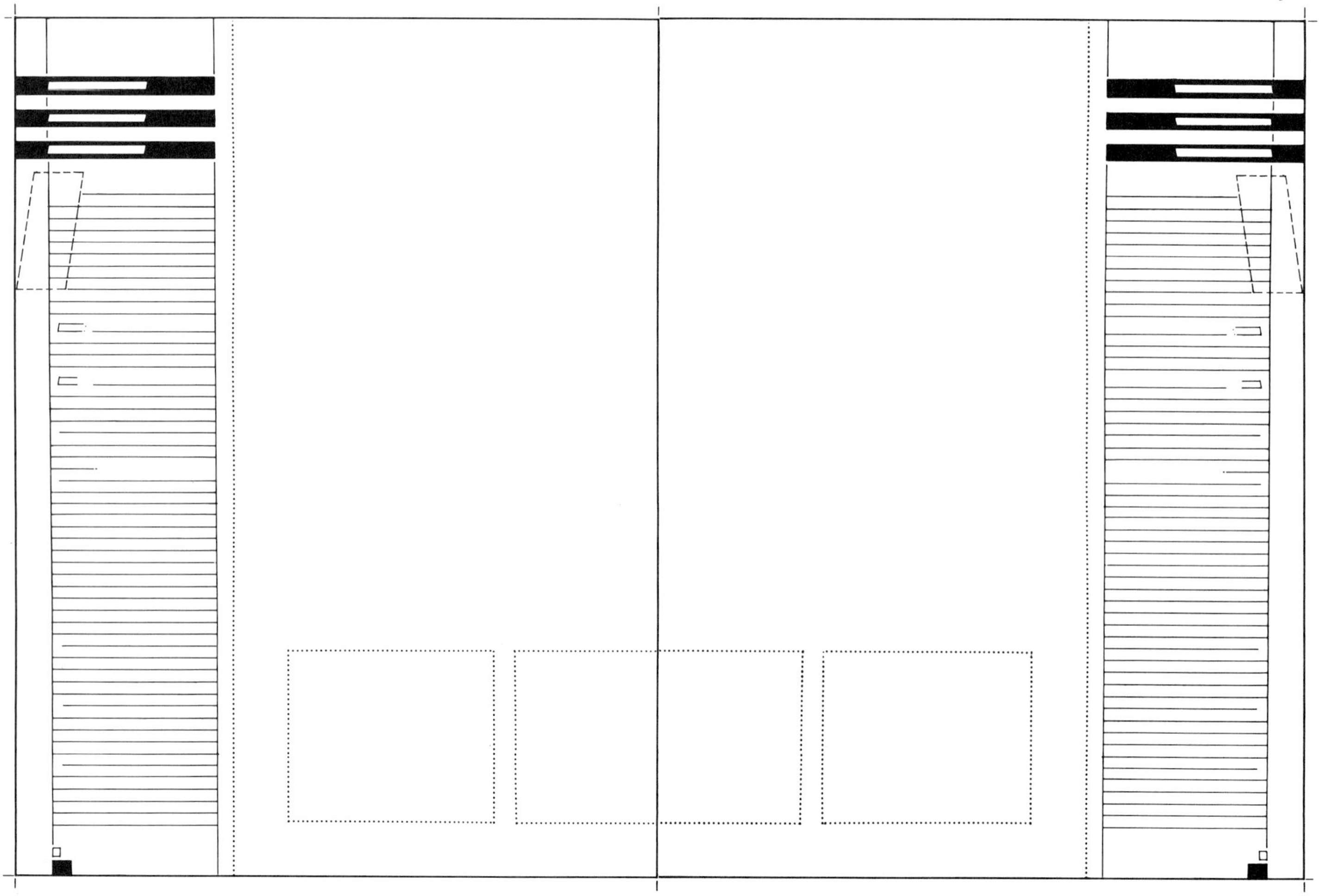

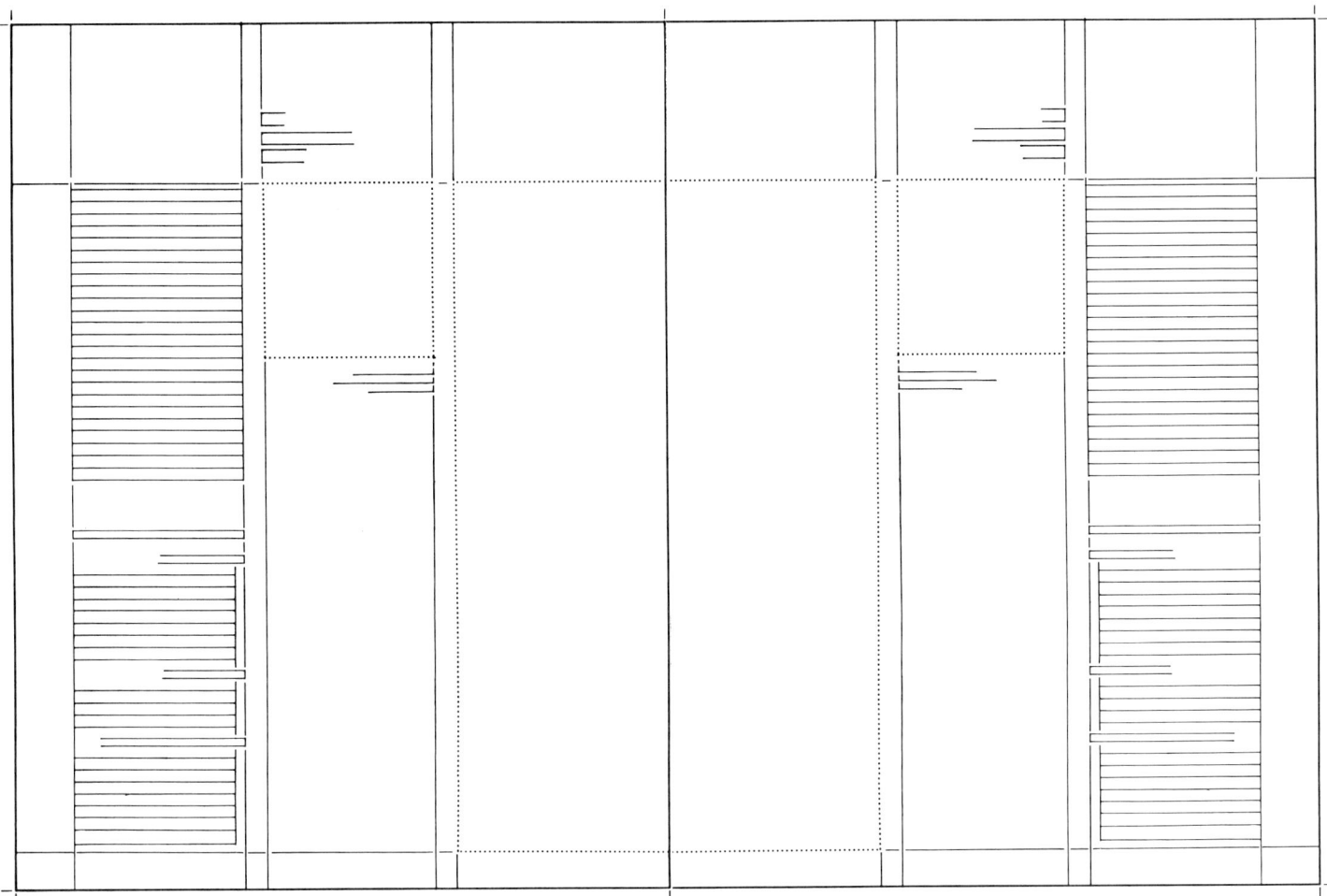

**Case Study 7** △

Enlarge to 230%

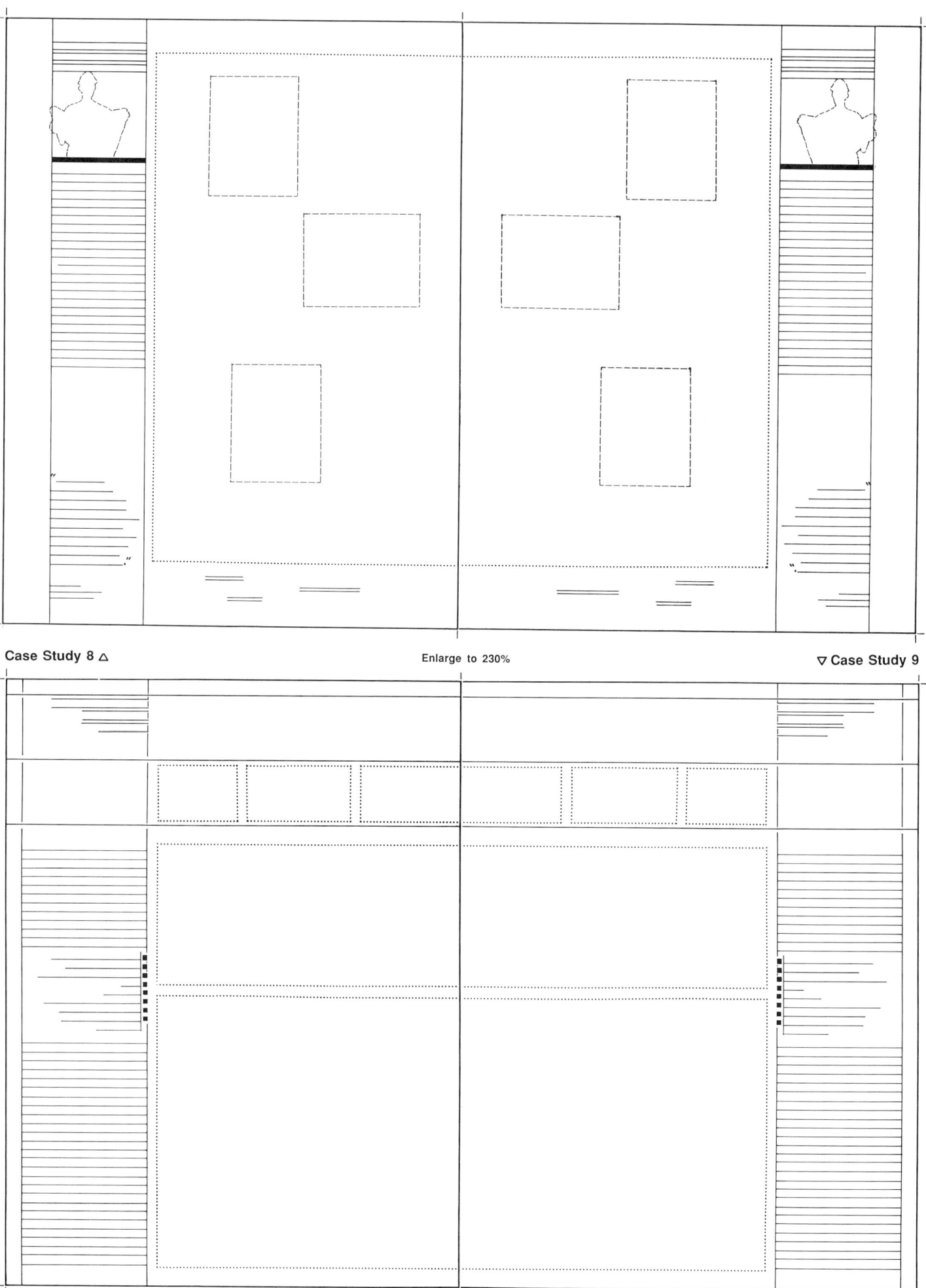

Case Study 8 △　　　Enlarge to 230%　　　▽ Case Study 9

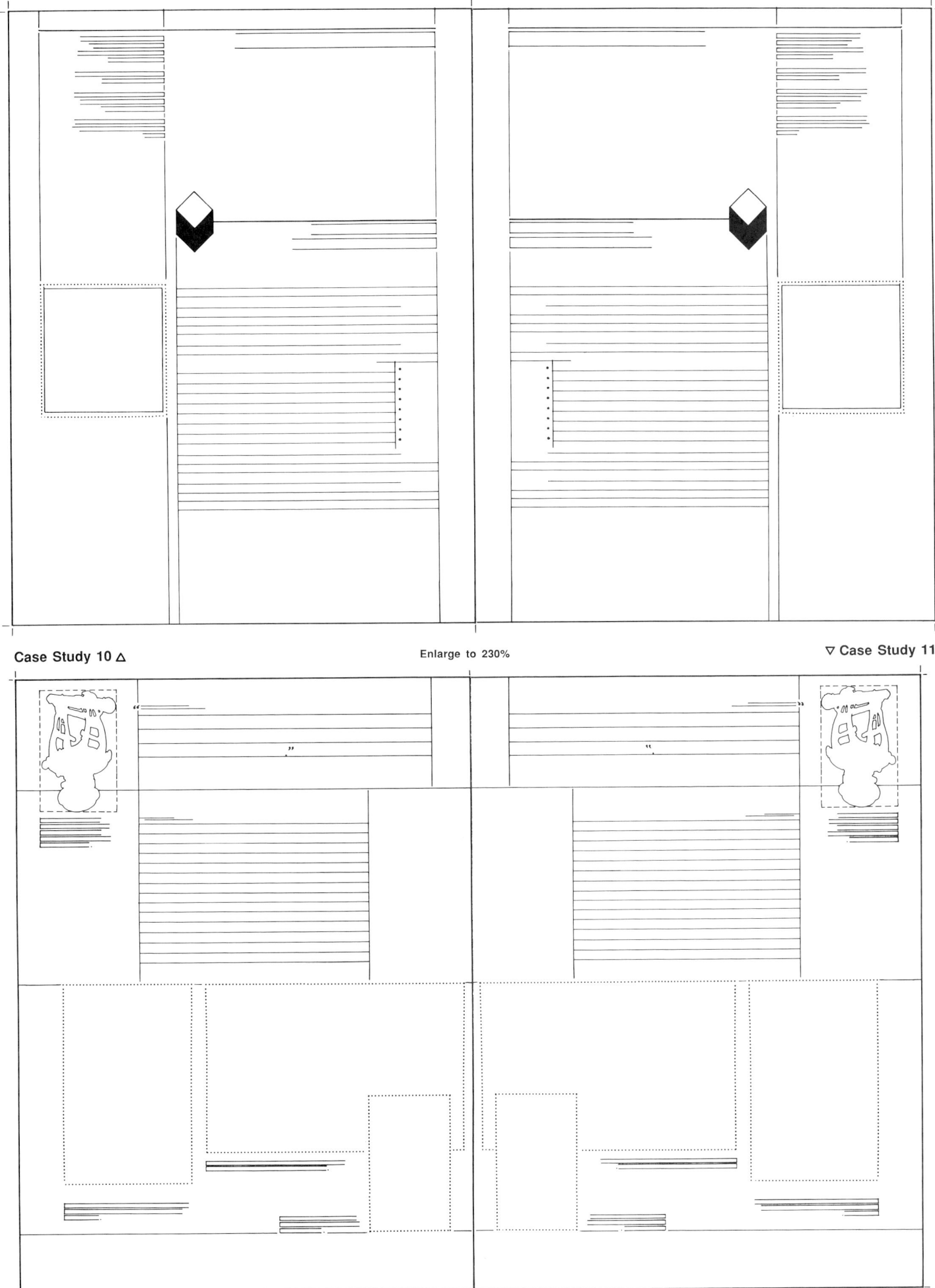

Case Study 10 △  Enlarge to 230%  ▽ Case Study 11

Appendix 197

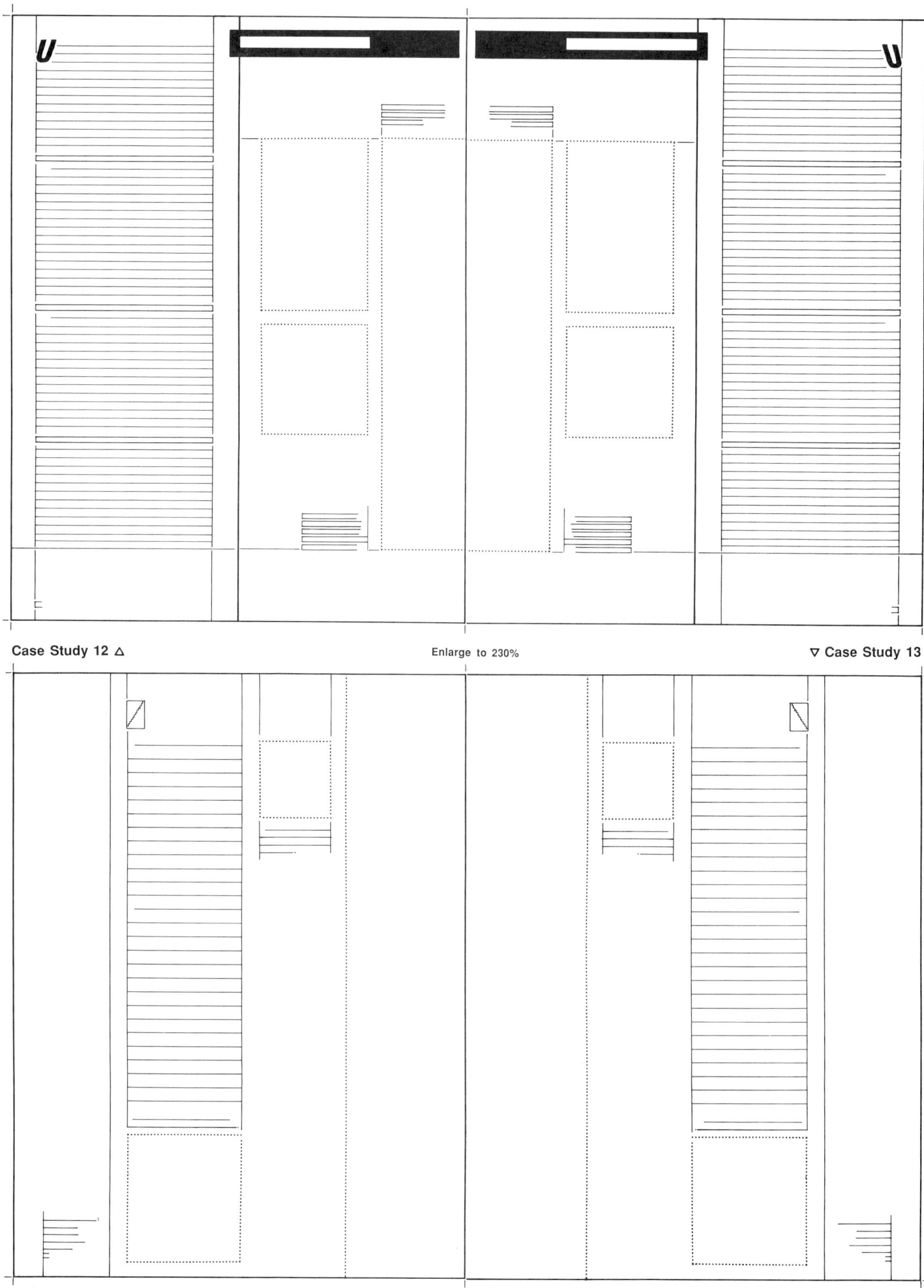

Case Study 12 △ Enlarge to 230% ▽ Case Study 13

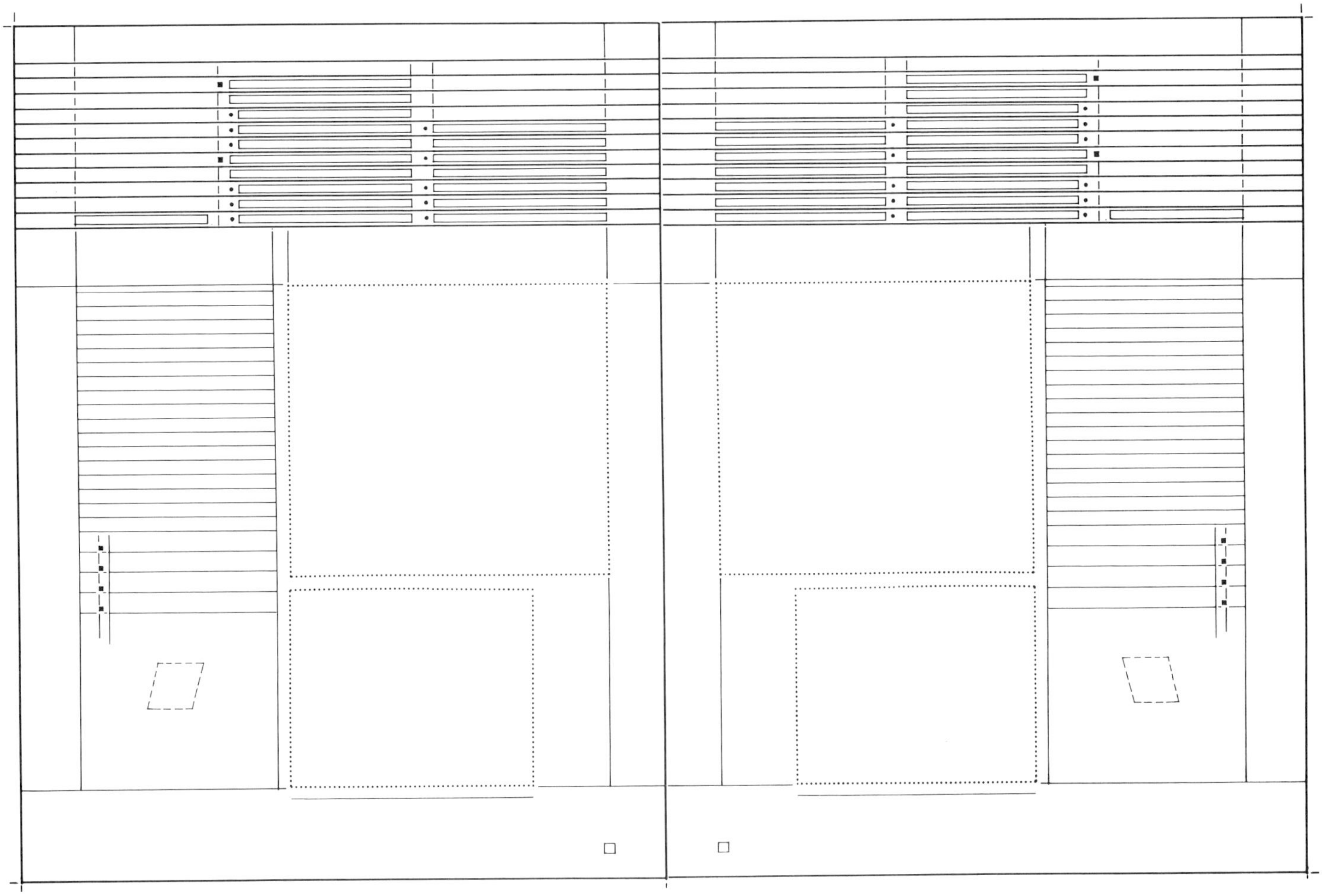

Case Study 14 △  Enlarge to 230%  ▽ Case Study 15

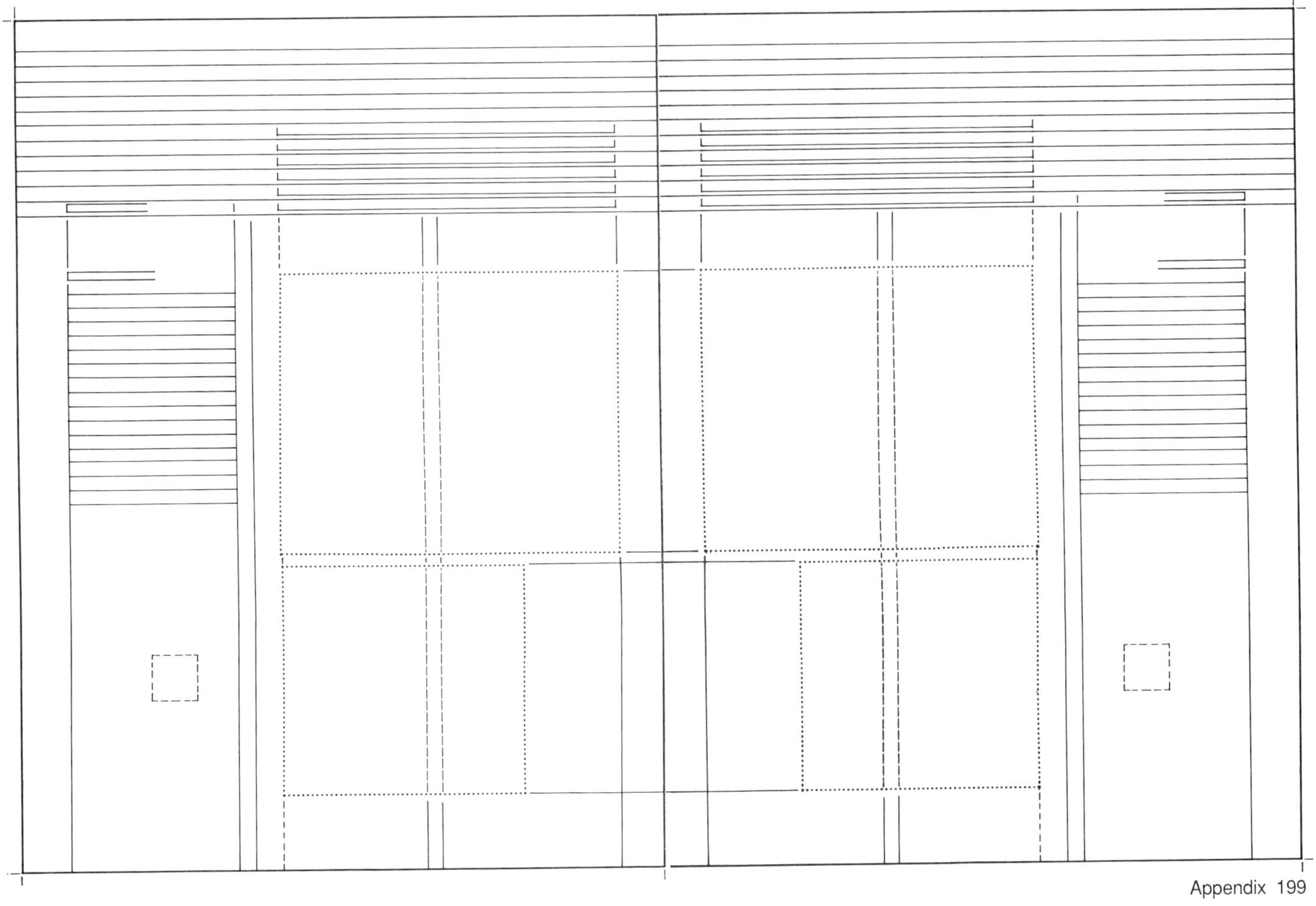

Appendix 199

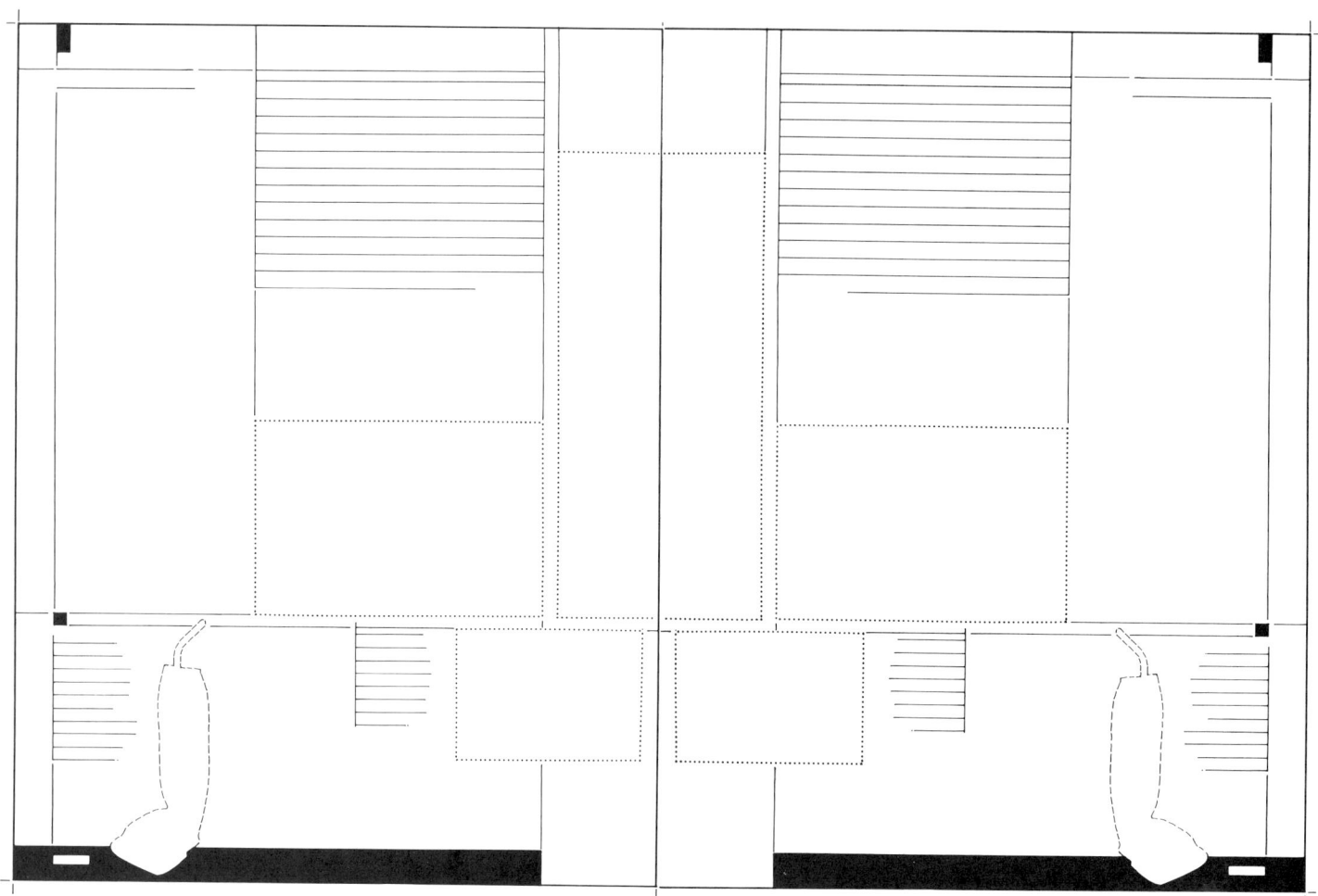

Case Study 16 △   Enlarge to 230%

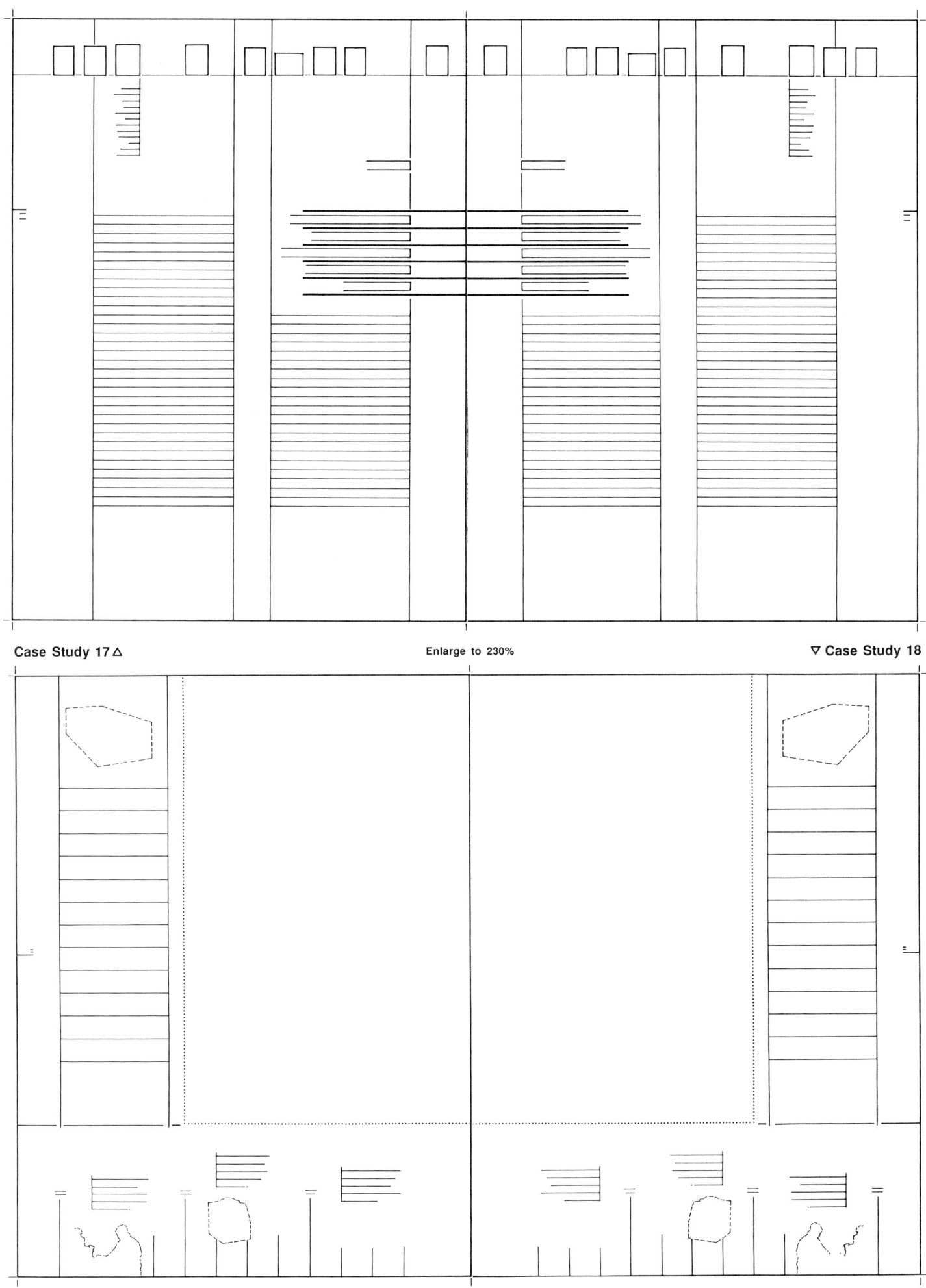

Case Study 17 △   Enlarge to 230%   ▽ Case Study 18

Appendix 201

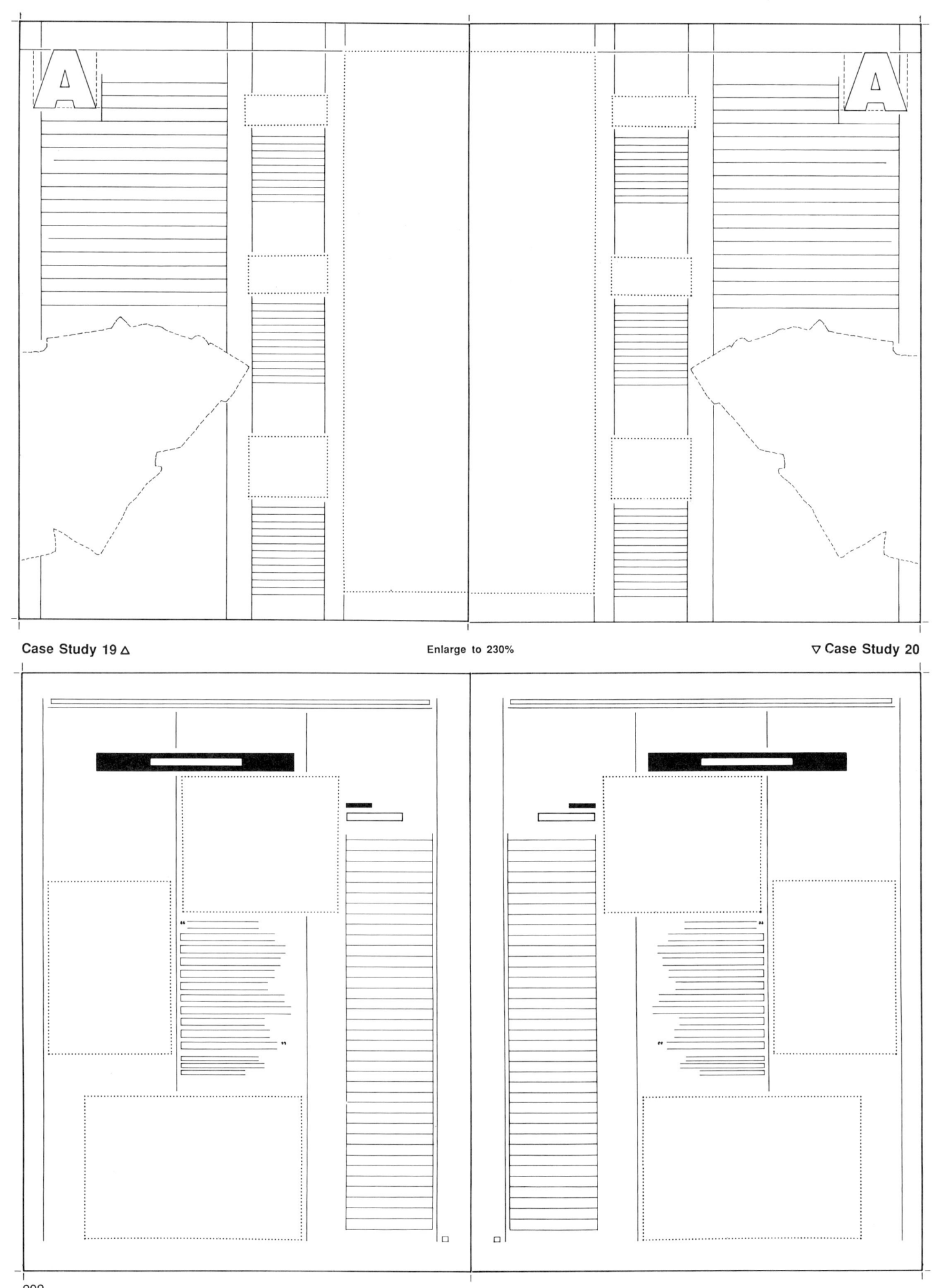

Case Study 19  Enlarge to 230%  Case Study 20

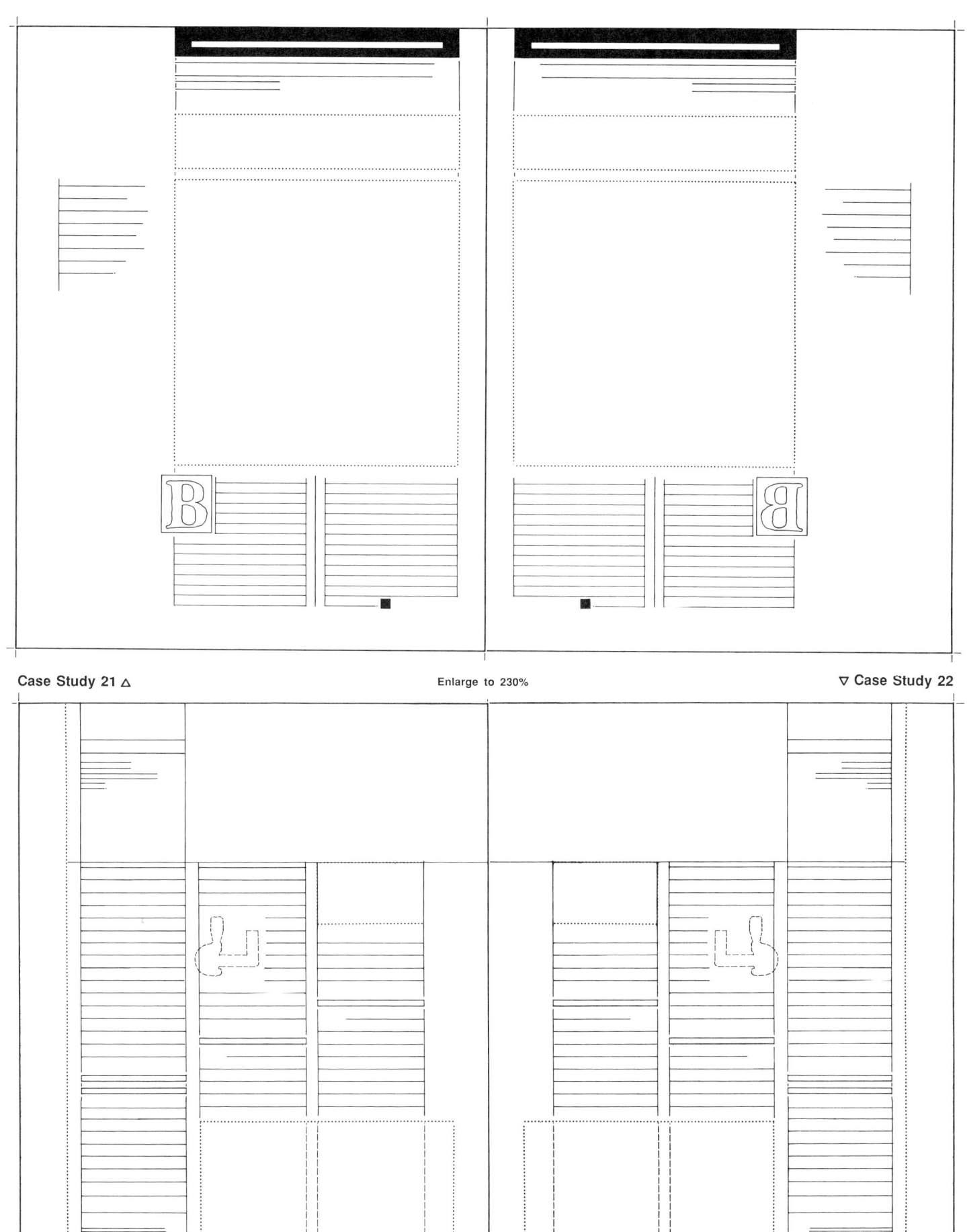

Case Study 21 △   Enlarge to 230%   ▽ Case Study 22

Appendix 203

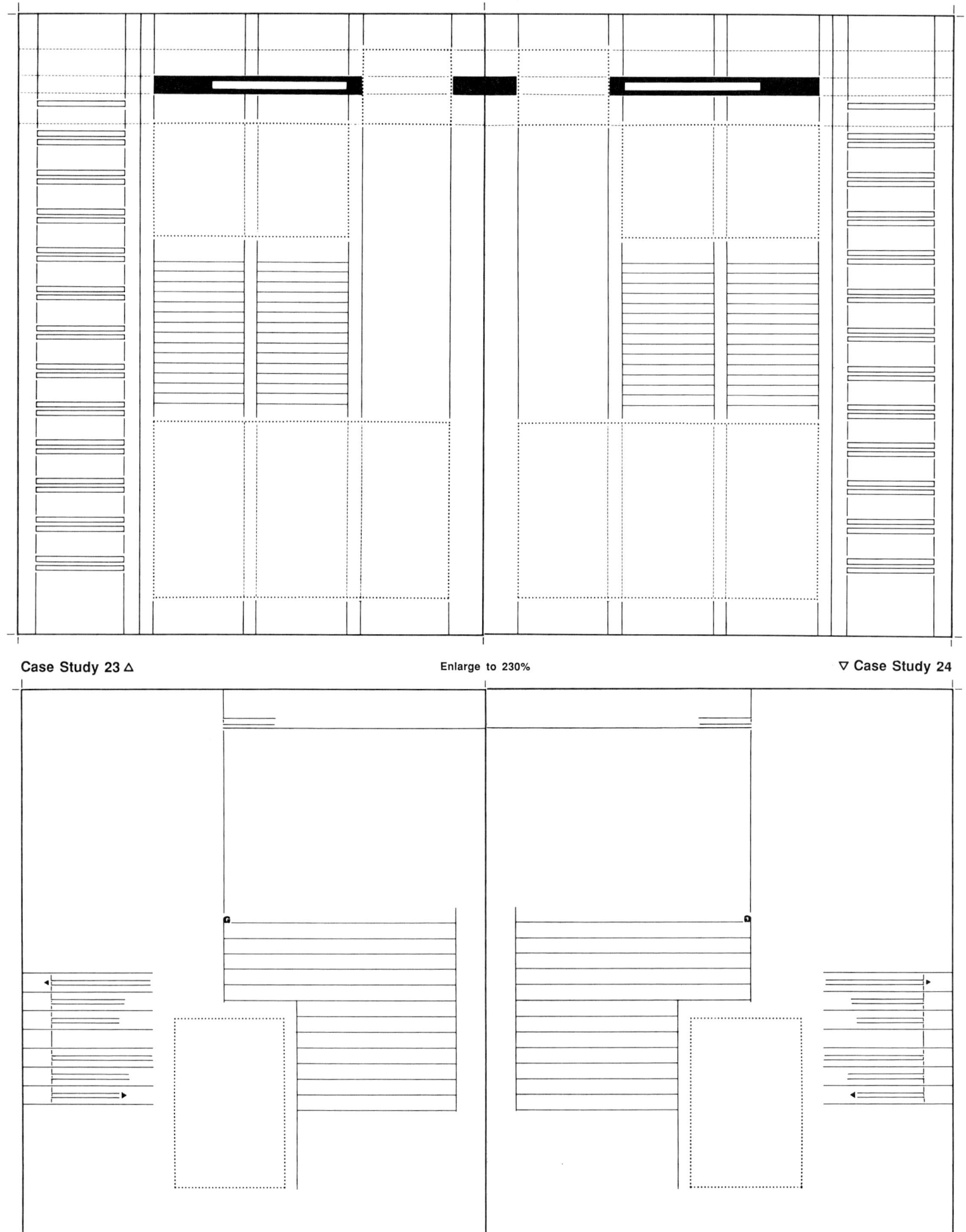

Case Study 23 △   Enlarge to 230%   ▽ Case Study 24

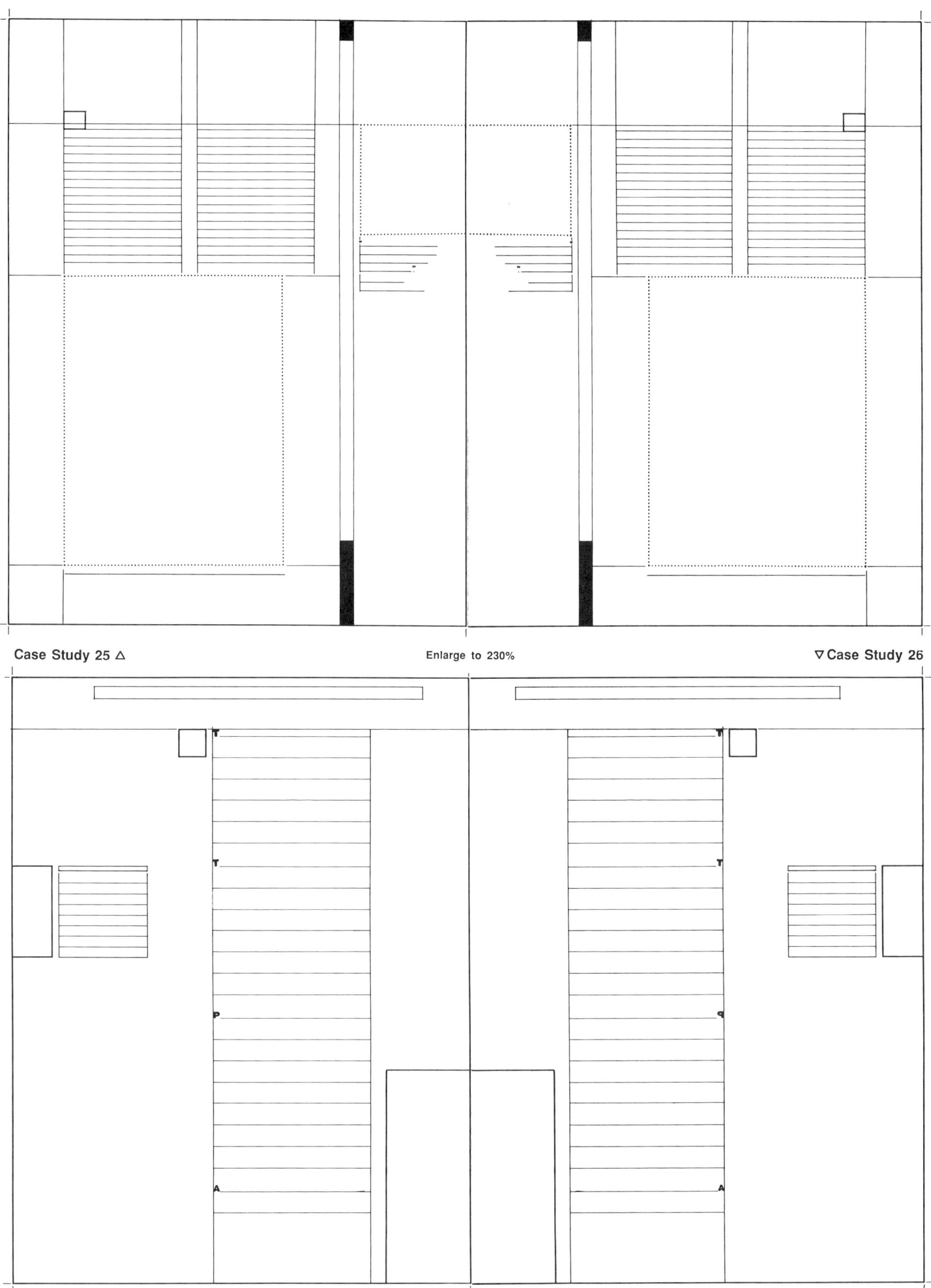

Case Study 25 △   Enlarge to 230%   ▽ Case Study 26

Appendix 205

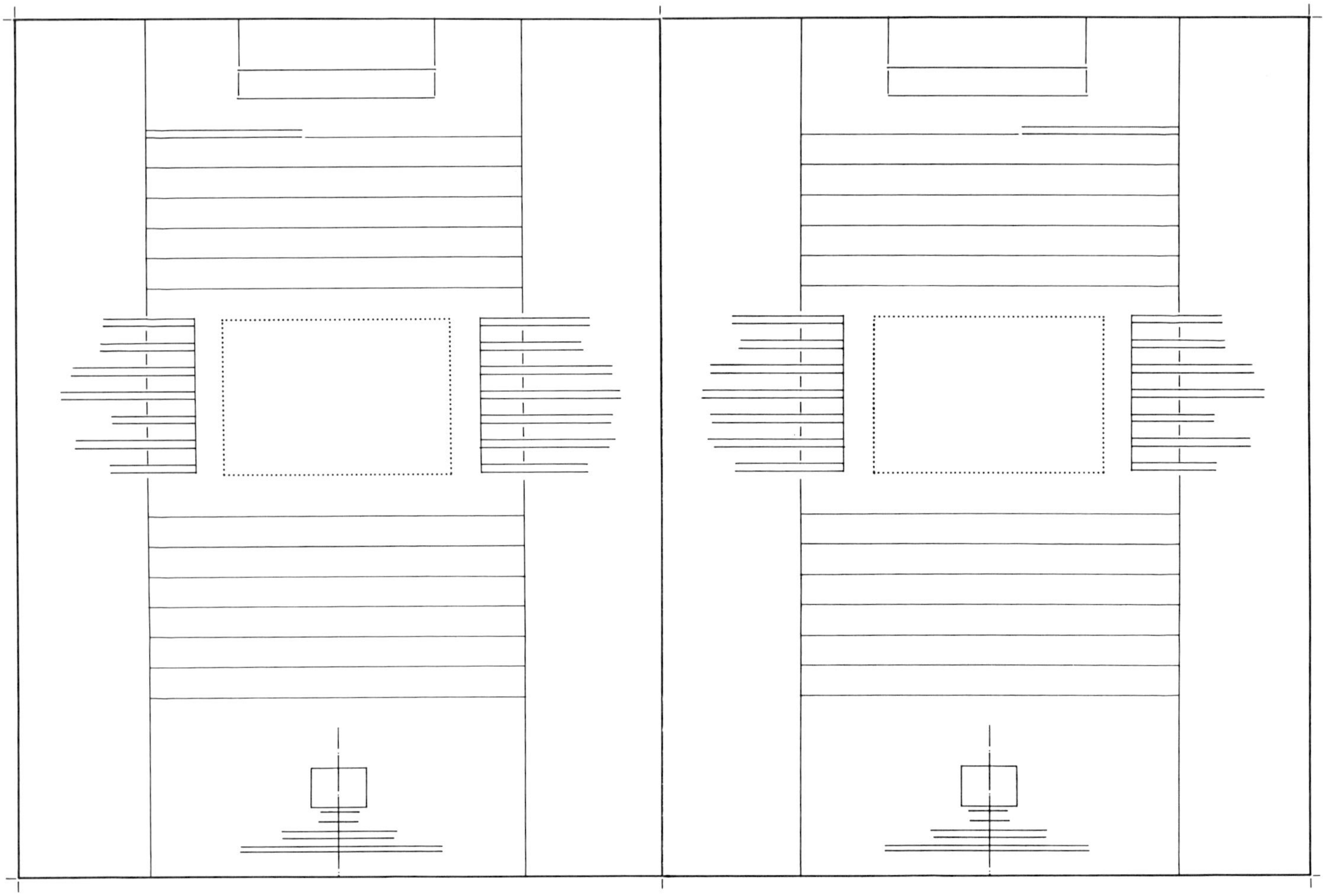

Case Study 27 △   Enlarge to 230%   ▽ Case Study 28

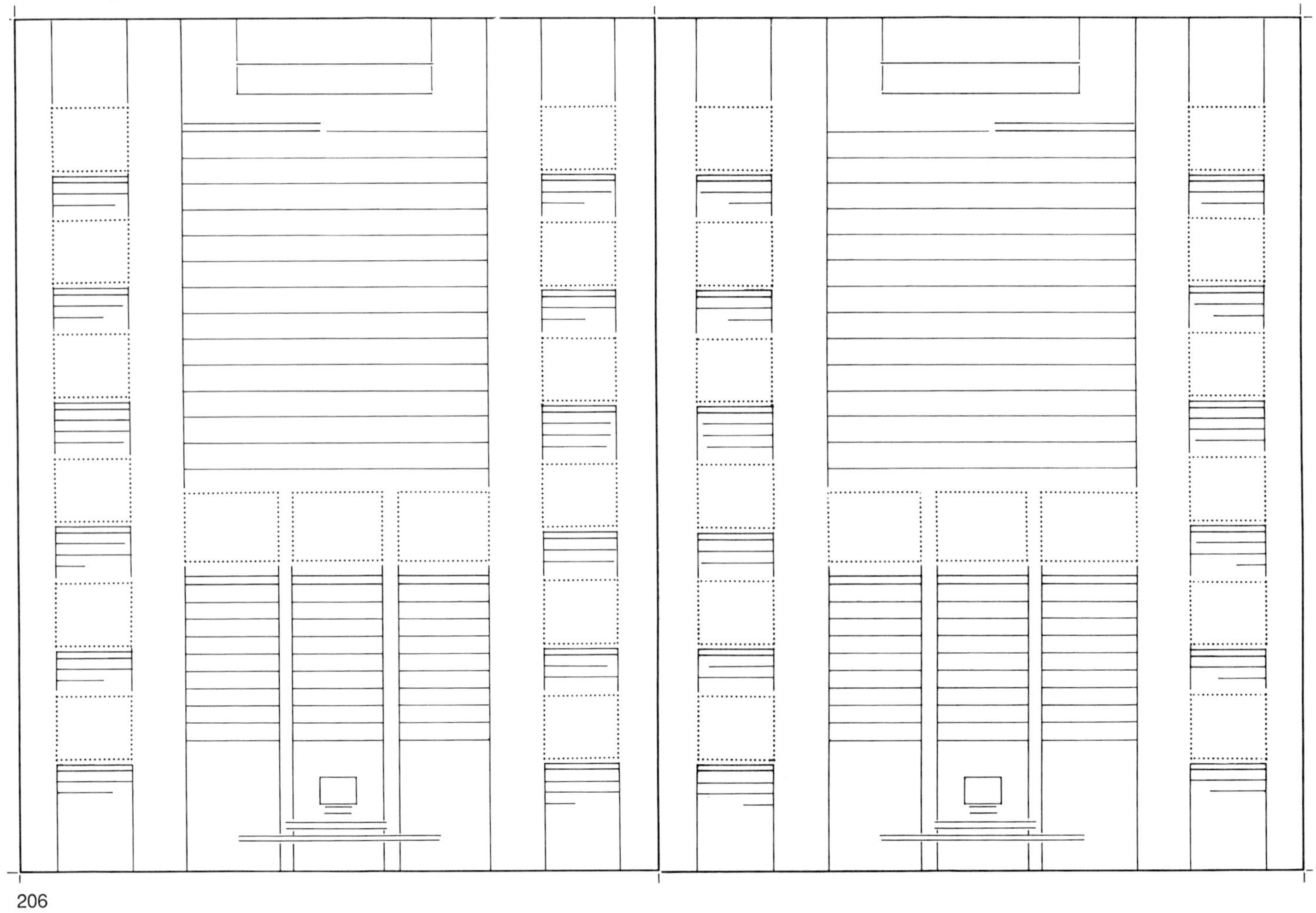

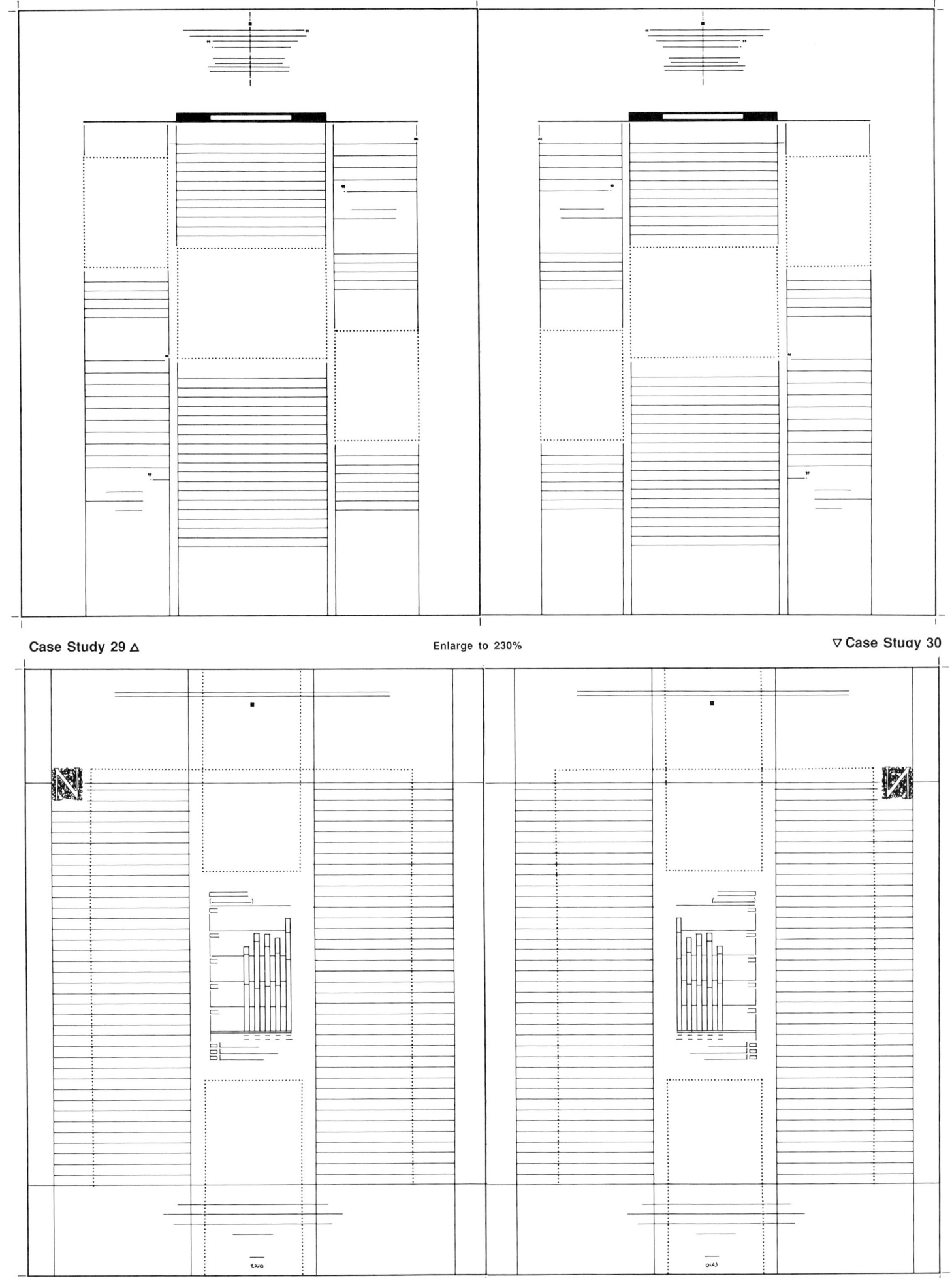

Case Study 29　　Enlarge to 230%　　Case Study 30

Appendix 207

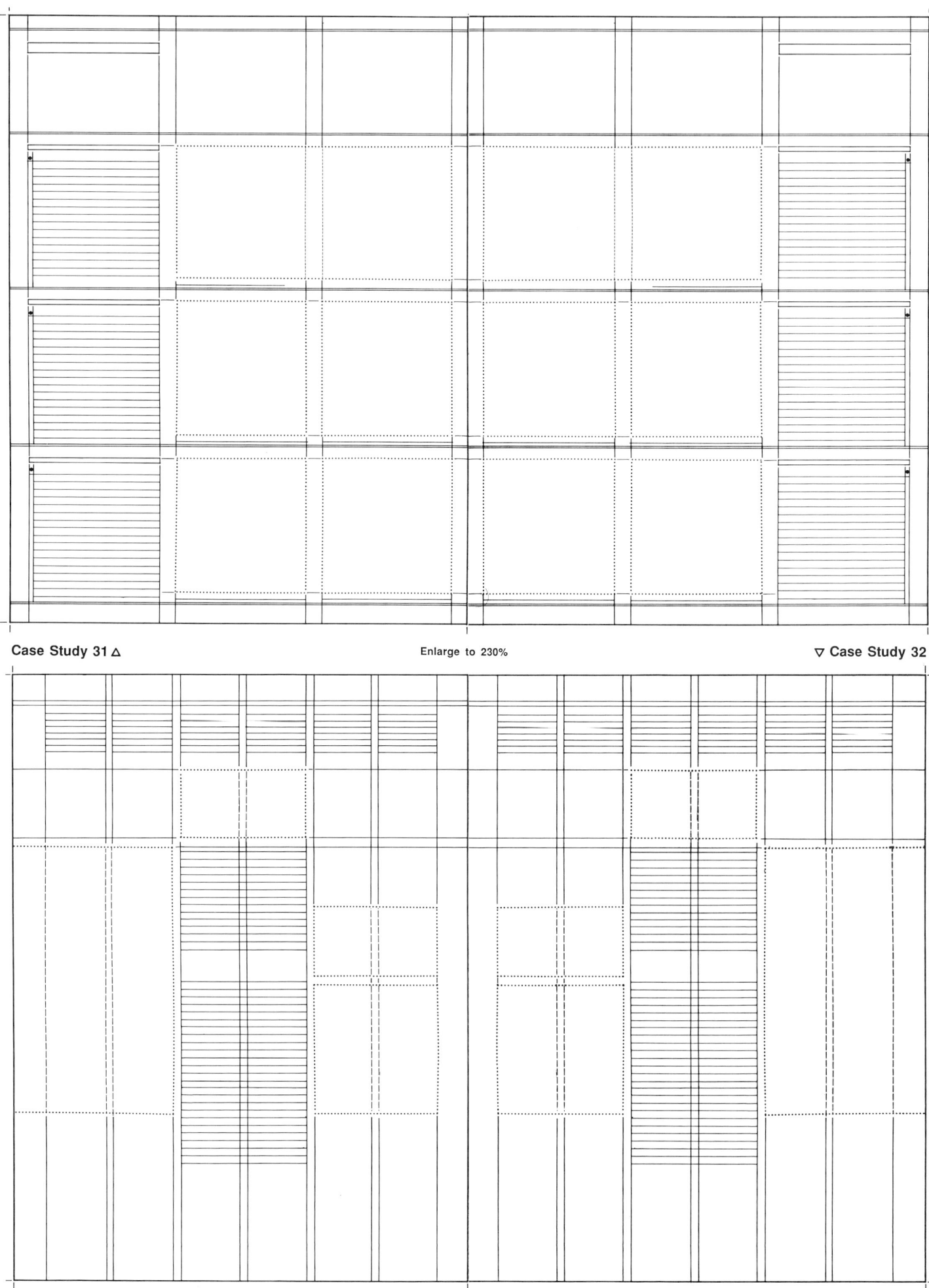

Case Study 31 △   Enlarge to 230%   ▽ Case Study 32

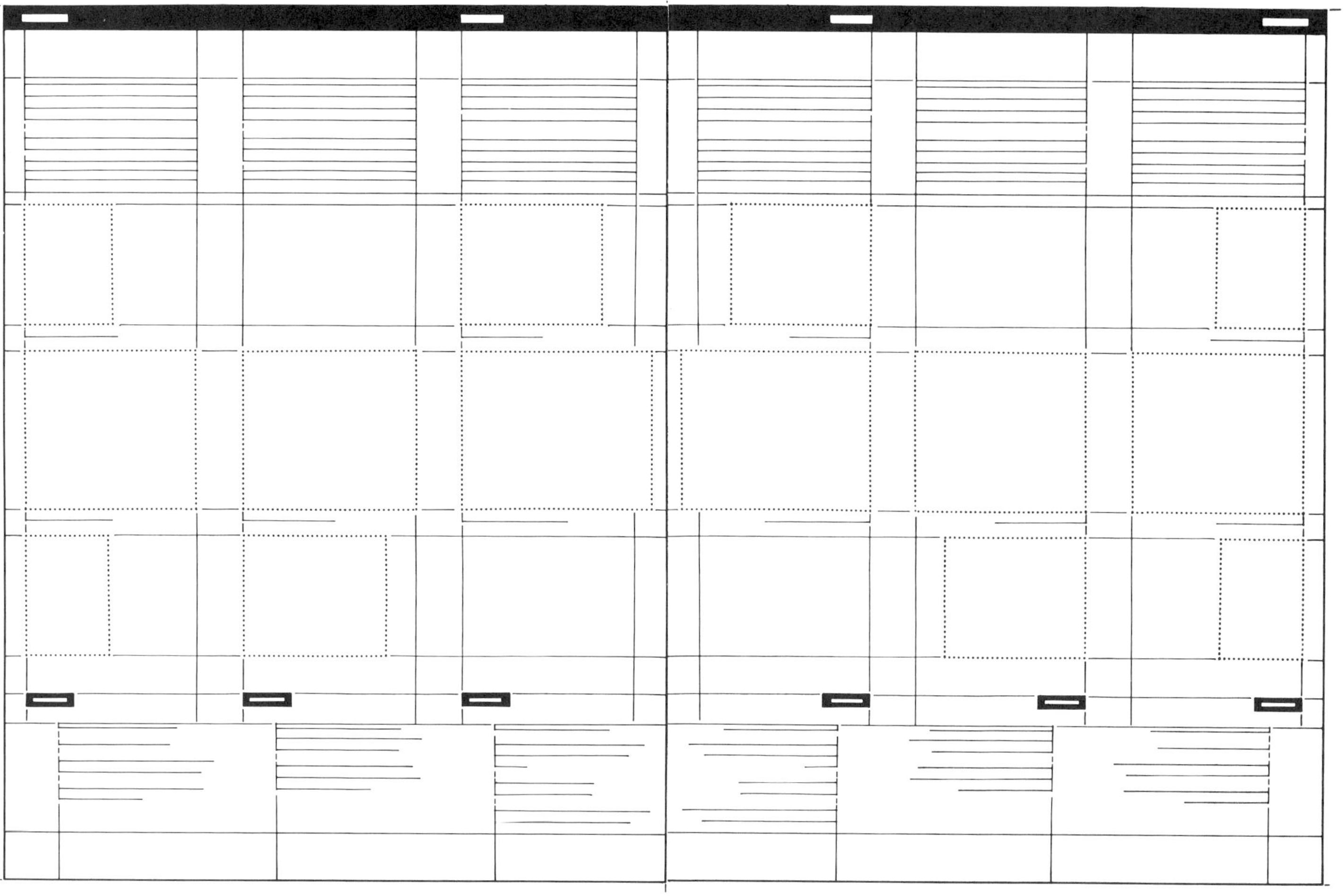

Case Study 33 △   Enlarge to 230%   ▽ Case Study 34

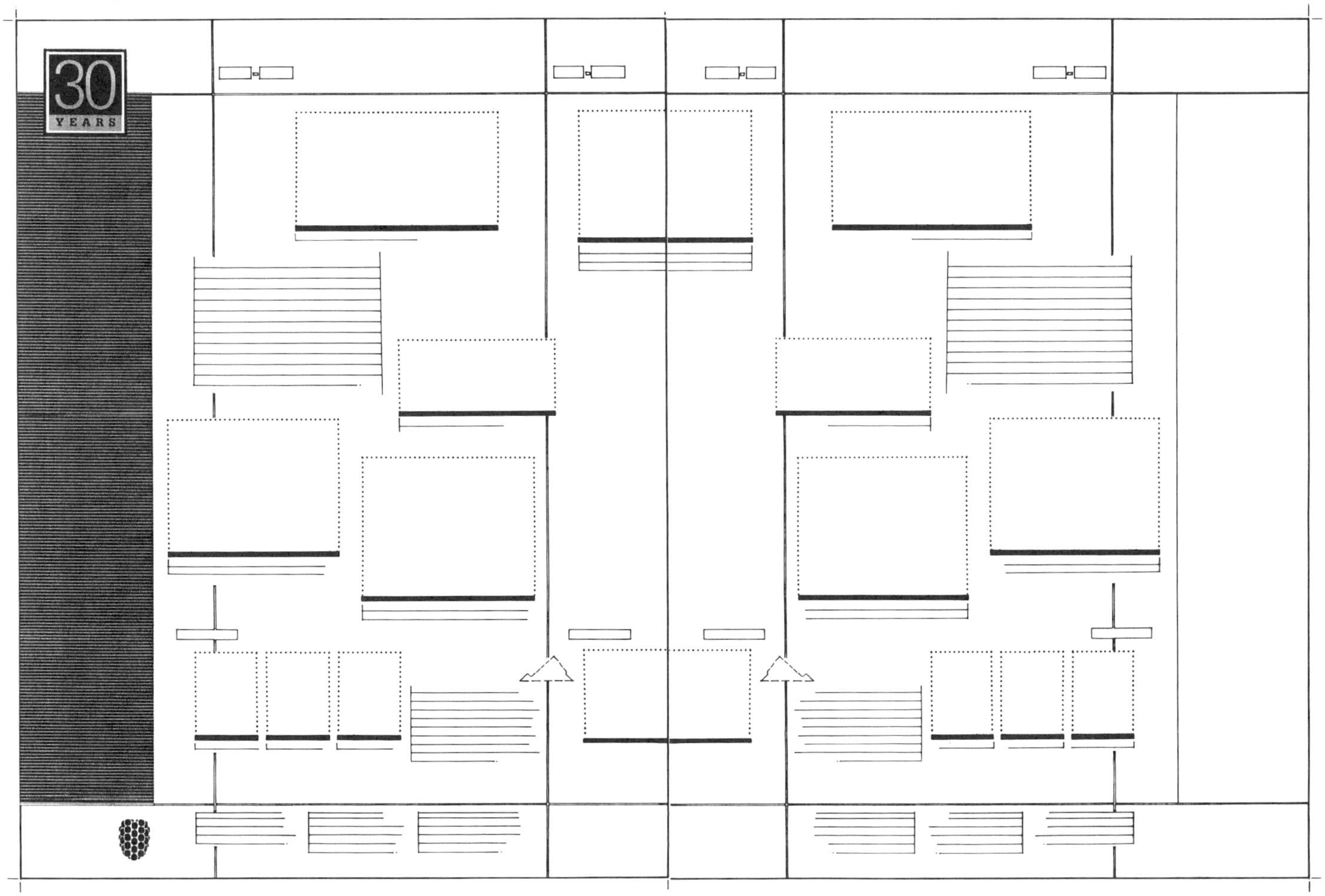

Appendix 209

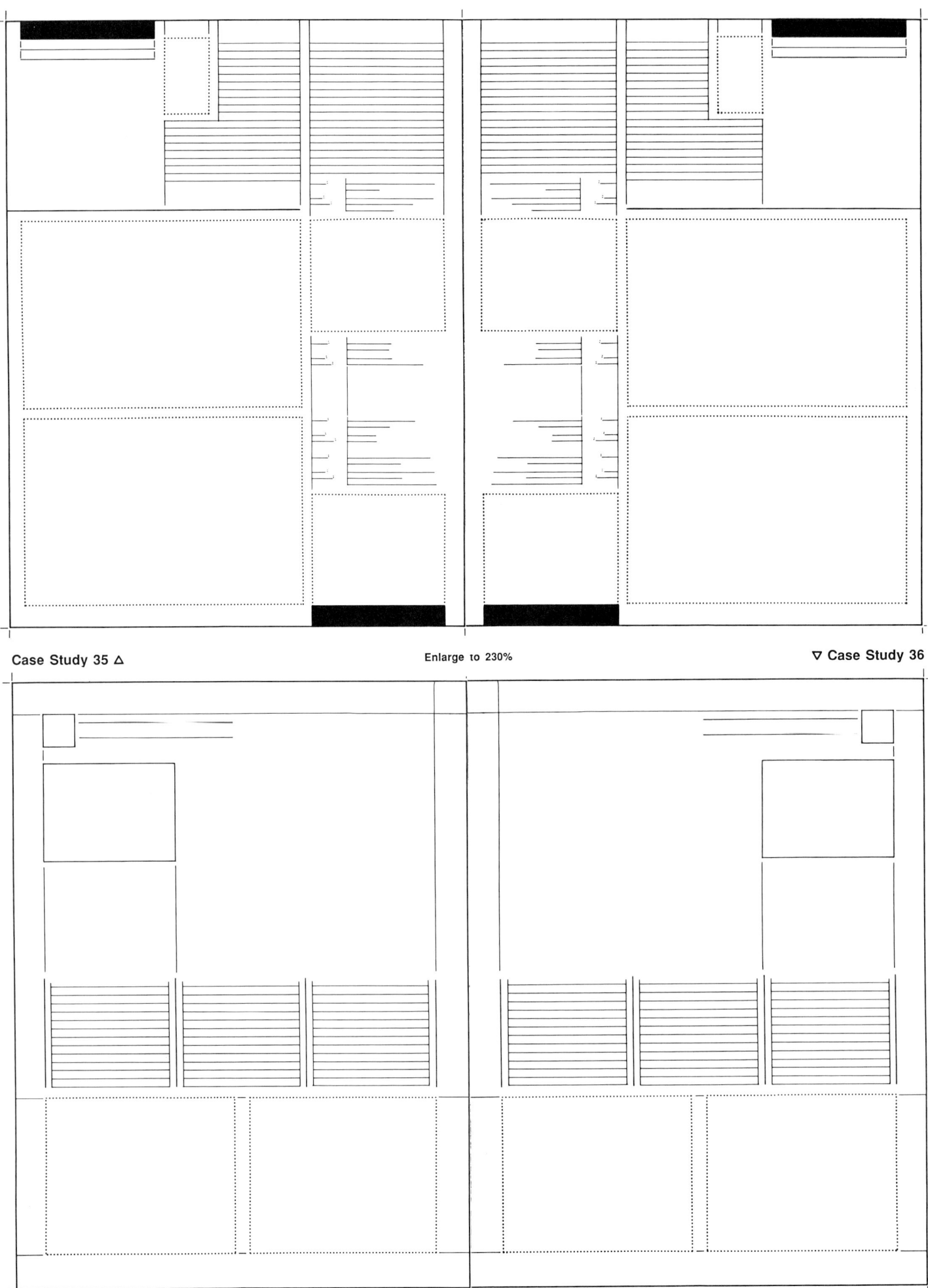

Case Study 35 △   Enlarge to 230%   ▽ Case Study 36

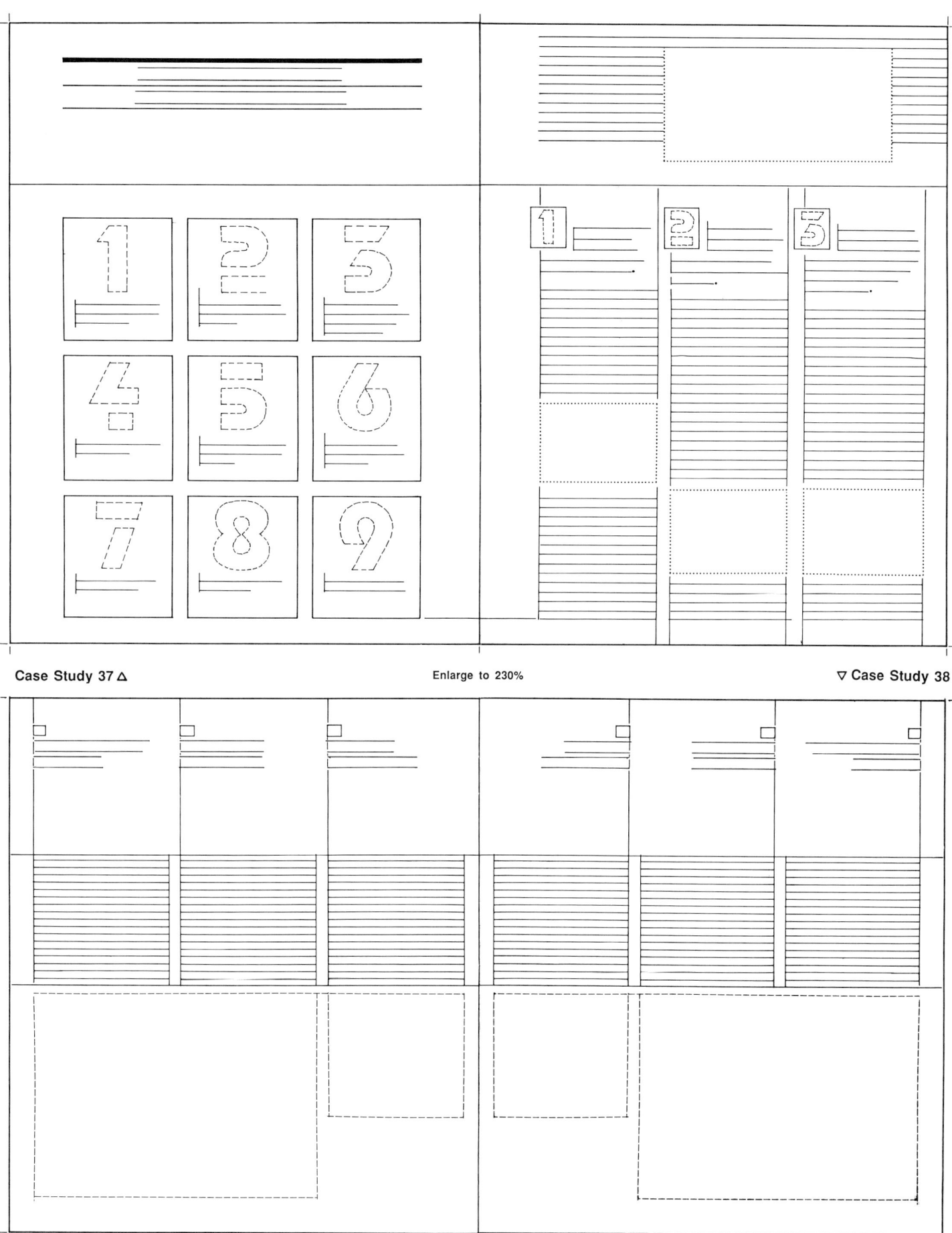

Case Study 37 △   Enlarge to 230%   ▽ Case Study 38

Appendix 211

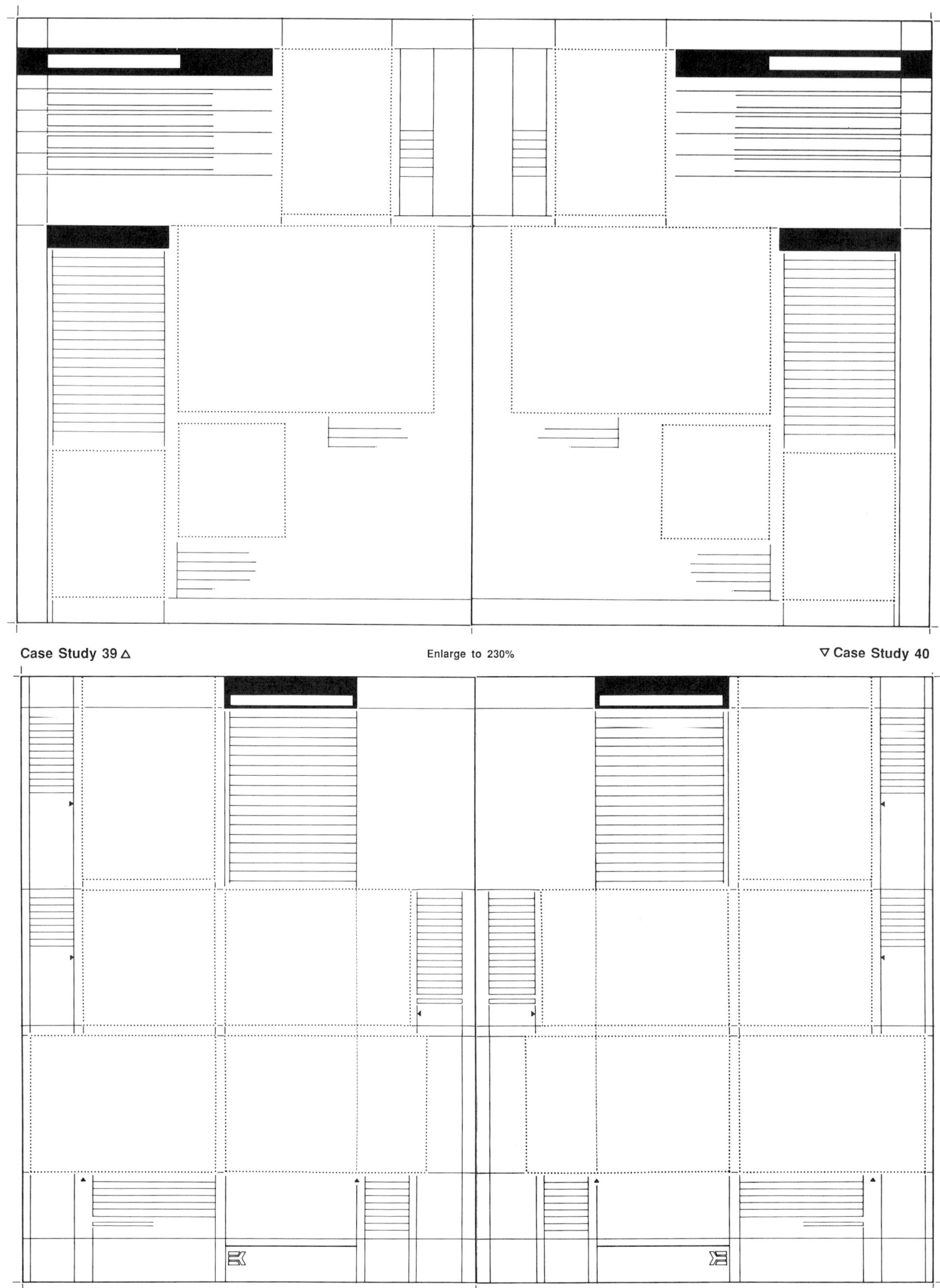

Case Study 39 △     Enlarge to 230%     ▽ Case Study 40

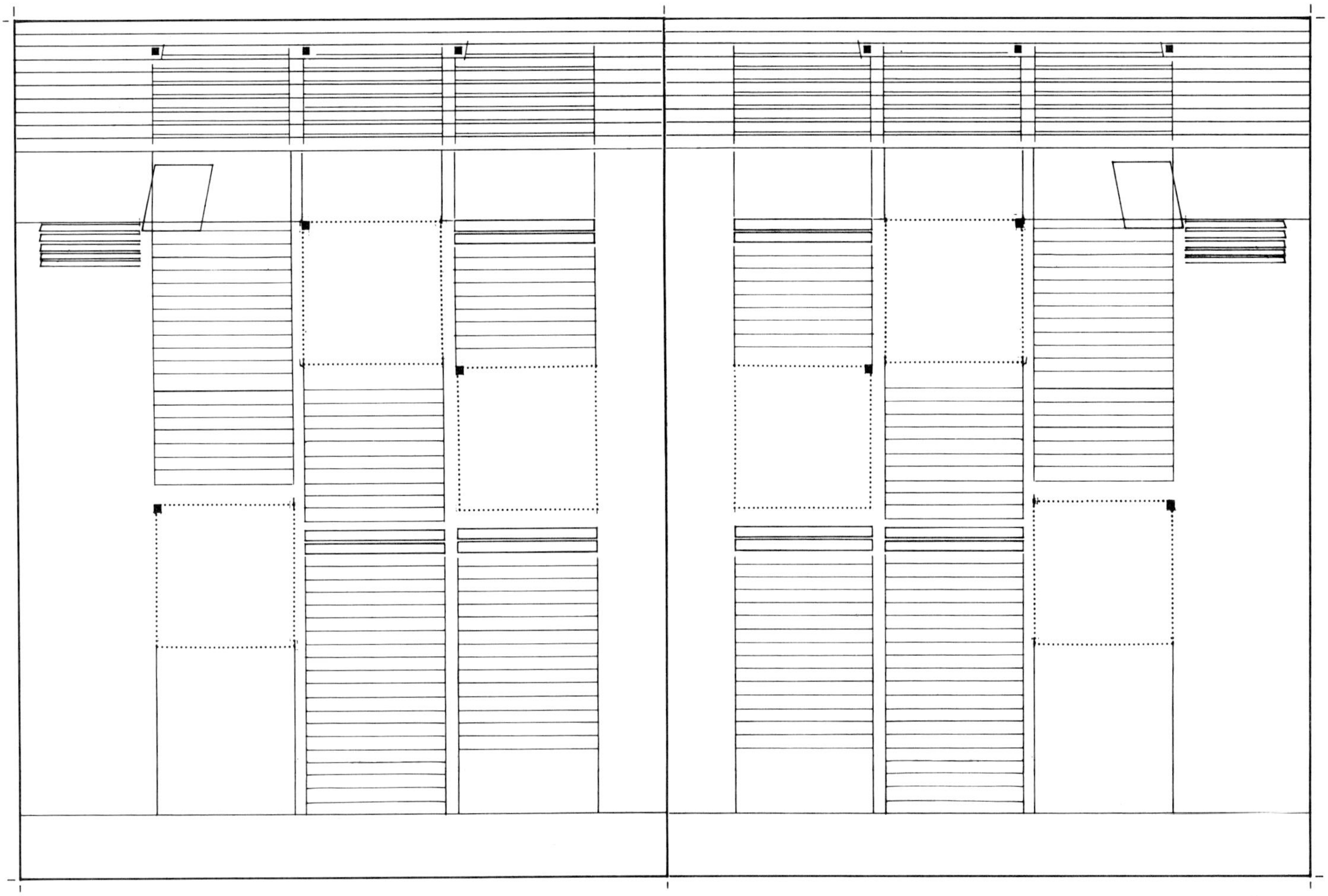

Case Study 41 △   Enlarge to 230%   ▽ Case Study 42

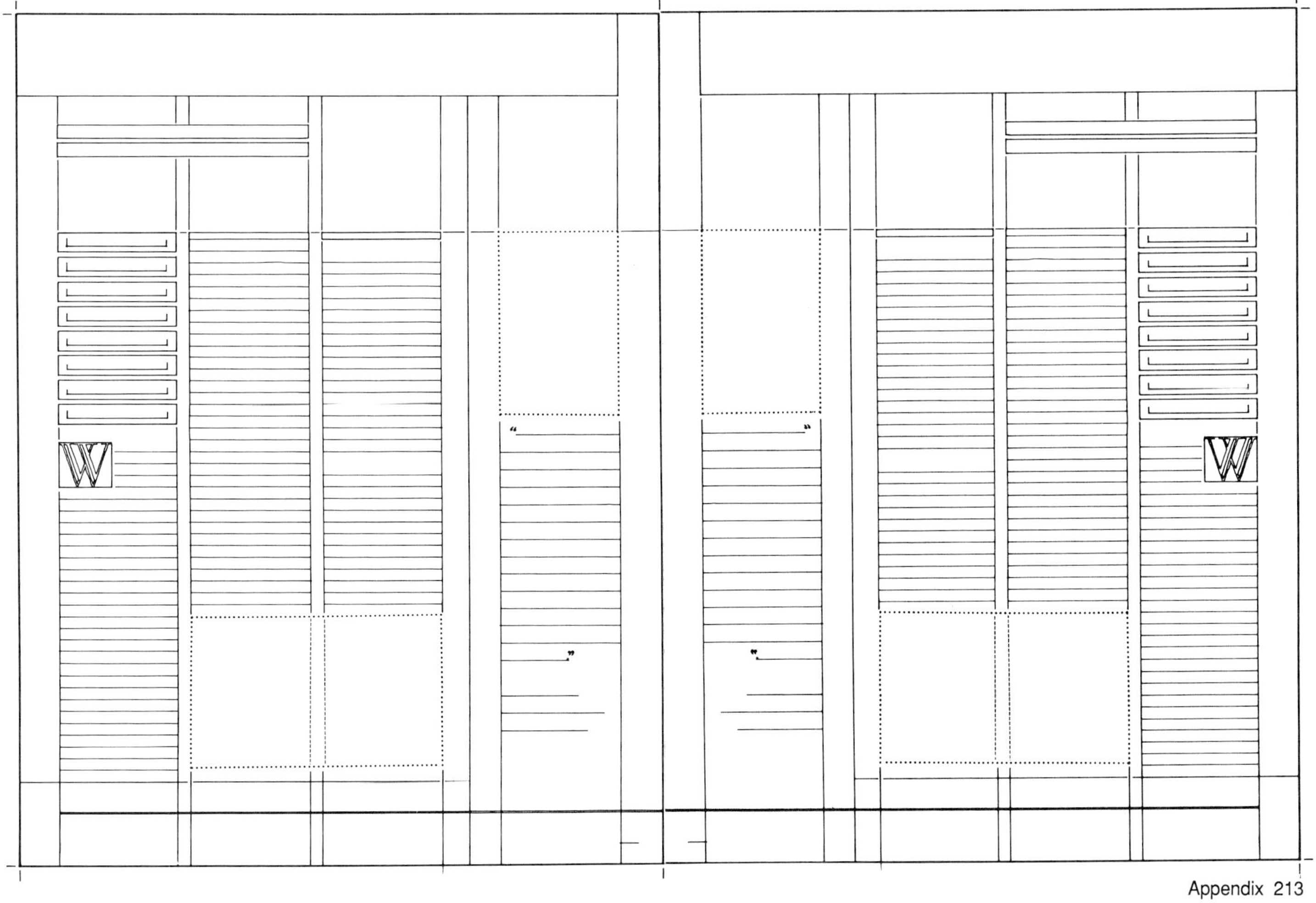

Appendix 213

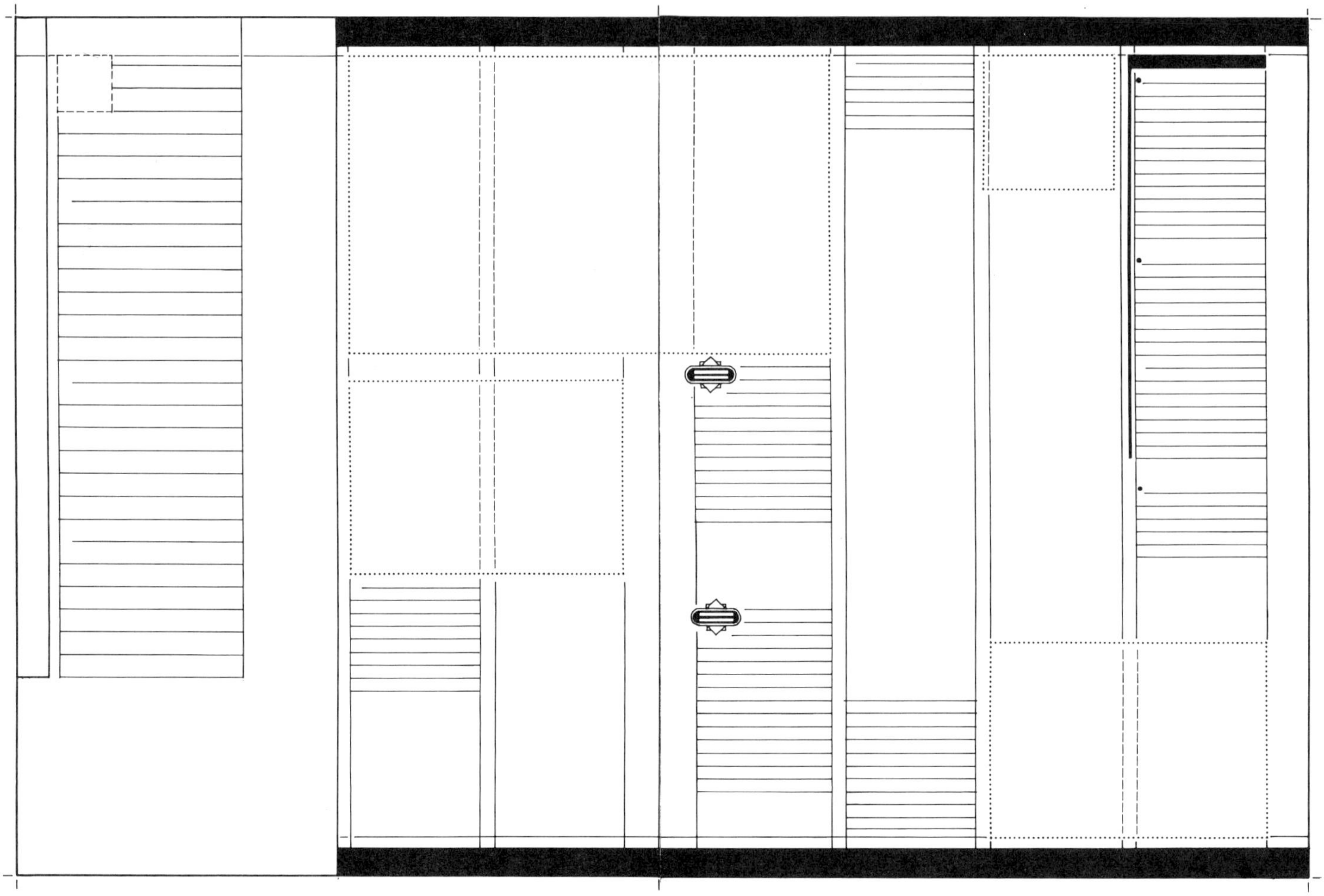

**Case Study 43** △▽   Enlarge to 230%

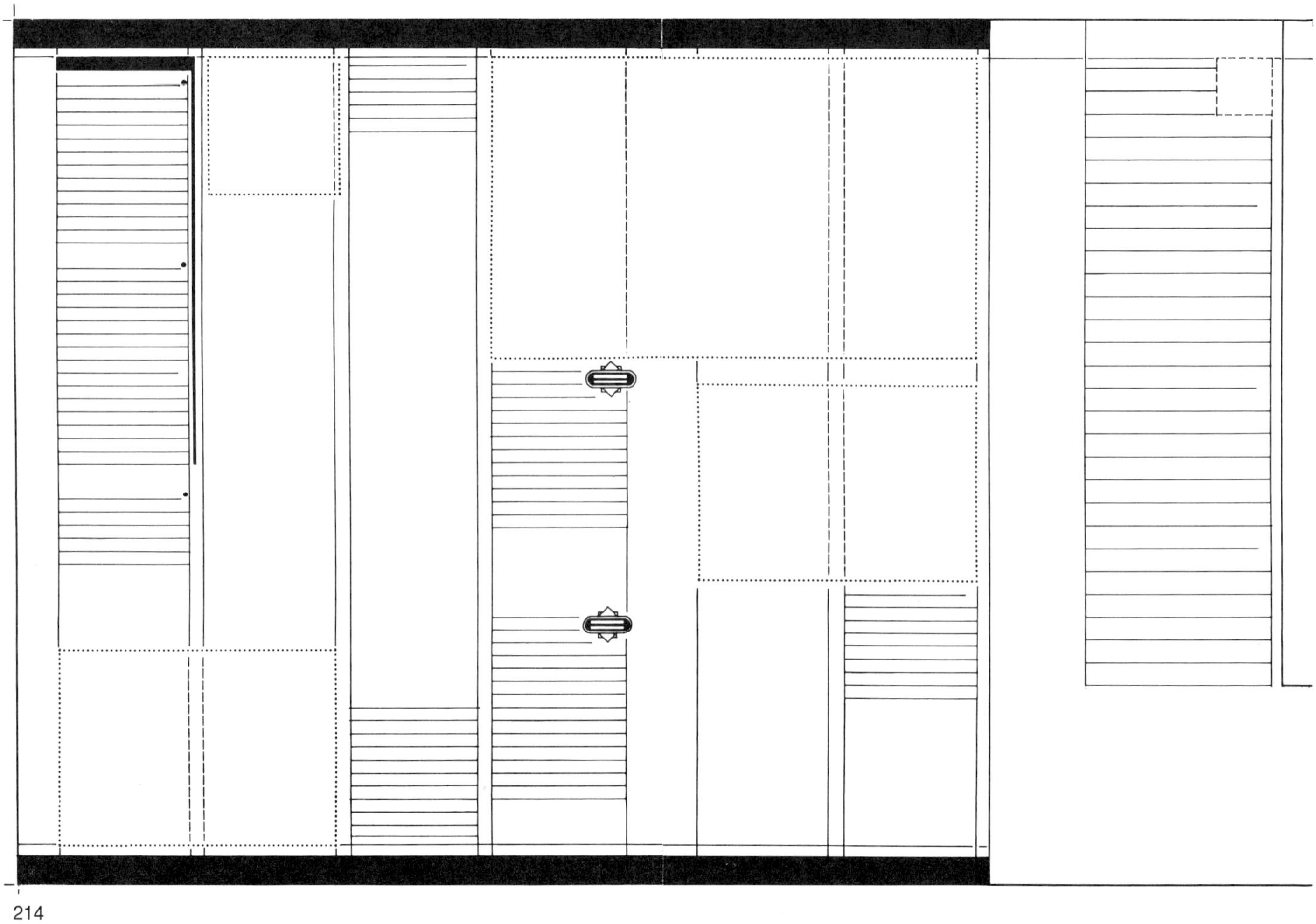

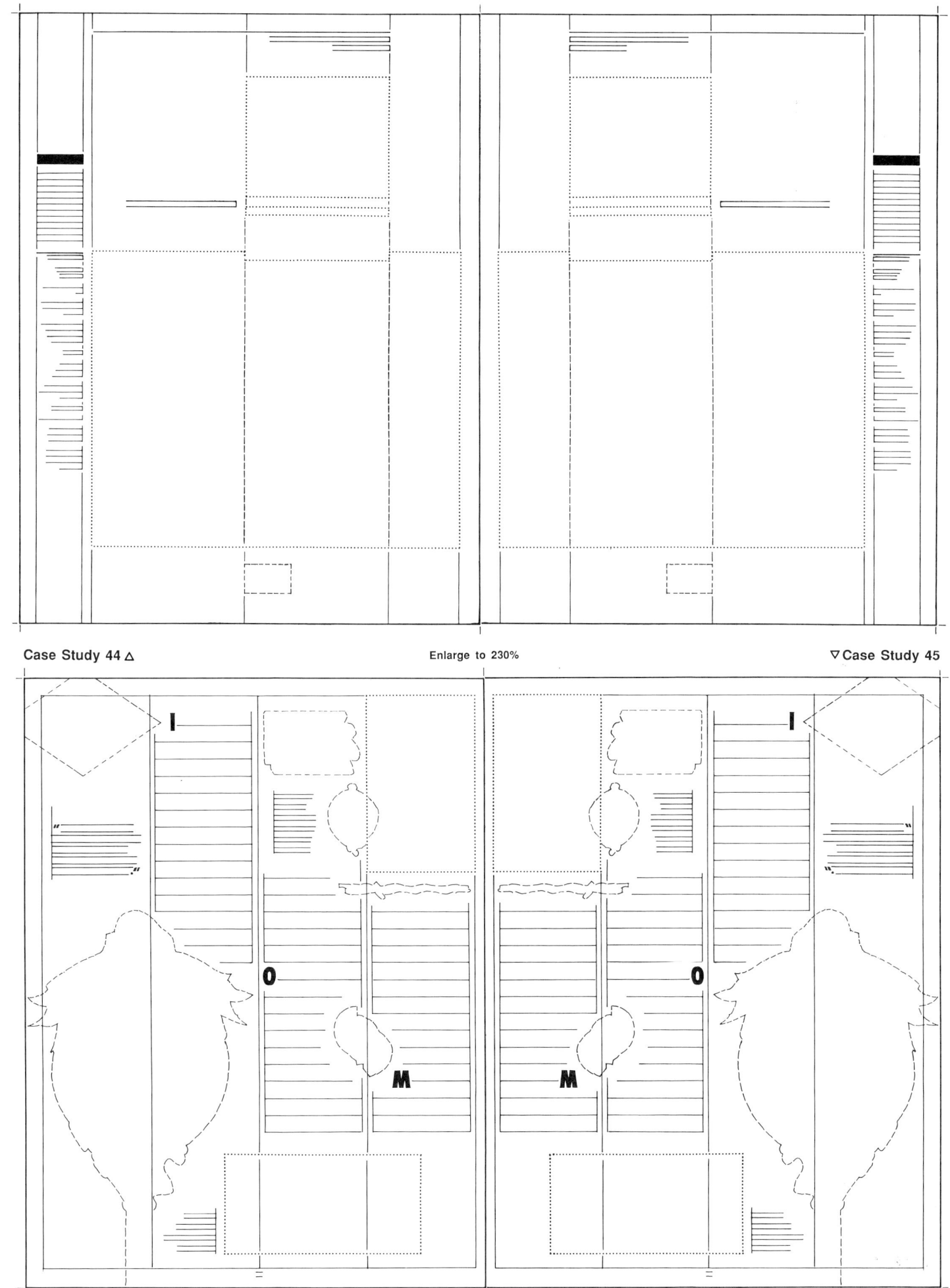

Case Study 44 △  Enlarge to 230%  ▽ Case Study 45

Appendix 215

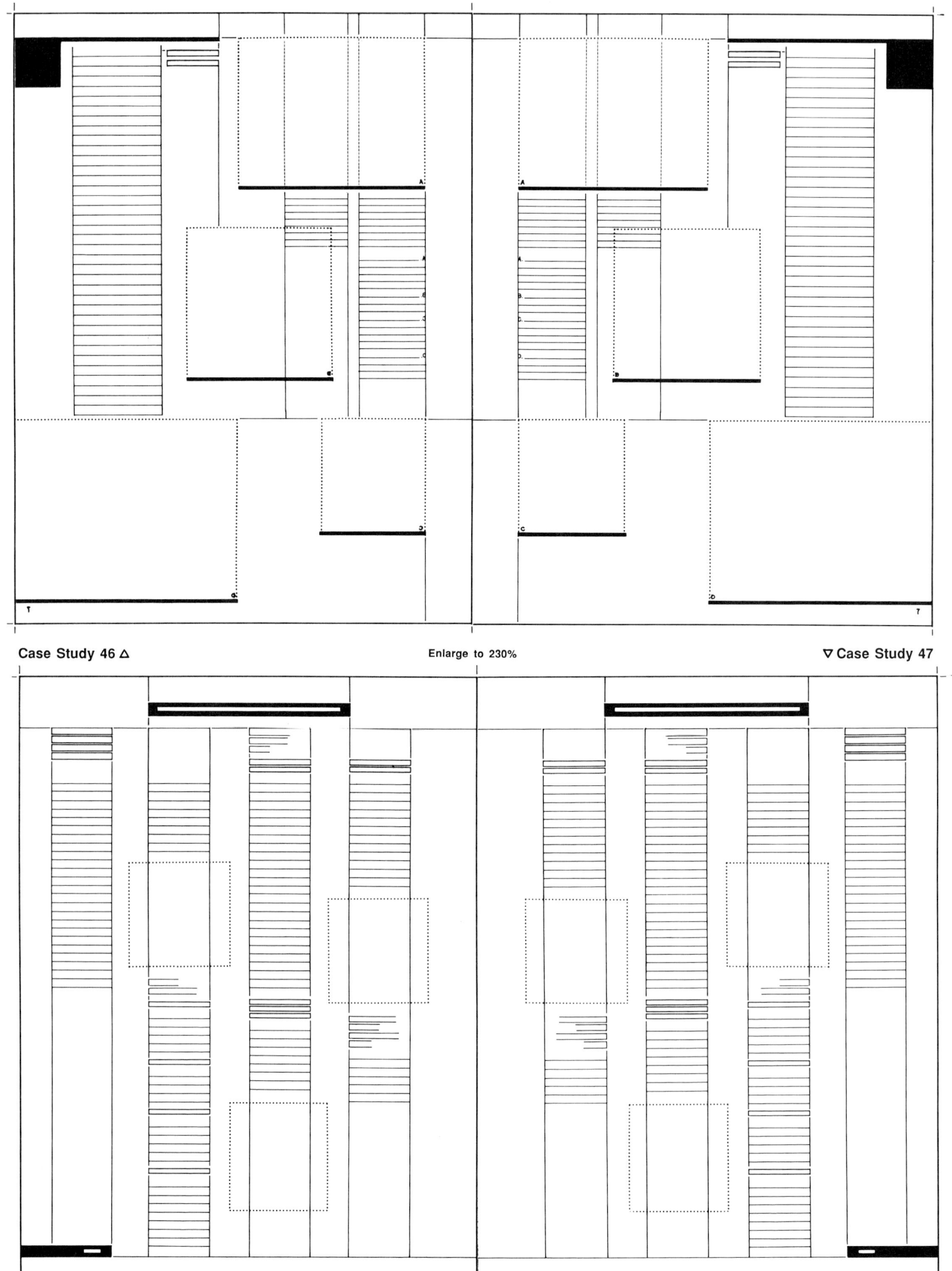

Case Study 46 △  Enlarge to 230%  ▽ Case Study 47

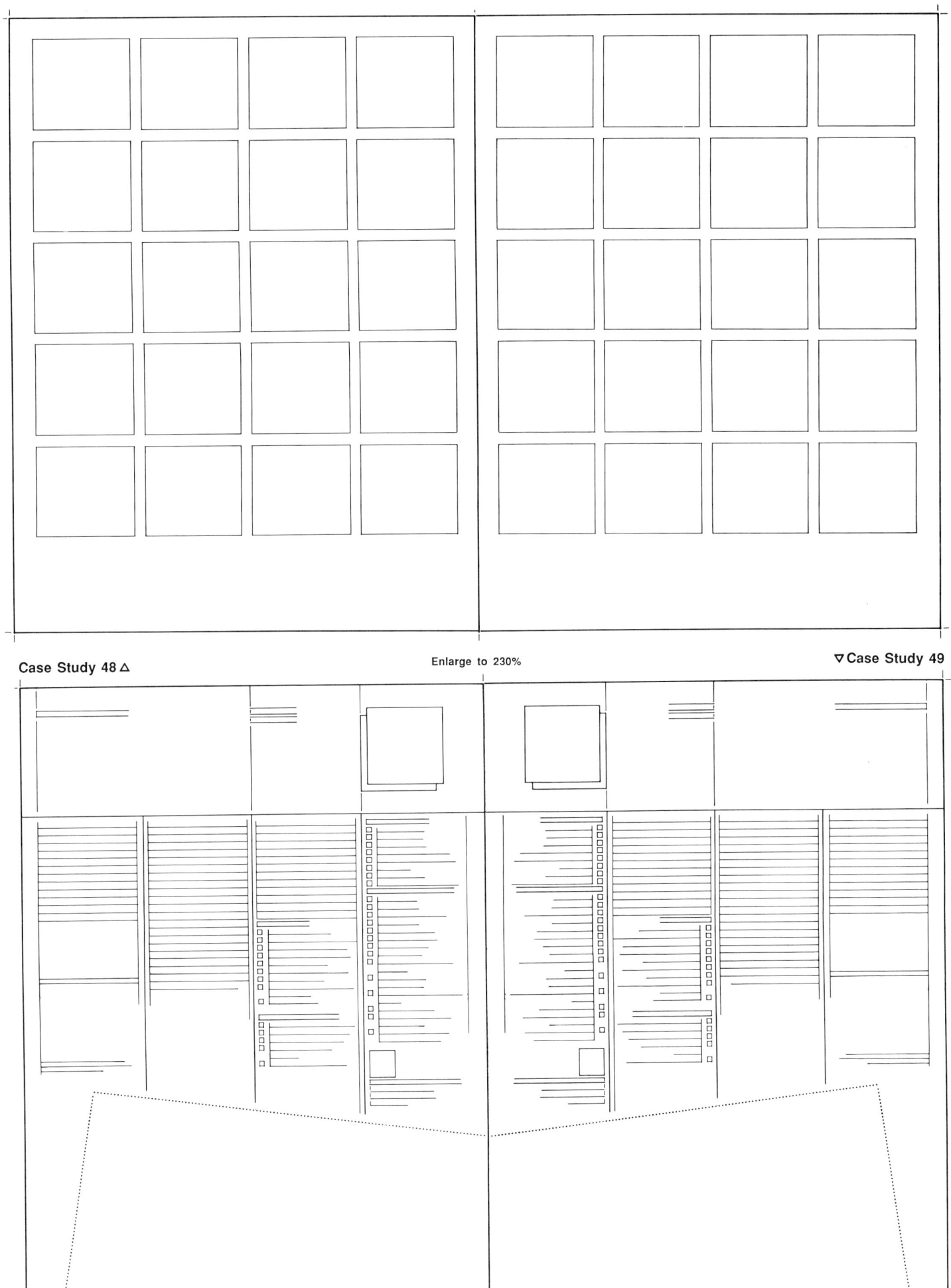

Case Study 48  Enlarge to 230%  ▽Case Study 49

Appendix 217

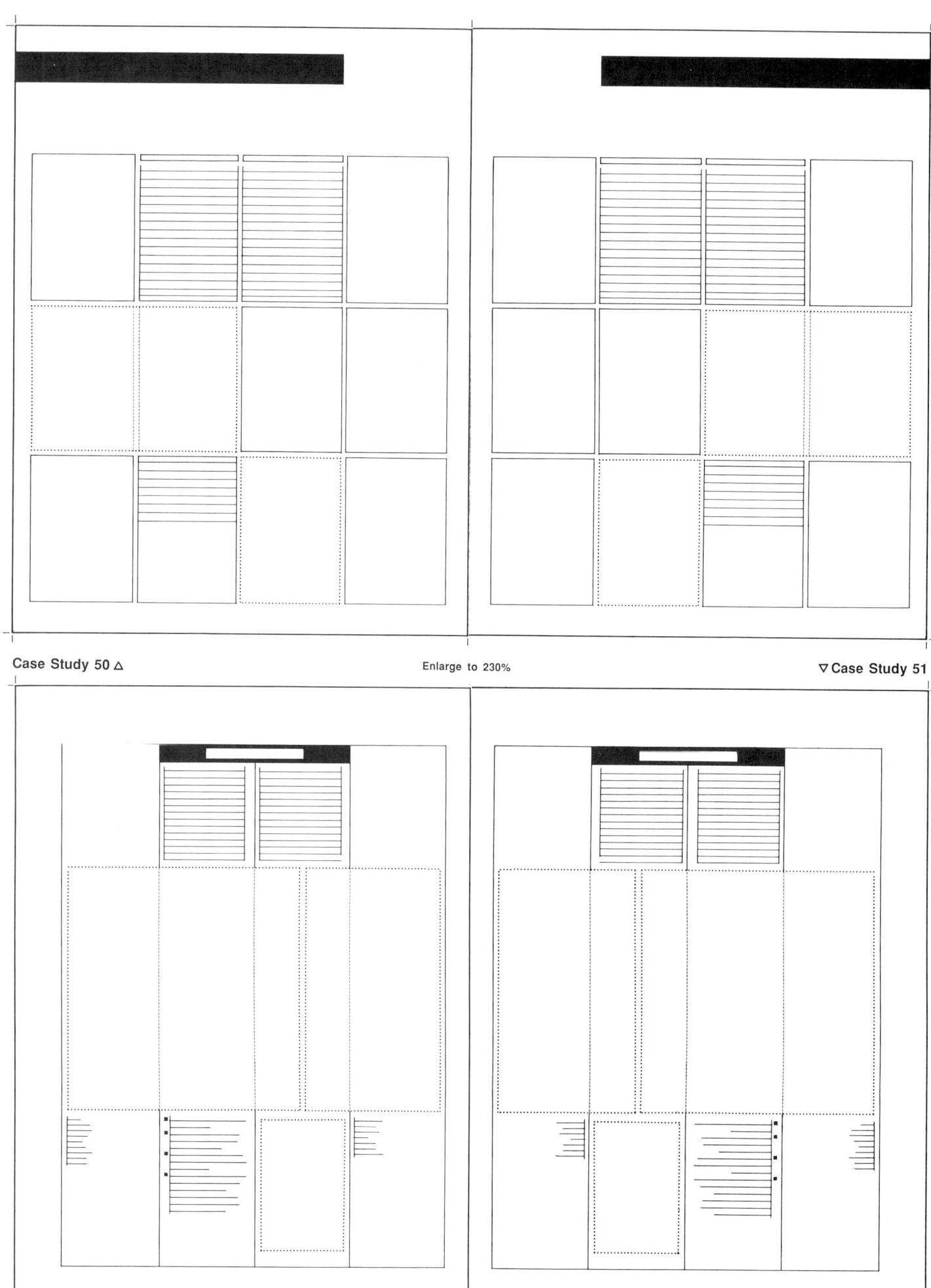

Case Study 50 △   Enlarge to 230%   ▽ Case Study 51

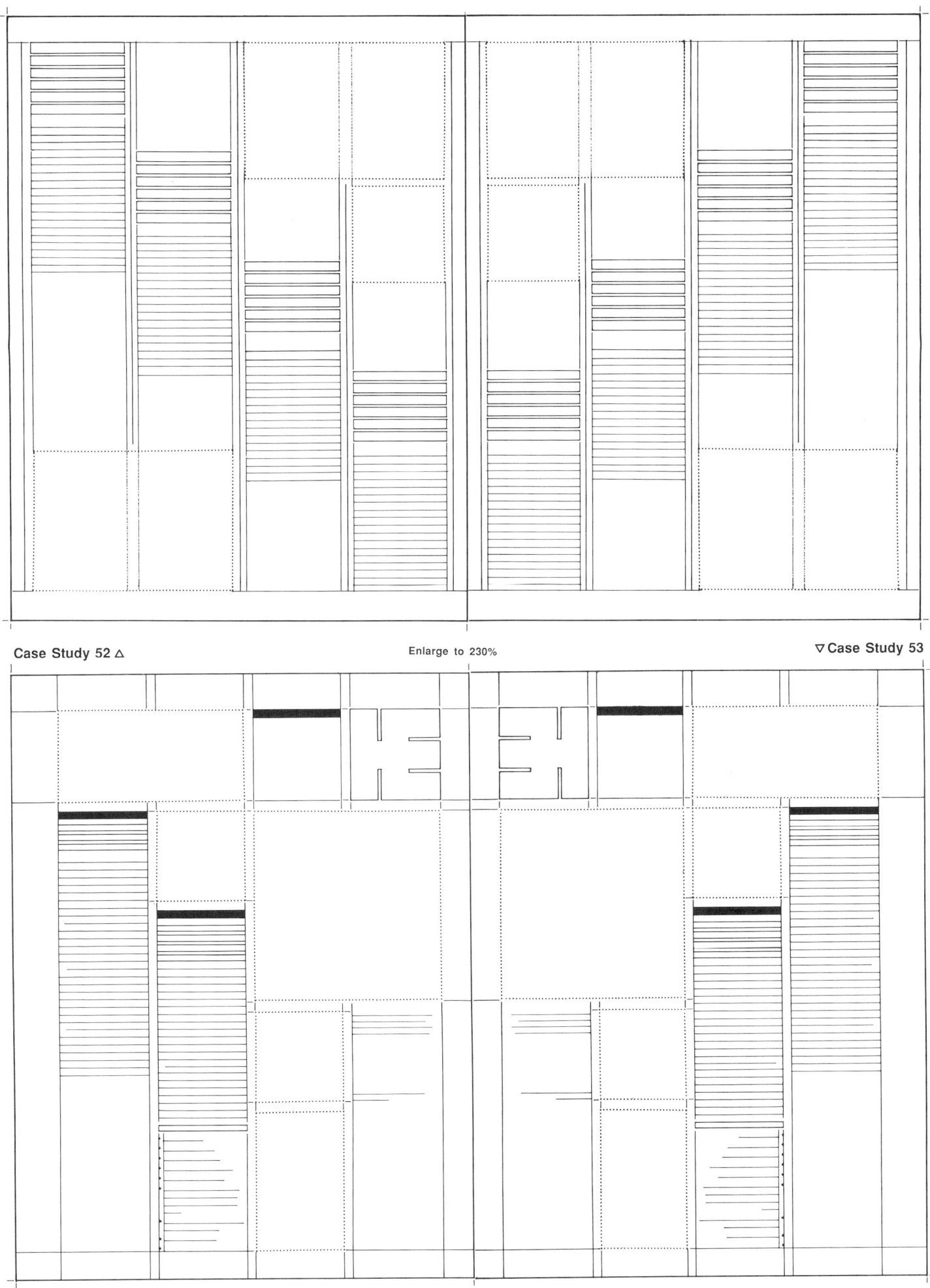

Case Study 52 △   Enlarge to 230%   ▽ Case Study 53

Appendix 219

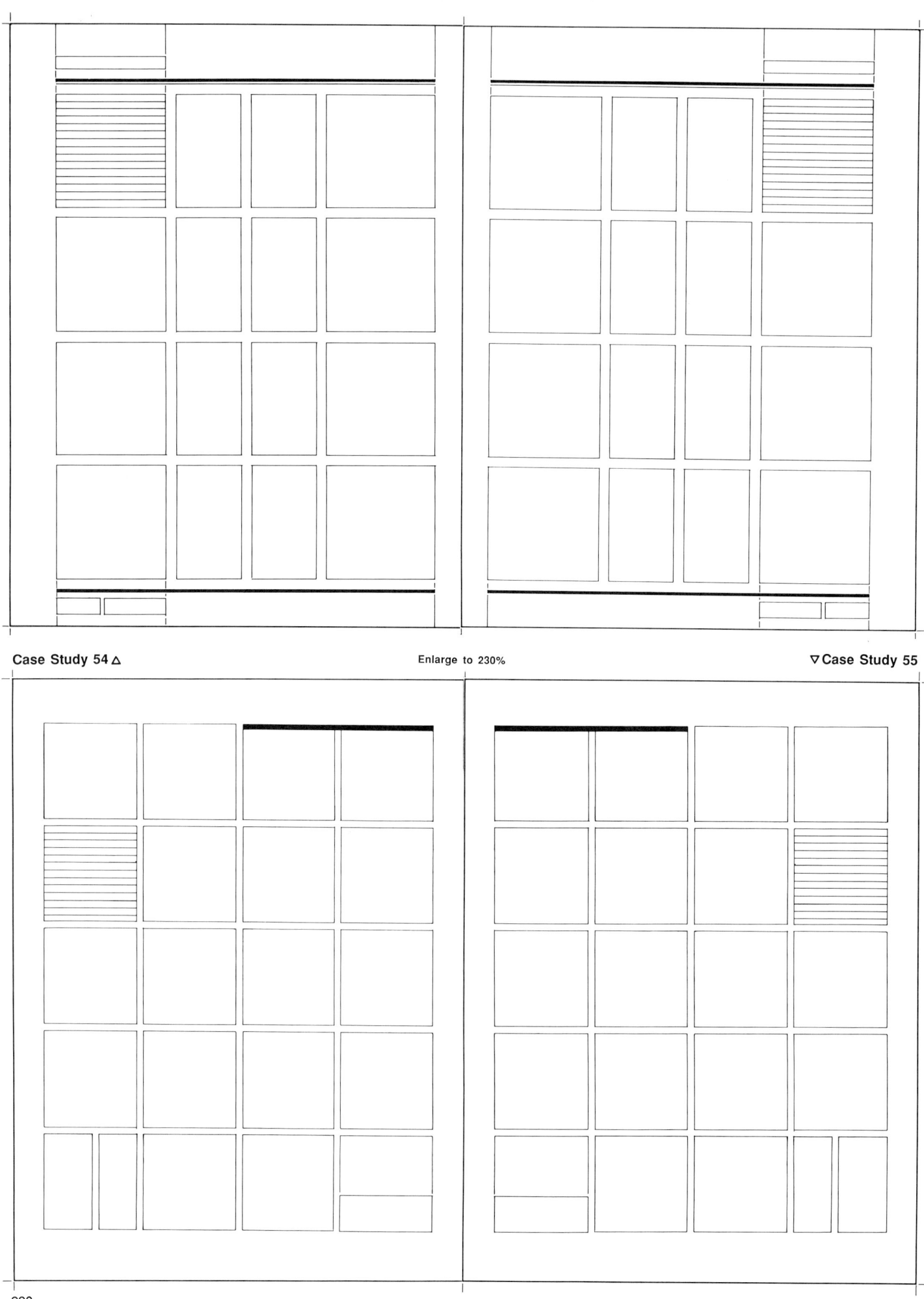

Case Study 54 △   Enlarge to 230%   ▽ Case Study 55

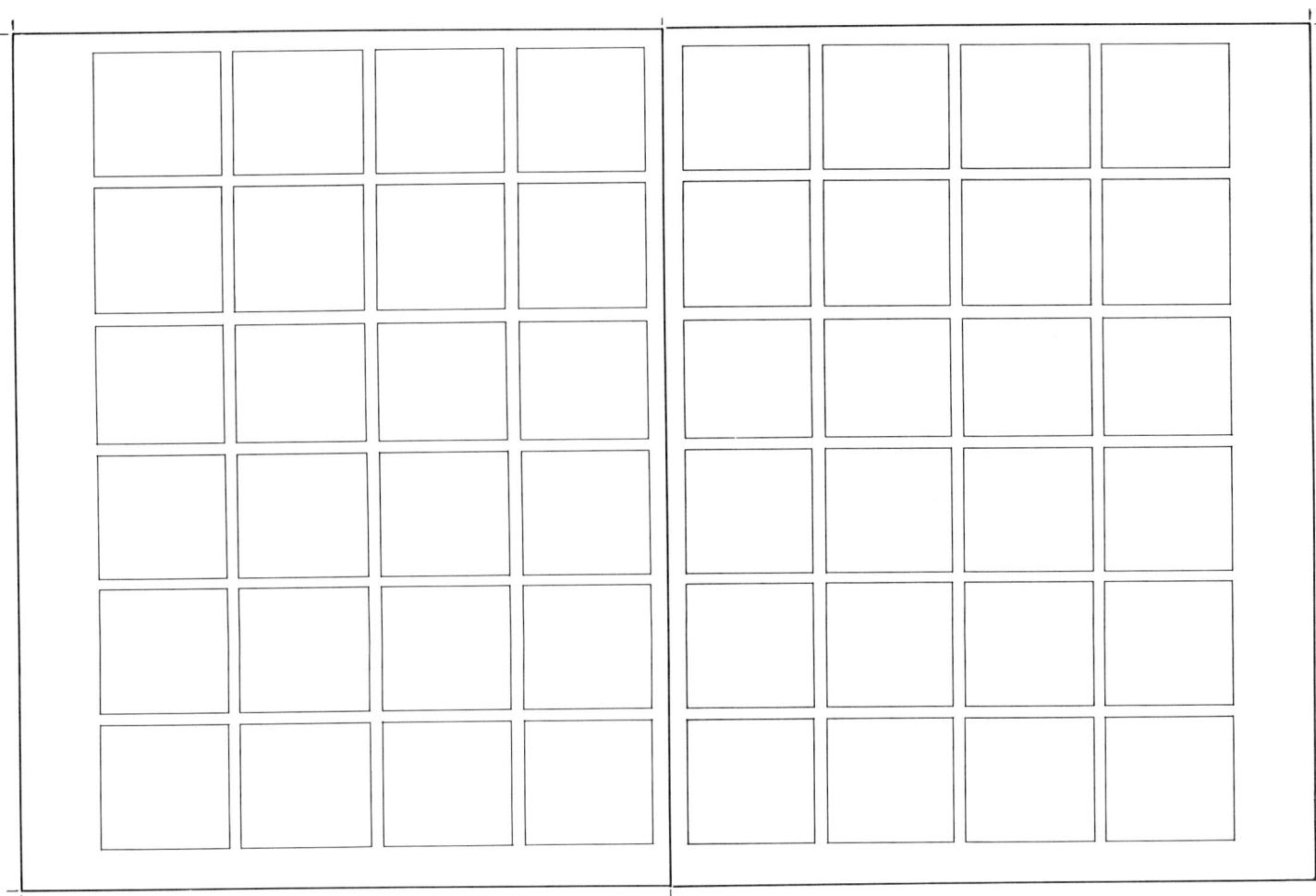

**Case Study 56** △  Enlarge to 230%

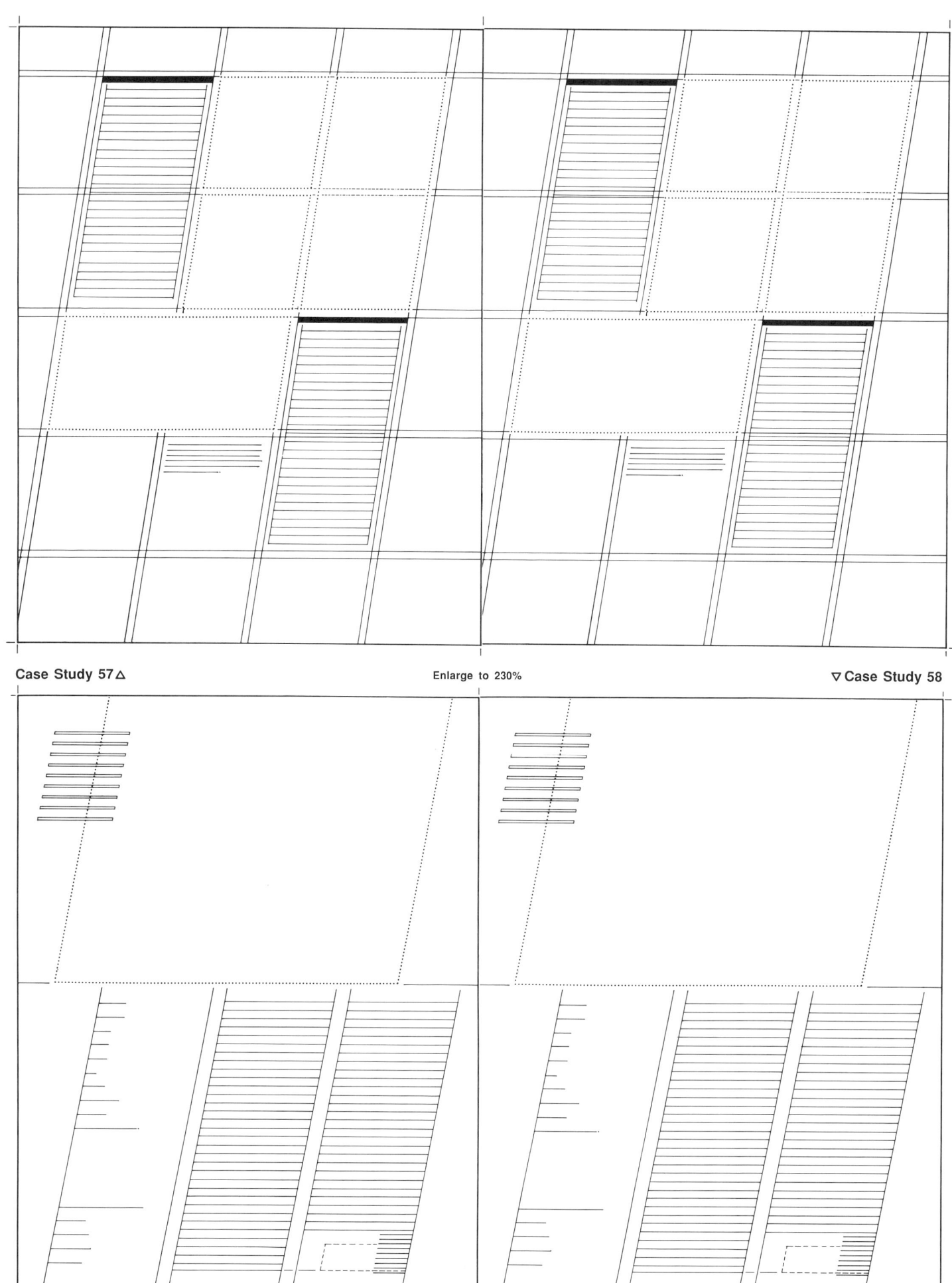

Case Study 57 △  Enlarge to 230%  ▽ Case Study 58

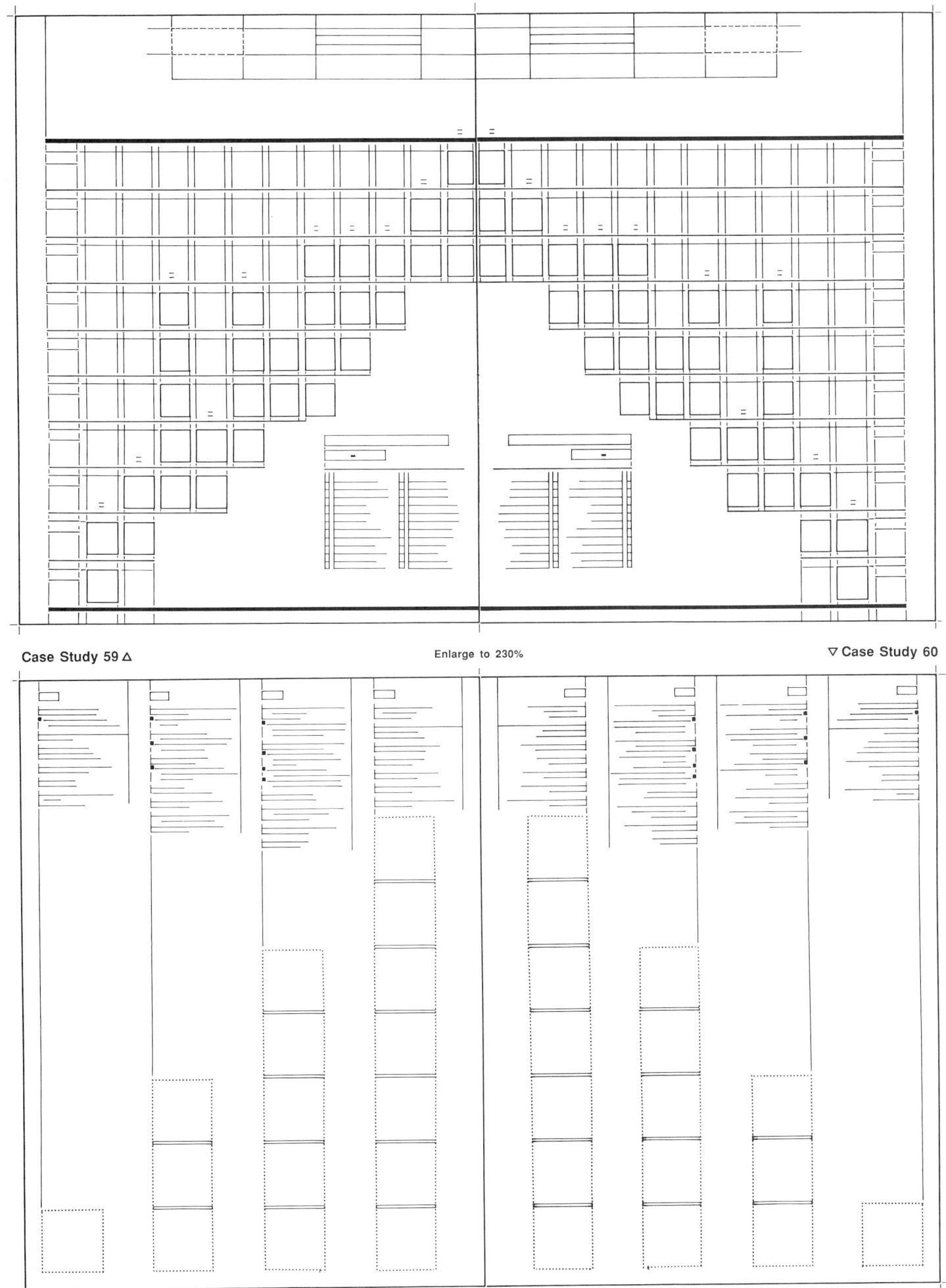

Case Study 59 △   Enlarge to 230%   ▽ Case Study 60

Appendix 223

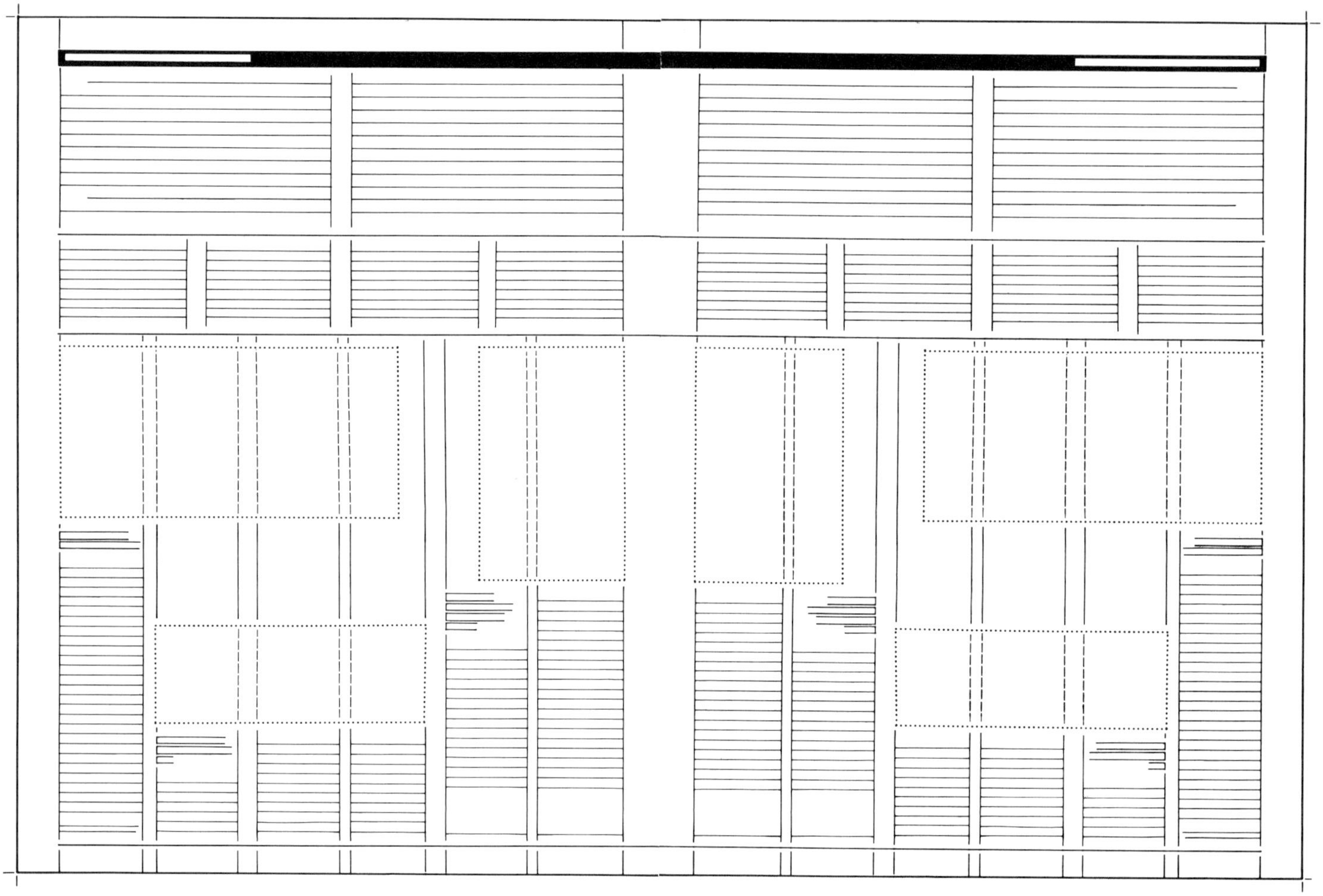

Case Study 61 △                                                                 ▽ Case Study 62

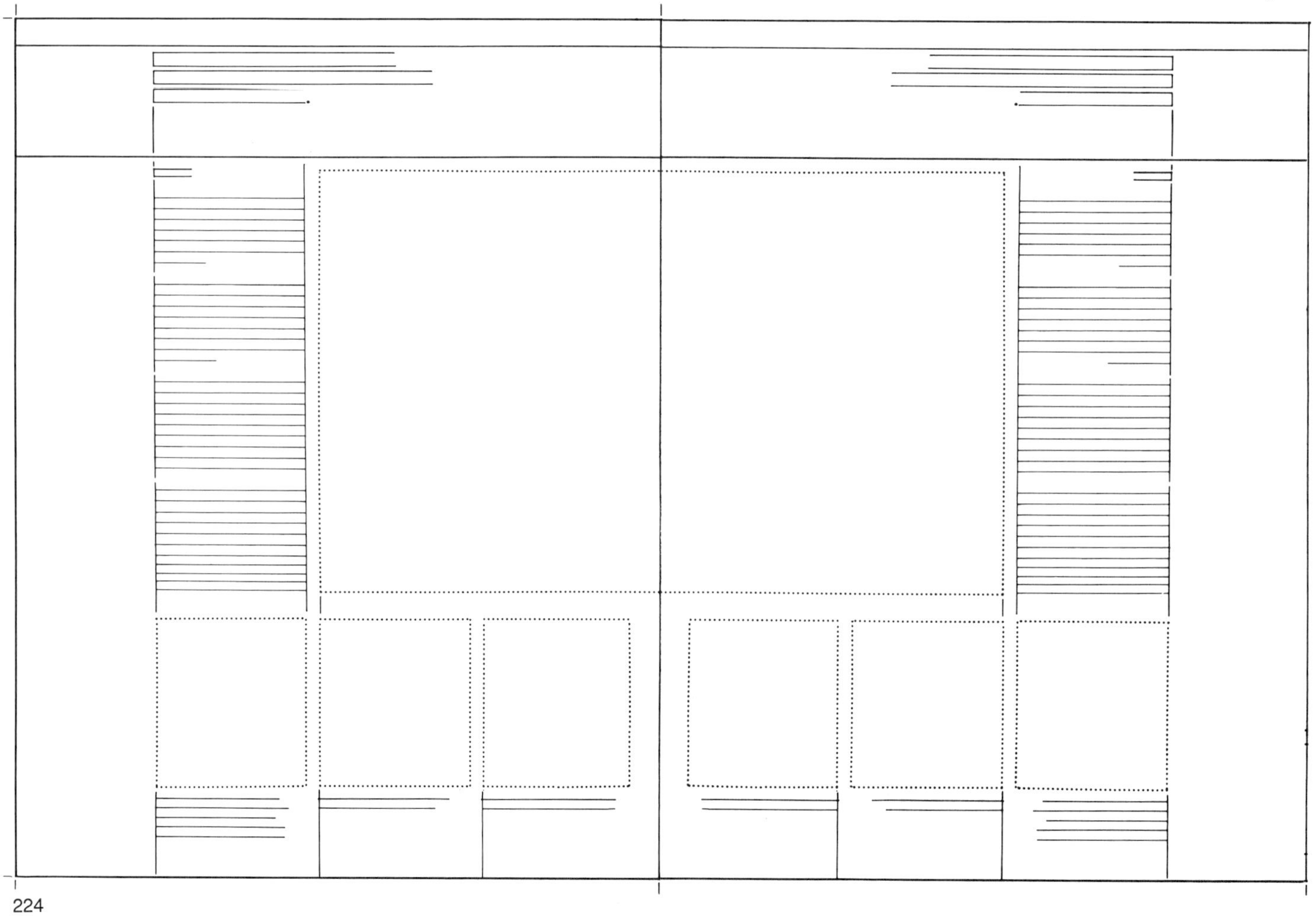

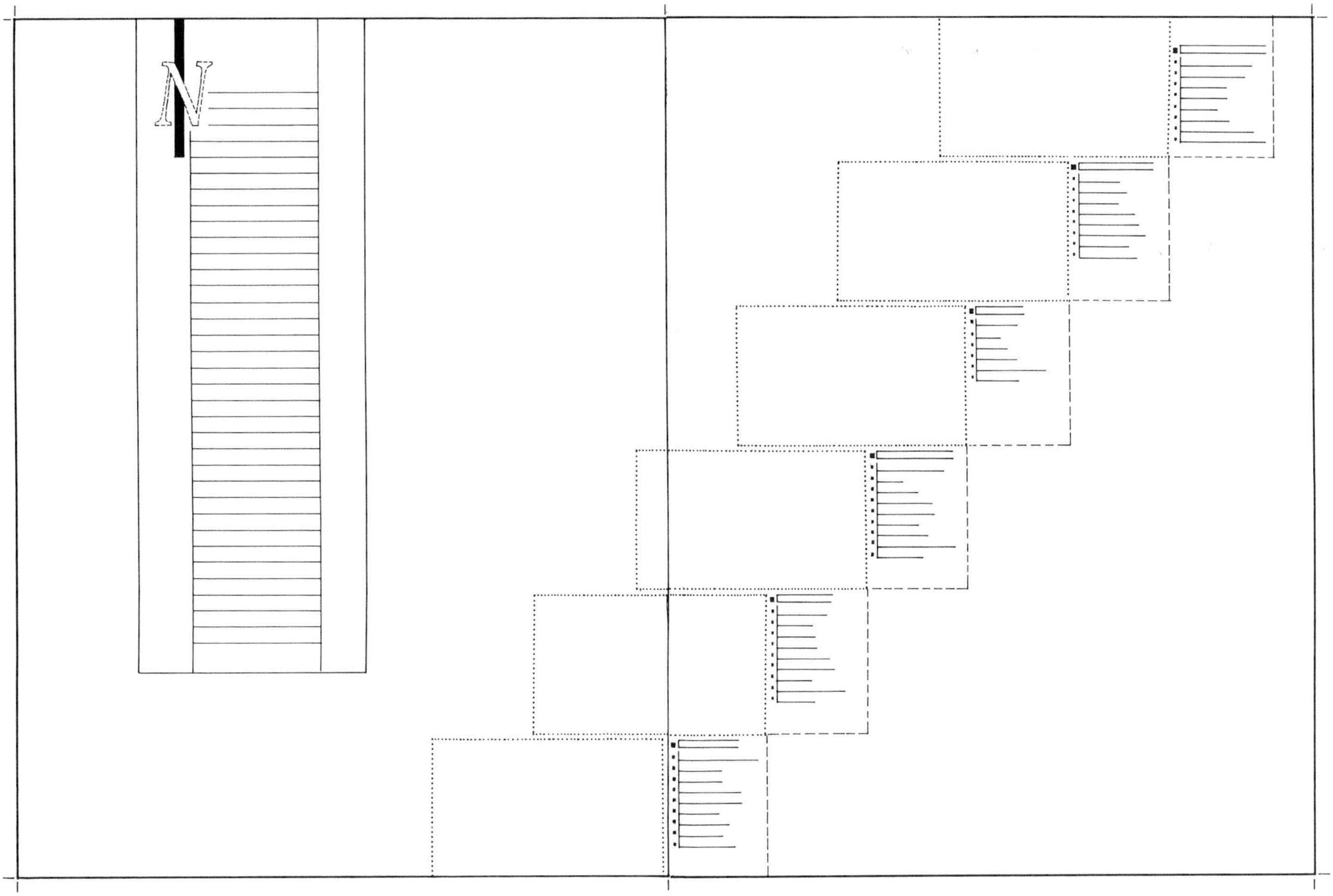

Case Study 63 △                    Enlarge to 230%                    ▽ Case Study 64

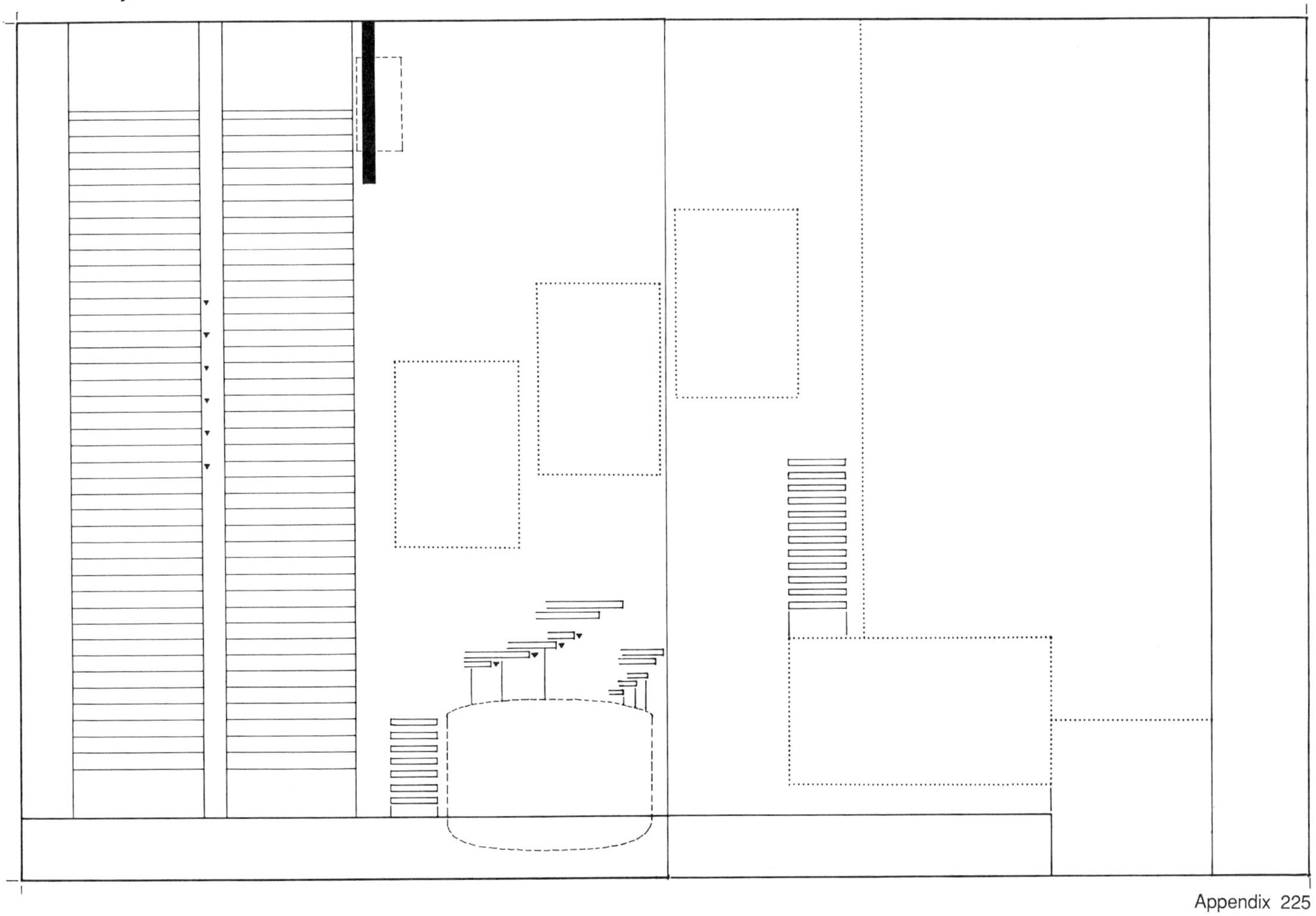

Appendix 225

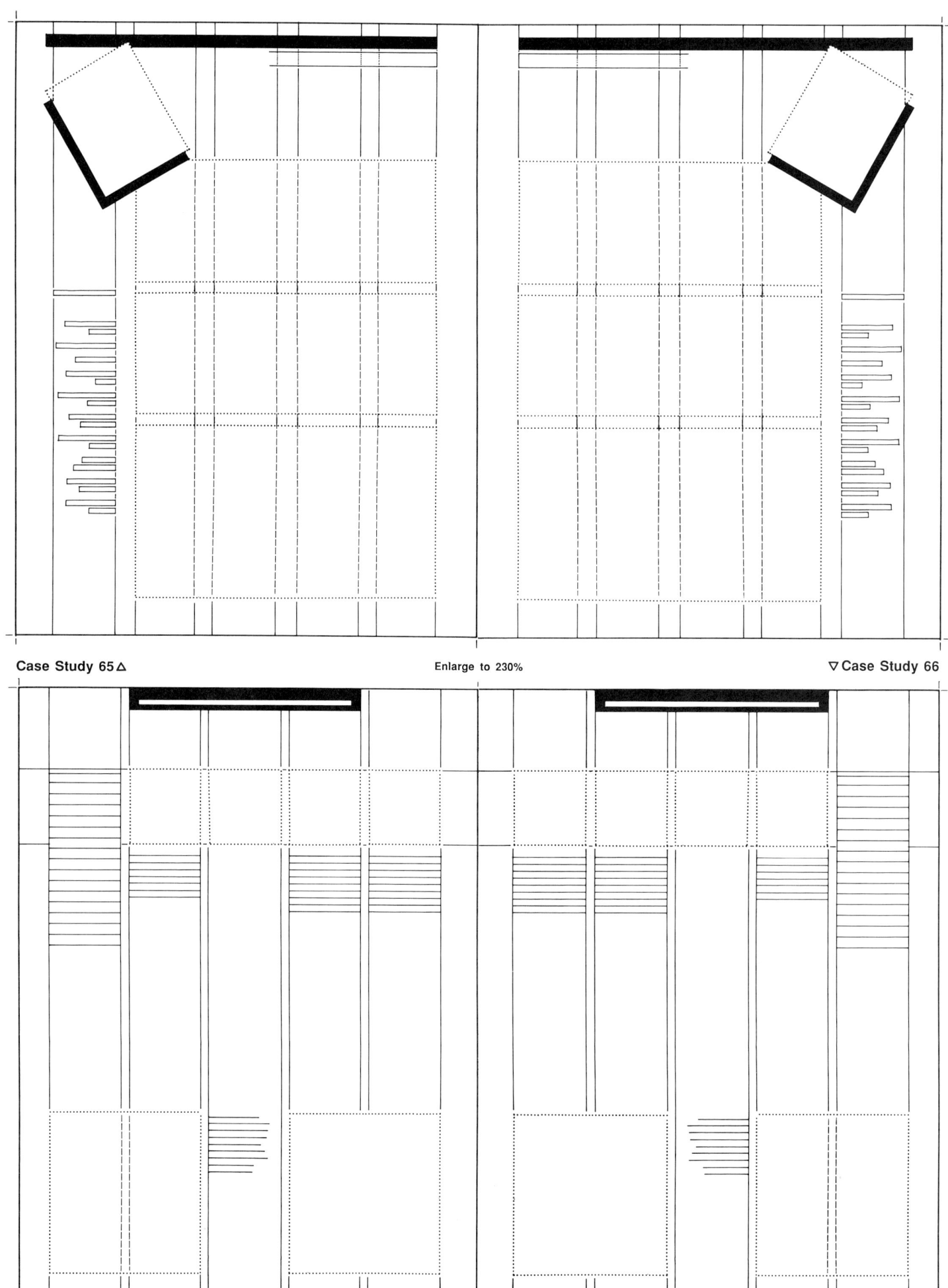

Case Study 65  Enlarge to 230%  Case Study 66

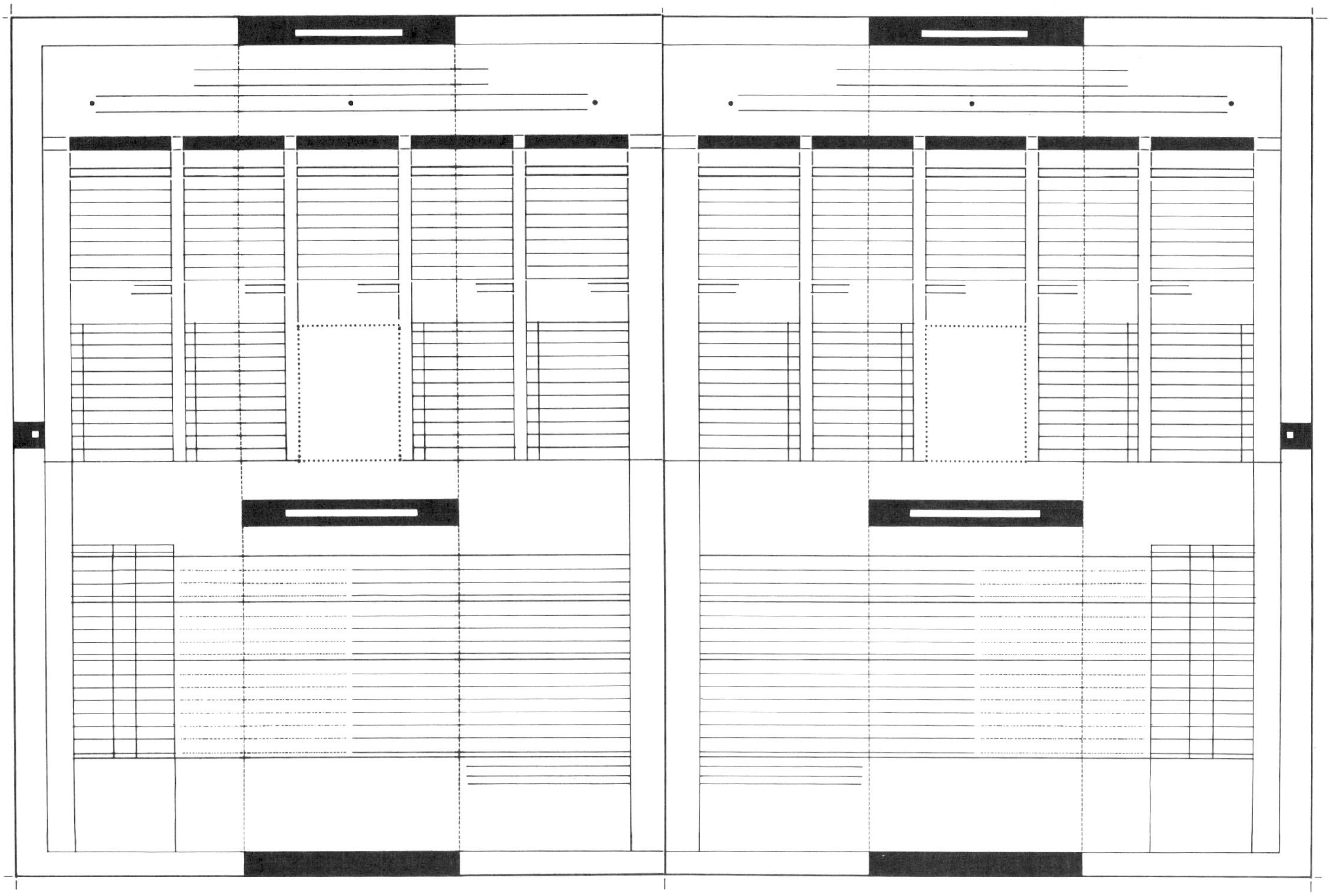

Case Study 67 △  Enlarge to 230%  ▽ Case Study 68

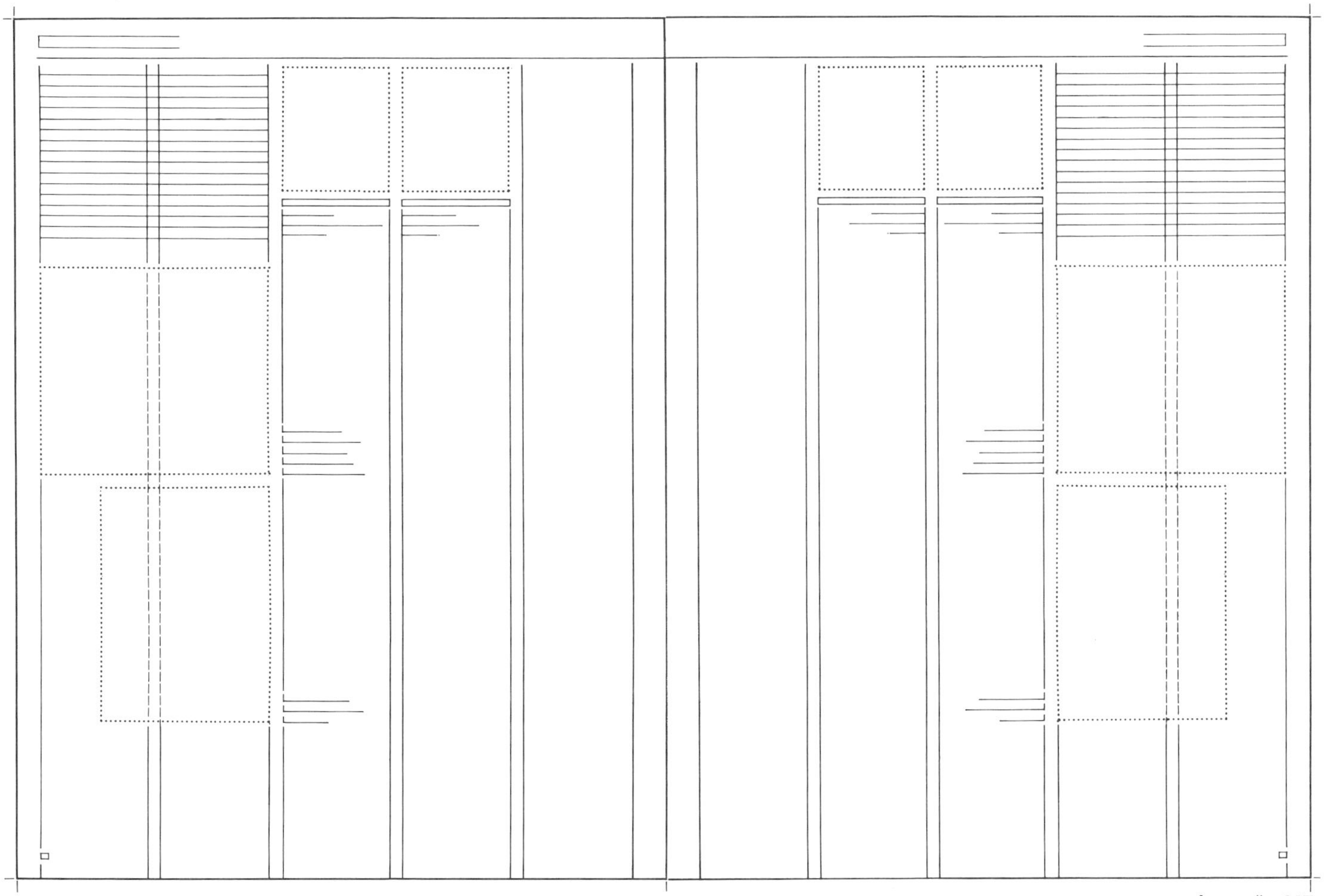

Appendix 227

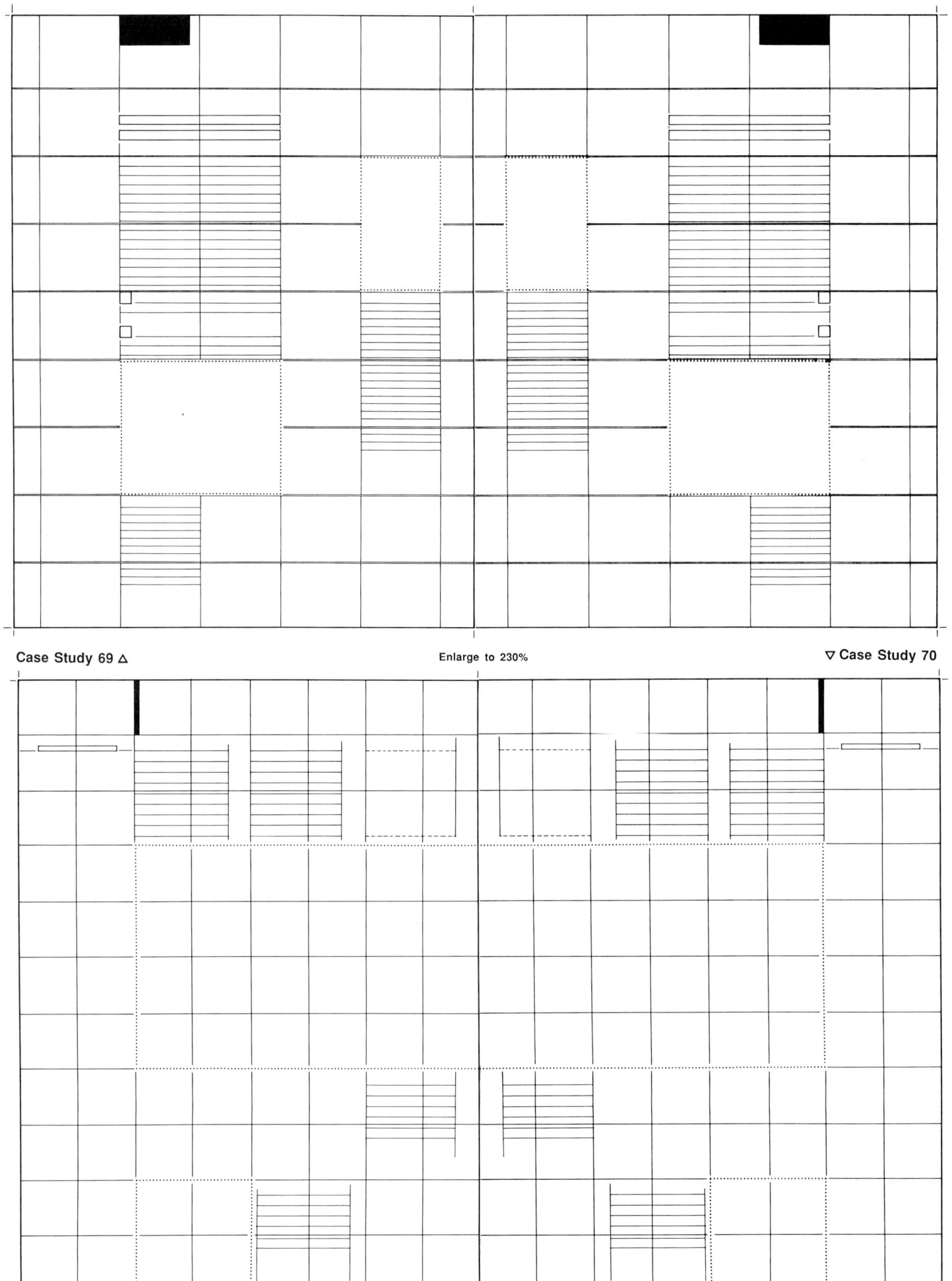

Case Study 69  △      Enlarge to 230%      ▽ Case Study 70

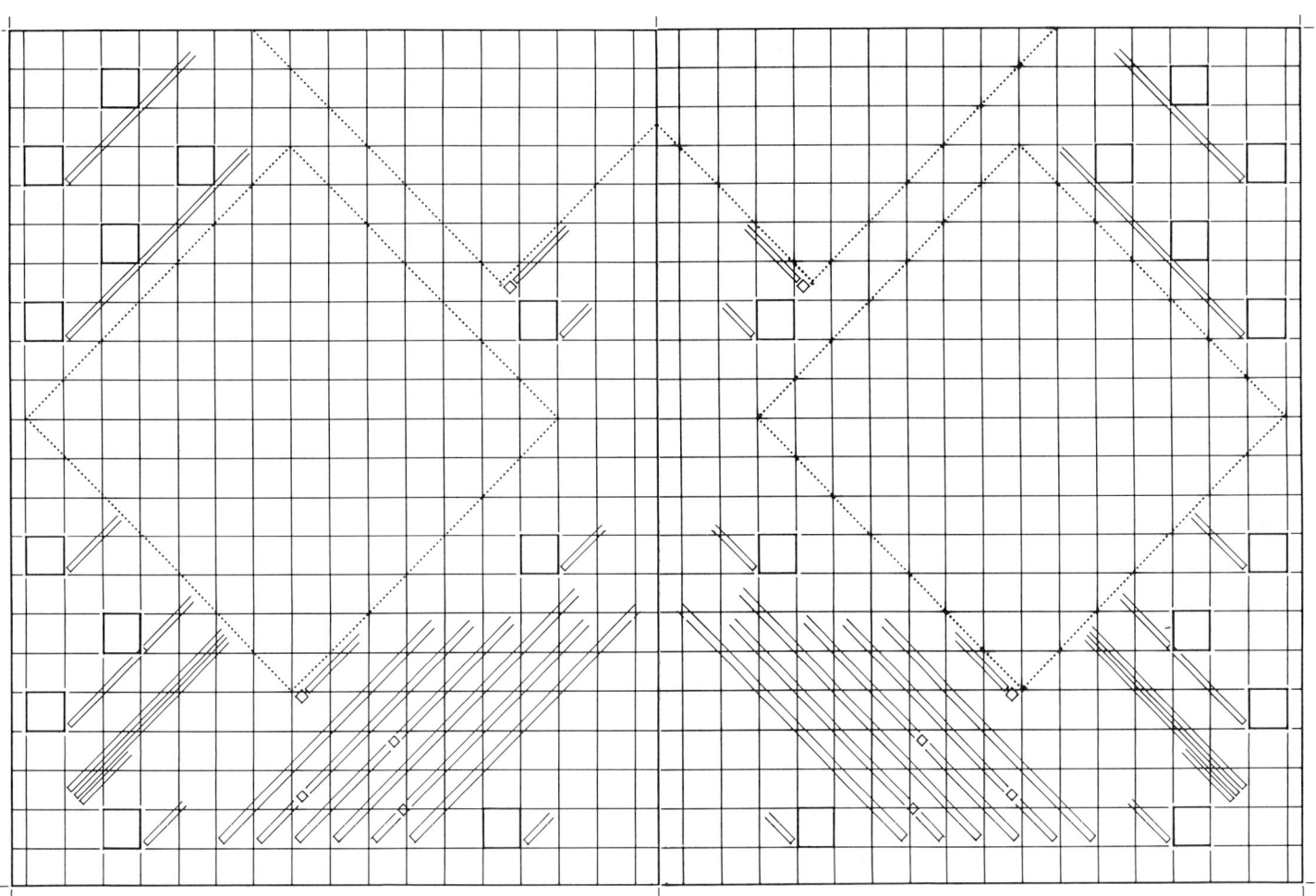

**Case Study 71** △   Enlarge to 230%

Appendix 229

# Design Credits

There are three distinct types of materials in this book. The first type is found in the chapters on Corporate ID and Anatomy of Formats. Here, the examples are composed of miniature reproductions of existing print material. The second type is found in the chapter on generic grids. These grids were originated in sketch form by the author, and constructed on a CADD system by the author's son, Ernest III. The third type is the Case Study Grids which were constructed on acetate over existing printed material. Thousands of brochures were evaluated for their appropriateness to analysis of the grid system used in their design. Many of these brochures, newsletters and annual reports were award winning designs. The sources of these materials are listed below.

## Corporate ID

10  ID Manual  Einhorn Yaffee Prescott

## Anatomy Of Formats

23  Collateral  Kirkham Michael & Associates
24  Advertisements  Lehrer McGovern (t.)/Turner Construction (m.)/Oliver Design (b.)
25  Advertisement  /Morris Aubrey (t.l.)/SHWC (t.r.)/ Tarlton Construction Co. (b.)
26  Announcements  Harley Ellington Pierce & Yee Assoc. (t.)/RTKL Associates Inc. (b.)
27  Announcements  Sandy & Babcock
28  Announcements  ISD Inc. (t.)/The Hillier Group (m)/Bobrow Thomas Associates (b.)
29  Announcements  Kay Lentz Marketing "Doors Open" (t.)/Troast "Blocks" (m.)/ Harriman Associates "Moving Pictures" (b.)
30  Accordion Folders  The Office of Pierce Goodwin Alexander (t.)/Kojian (b.)
31  Accordion Folders  Wallace McHarg Roberts+Todd (t.)/Fletcher-Thompson, Inc. (b.)
32  Fliers  TAC, The Architects Collaborative (.t)/Miles Treaster and Associates (b.)
33  Fliers  Scholastic Structures/Arthrur L. Spaet and Associates (b.)
50  Covers/Insides  Wallach Glass Studio (t.)/Space Design International, Inc. (m.)/ Einhorn Yaffee Prescott (b.)
51  Covers/Insides  Michael Brandman Associates (t.)/Bullfield Volkman Stockwell (m.)/ Kober Bulluschi Associates, Inc. (b.)
54  Proposals  Arthur L. Spaet and Associates
62  Organization Chart  Kirkham Michael and Associates
70  Templates  Image Express
72  Electronic Formats  Creighton C. Nolte, AIA
73  CADD Formats  Creighton C. Nolte, AIA

## Generic Grids

76  Generic Grids Drawn on a CADD System  Ernest Burden III

## Case Study Formats

116  Case Study 1   James M. Montgomery / Corp. Brochure
117  Case Study 2   Parsons Brinckerhoff / Tudor/Specialty Brochure
118  Case Study 3   Dura Art Stone / Product Catalog
119  Case Study 4   Hill Pinkert Architects, Inc. / Corp. Brochure
120  Case Study 5   Civale & Trovato / Corp. Brochure
120  Case Study 6   U.S. Gypsum Co., Ultrawall / Product Catalog
121  Case Study 7   Cole Associates / Corp. Brochure
122  Case Study 8   The Callison Partnership / Corp. Brochure
123  Case Study 9   Gibbs & Hill / Specialty Brochure
124  Case Study 10  Sverdrup Corporation / Specialty Brochure
125  Case Study 11  Mid-State Assoc., Inc. / Specialty Brochure
126  Case Study 12  Humphrey's & Harding / Corp. Brochure
127  Case Study 13  WilsonArt /  Product Brochure
128  Case Study 14  The Earth Technology Corp. / Specialty Brochure
129  Case Study 15  The Earth Technology Corp. / Specialty Brochure
130  Case Study 16  Henry Dreyfuss Associates / Corp. Brochure
131  Case Study 17  Merideth Corporation / Annual Report
131  Case Study 18  Hewlett Packard / Annual Report
132  Case Study 19  Schwartz / Silver Architects/Corp. Brochure
133  Case Study 20  RTKL Associates, Inc. / Planning Brochure
134  Case Study 21  National Concrete Masonry Assoc. / Magazine
135  Case Study 22  Alply, Inc. /Techwall/Brochure
136  Case Study 23  Pare Engineering Corp. / Service Brochure
137  Case Study 24  The Kling-Lindquist Partnership / Corp. Brochure
138  Case Study 25  Dimeo Construction Co. / Corp. Brochure
139  Case Study 26  CMA Enterprises Ltd. / Corp. Brochure
140  Case Study 27  ASG / Product Folder
141  Case Study 28  ASG / Product Folder
142  Case Study 29  EDAW, Inc ./ General Brochure
143  Case Study 30  Homestead Financial Corp. / Annual Report
144  Case Study 31  Haines Lundberg Waehler / Planning Brochure
145  Case Study 32  Strang Partners/Services Brochure
146  Case Study 33  Gensler and Associates Architects/Anniversary Brochure
147  Case Study 34  Dewberry & Davis / Anniversary Foldout
148  Case Study 35  M.A. Mortenson Co. / Corp. Brochure
149  Case Study 36  Herbert Construction Co., Inc. / Corp. Brochure
150  Case Study 37  Seal Dry / Product Brochure

| | | |
|---|---|---|
| 163 | Case Study 49 | Van Dell and Assoc. / Services Brochure |
| 164 | Case Study 50 | STO Industries, Inc. / Products Catalog |
| 165 | Case Study 51 | National Products, Inc. / Products Catalog |
| 166 | Case Study 52 | Wright-Pierce / Specialty Brochure |
| 167 | Case Study 53 | Havens and Emerson / Specialty Brochure |
| 168 | Case Study 54 | Office Design Associates / Project Pages |
| 169 | Case Study 55 | Oliver Design Group / Project Pages |
| 170-1 | Case Study 56 | Omega Lighting / Product Catolog |
| 172 | Case Study 57 | Sugimora & Assoc. / General Brochure |
| 173 | Case Study 58 | Graftel Systems / Advertisement |
| 174 | Case Study 59 | Haines Lundberg Waehler / Partners Folder |
| 175 | Case Study 60 | Thompson Ventulett Stainback / Anniversary Folder |
| 176 | Case Study 61 | EDAW, Inc. / Newsletter |
| 177 | Case Study 62 | Syska & Hennessy / Corp. Brochure |
| 178 | Case Study 63 | CH2M Hill / Specialty Brochure |
| 179 | Case Study 64 | CH2M Hill / Specialty Brochure |
| 180 | Case Study 65 | Einhorn Yaffee Prescott / Corp. Brochure |
| 181 | Case Study 66 | The Callison Partnership / Corp. Brohcure |
| 182 | Case Study 67 | Mobil Corp. / Annual Report |
| 183 | Case Study 68 | Syska & Hennessy / Brochure |
| 184 | Case Study 69 | Opus Corp. / Corp. Brochure |
| 185 | Case Study 70 | Seitz Yamamoto Moss, Inc. / Corp. Brochure |
| 186 | Case Study 71 | Classiquez / Products Catalog |
| 187 | Case Study | Stone & Webster / Services Folder |
| 187 | Case Study | Davy McKee Corp. / Services Folder |
| 151 | Case Study 38 | Wolff Zimmer Gunsul Frasca / Specialty Brochure |
| 152 | Case Study 39 | Sverdrup Corp. / Specialty Brochure |
| 153 | Case Study 40 | The Edwards and Kelcey Organization / Services Folder |
| 154 | Case Study 41 | Post, Buckley, Schuh & Jernigan / General Brochure |
| 155 | Case Study 42 | Dewberry & Davis / Magazine |
| 156 | Case Study 43 | CH2M Hill / Anniversary Brochure |
| 157 | Case Study 44 | The NBBJ Group / Anniversary Brochure |
| 158 | Case Study 45 | RTKL Associates, Inc. / Specialty Brochure |
| 159 | Case Study | Weck Glass Blocks / Product Brochure |
| 160 | Case Study 46 | American Seating / Product Brochure |
| 161 | Case Study 47 | Armstrong / Product Catalog |
| 162 | Case Study 48 | Ruth, Shives & Williams, Inc. / Services Folder |

# Bibliography

There are many books on typography and graphic design which are useful resources. Many complement the material in this book. A few contain grids. While it is not possible to review each one, they are categorized her according to their main focus.

## General / DTP Publishing

Collier, David and Flood, Kay:
LAYOUT FOR DESKTOP DESIGNS
Northlight Books, 1989

Gosney, Michael / Linnea Dayton:
VERBUM BOOK OF ELECTRONIC
PAGE DESIGN
Van Nostrand Reinhold, 1983

Shushan, Ronnie / Wright, Don
Bimerle, Ricardo
DESKTOP PUBLISHING
BY DESIGN
Ventura Publisher Edition,
Microsoft Press, 1991

Meggs, Phillip B:
HISTORY OF GRAPHIC DESIGN
Van Nostrand Reinhold, 1983

Miles, John:
DESIGN FOR DESKTOP
PUBLISHING
Chronicle Books, 1987

White, Jan V.:
GRAPHIC DESIGN FOR THE
ELECTRONIC AGE
Watson-Guptil, 1988

## Grids and Layouts

Hurlburt, Allen:
THE GRID
Van Nostrand Reinhold 1978

Hurlburt, Allen:
LAYOUT: The Design Of The
Printed Page
Watson-Guptil, 1989

Parker, Roger C.:
THE MAKE-OVER BOOK
101 Design Solutions
Ventana Press, 1989

Swann, Allen:
How To Understand And Use
GRIDS
Northlight Books, 1898

Swann, Allen:
How To Understand And Use
DESIGN AND LAYOUTS
Northlight Books, 1897

West, Suzanne:
WORKING WITH STYLE
Watson-Guptil, 1989

## Typography

Bauermeister, Benjamin
A MANUAL OF COMPARATIVE
TYPOGRAPHY, The Panose System
Van Nostrand Reinhold, 1988

Carter, Rob / Day, Ben /
Meggs, Phillip B.:
TYPOGRAPHIC DESIGN
Form And Communication
Van Nostrand Reinhold, 1985

King, Jean Callen / Esposito, Tony
DESIGNERS GUIDE TO TEXT
TYPE
Van Nostrand Reinhold, 1980

Rosen, Ben:
TYPE &!,:"?" 2e
Van Nostrand Rei Reinhold, 1989

THE TYPE SPECIMEN BOOK
544 Different Typefaces
Van Nostrand Reinhold, 1974

## Charts / Presentation / Production

HHolmes, Nigel:
DESIGNERS GUIDE TO
DESIGNERS GUIDE TO
CREATING CHARTS AND
DIAGRAMS
Watson-Guptil, 1988

Craig, James:
PRODUCTION FOR THE
GRAPHIC DESIGNER, 2e
Watson-Guptil, 1990

John, Lynn:
PREPARING YOUR DESIGN FOR
PRINT
Northlight Books, 1989

Raab, Margaret Y.:
THE PRESENTATION DESIGN
BOOK
Ventana Press, 1990

## Brochures

Gedney, Karen / Fultz, Patrick:
COMPLETE GUIDE TO
CREATING SUCCESSFUL
BROCHURES
Asher Gallant, 1988

Jones, Gerre
HOW TO PREPARE
PROFESSIONAL DESIGN
BROCHURES
McGraw—Hill, 1976

Travers, David
PROFESSIONAL DESIGN OFFICE
BROCHURES
2e, Arts and Architecture Press,'78

## DTP Manuals

Quark, Ventura, Pagemaker, Postscript, and others. Many software companies publish extensive guides for their programs, which contain valuable information on layouts and page composition. Some programs have "templates" or other internal grid systems.